THE IMMORTALISTS

Also by
DAVID M. FRIEDMAN

—

A Mind of Its Own:
A Cultural History of the Penis

An Imprint of HarperCollins*Publishers*

THE IMMORTALISTS

—

Charles Lindbergh,
Dr. Alexis Carrel,
and
Their Daring Quest
to Live Forever

David M. Friedman

HarperCollins books may be purchased for educational, business, or
sales promotional use. For information, please write: Special Mar-
kets Department, HarperCollins Publishers, 10 East 53rd Street,
New York, NY 10022.

Grateful acknowledgment is made for permission to quote from the
following: The writings of Charles A. Lindbergh in the Charles Au-
gustus Lindbergh Papers, Manuscripts and Archives, Yale Univer-
sity Library, granted by Yale University. Materials in the Alexis
Carrel Papers at Georgetown University, granted by the Special
Collections Division, Georgetown University Library. Materials in
the Rockefeller Archive Center, granted by the Rockefeller Archive
Center, Sleepy Hollow, New York. Lyrics from "Mister Charlie
Lindbergh," by Woody Guthrie, © 1977 by WOODY GUTHRIE
PUBLICATIONS, INC. All rights reserved. Used by permission.

FIRST EDITION

Designed by Barbara M. Bachman

LIBRARY OF CONGRESS
CATALOGING-IN-PUBLICATION DATA
is available upon request.

ISBN: 978-0-06-052815-7
ISBN-10: 0-06-052815-X

07 08 09 10 11 DIX/RRD 10 9 8 7 6 5 4 3 2 1

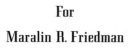

For
Maralin R. Friedman

CONTENTS

—

1	I Will Show You What I'm Doing Here	1
2	If Man Could Learn to Fly	17
3	A Student Who May Amount to Something	25
4	Isn't He in the Crib?	41
5	The Chamber of Life	54
6	Every Act of His Is Not a Fluke	75
7	Men of Genius Are Not Tall	86
8	A Tiny Puff of Smoke	95
9	The Most Interesting Place in the World Today	103
10	The Exploration of This Realm Is a Great New Adventure	118
11	The Reaction of a Man Would Probably Be Similar	128
12	Two Men Sitting on Two Rocks	137
13	By Order of the Führer	151
14	I'll Take a Rain Check	157

15   Not Merely a Schoolboy Hero,
       but a Schoolboy   171

16   For We All Know What Awaits Us   177

17   There Is Much I Do Not Like That
       Is Happening in the World   186

18   The Tissues, the Blood, and the
       Mind of Man   201

19   The World Was Never Clearer   220

20   The Grandeur of His Life   233

21   I Felt the Godlike Power   247

22   Only by Dying   258

23   No, Science Has Abandoned *Me*   270

EPILOGUE   All of Us Are Following   283

ACKNOWLEDGMENTS   285

NOTES   287

INDEX   321

THE IMMORTALISTS

I WILL SHOW YOU WHAT
I'M DOING HERE

CHARLES LINDBERGH'S FAMILIARITY WITH PRYING GAZES began on May 21, 1927, the day he became the most famous man in the world. That status was conferred on the unsuspecting twenty-five-year-old, literally overnight, when he was the first aviator to fly without stopping from New York to Paris, a feat that many people—even many aviators—had thought impossible. Making Lindbergh's triumph all the more newsworthy was that he flew without a copilot, a radio, or even a front window for thirty-three and a half hours in a single-engine airplane made from wood, canvas, and piano wire. The *New York Times* showed its awe by devoting its first five pages to the dimple-chinned American's landing at Le Bourget, the dusty airfield where 100,000 Frenchmen—nearly all of them chanting "LAN-BAIRGH! LAN-BAIRGH!"—were so eager to stare at the spent pilot that they almost trampled him to death after he climbed out of his cockpit. Maintenance crews swept up a ton of personal items lost or abandoned in that lovefest, including a sable coat and six sets of false teeth.

In the week that followed, Lindbergh, in new clothes made by a Paris tailor, was seen by the president and premier of France, the French chamber of deputies, and a million more Frenchmen who lined a parade route along the Champs-Élysées. After leaving France he was presented to the king of

Belgium, the king of England (who asked, "How did you pee?"), and the prince of Wales (the future duke of Windsor), whom Lindbergh quickly replaced as the most photographed person on earth. Returning home, he was gawked at by 300,000 Americans at the Washington Monument, where President Coolidge pinned the Distinguished Flying Cross on his chest. Four million New Yorkers showered him with cheers and paper scraps, as 10,000 schoolchildren sang "Hail the Conquering Hero Comes," in the largest ticker-tape parade the world had ever seen. Taking a three-month "victory lap," Lindbergh flew his plane, the *Spirit of St. Louis*, to every state in the union; rode in 1,300 miles of motorcades; gave 147 speeches; and was seen, in the flesh, by 30 million people—one out of every four Americans then living. Still, none of this prepared Lindbergh for the way he was stared at on November 28, 1930, the day the world's most famous man was introduced to the person some considered the world's smartest.

Ironically, that day began with steps aimed at preventing Lindbergh from being stared at. Before entering his black Franklin sedan outside his rented home in Princeton, New Jersey, Lindbergh slipped a pair of lensless eyeglasses over his famously blue eyes and a fedora hat over his equally famous blond hair. Much to his pleasure, he'd found this simple disguise was usually enough to afford him some privacy in public. Privacy was always important to Lindbergh, but it was crucial this morning because he wanted to think quietly in his car as he made the two-hour drive into New York, without interruptions from starstruck toll takers or fellow motorists. What Lindbergh wanted to think about was the list of questions he planned to ask the man he was driving to meet, a man whose name he'd only recently heard for the first time.

He heard it from Dr. Paluel Flagg, the anesthetist who attended Lindbergh's wife, the former Anne Morrow, when she gave birth to the Lindberghs' first child, Charles Jr., on June 22, 1930. Lindbergh had met his wife in December 1927 when he flew to Mexico City, an event that caused nearly as much pandemonium as his landing in Paris. Anne, then an introverted twenty-one-year-old college student, was worried at first that her Christmas holiday with her family at the U.S. embassy—her father, Dwight, was the American ambassador—would be spoiled by the presence of someone she called "a sort of baseball player."

But that anxiety vanished when the tall, handsome aviator took the ambassador's middle daughter on her first airplane flight, a thrill that Anne, a short brunette unsure of her own attractiveness, described in her diary in near-orgasmic terms. Lindbergh "moved so very little" in the cockpit, "yet you felt the harmony of it," she wrote. "It was a complete and intense experience." The pilot and his sated passenger were married on May 27, 1929, in a secret ceremony at Next Day Hill, the Morrow family mansion set on a verdant fifty-acre estate overlooking the Hudson River in Englewood, New Jersey. Most of the two dozen guests thought they'd been invited to play bridge.

When Anne went into labor in the same mansion the next summer, her husband, waiting in the next room, struck up a conversation with Dr. Flagg. The topic was Anne's older sister Elisabeth, whose health had deteriorated dramatically after a bout of rheumatic fever damaged her heart's mitral valve, the valve that regulates the flow of blood from the left atrium into the left ventricle, the heart's main pumping chamber.

Lindbergh, who knew a fair bit about valves in machines, was puzzled that a mere valve in the heart—the body's engine, as he saw it—could cause so much trouble in an otherwise vibrant woman in her mid-twenties. He was similarly vexed to learn that not one of the doctors consulted by Elisabeth's family, several of whom Lindbergh had questioned personally, had any ideas on how to proceed. Lindbergh had several: remove and replace the broken valve, as he would do in an airplane engine; replace the entire heart with a mechanical pump—an "artificial heart," he called it—just as Lindbergh would replace a failed airplane motor; or insert a temporary blood pump, remove the heart, fix it, then put it back.

Lindbergh spoke to Flagg about Elisabeth Morrow's situation because he noticed that the anesthetist had brought with him a machine he invented to give artificial respiration to newborns, in case a breathing emergency arose with the Lindberghs' baby. Fascinated by all things mechanical, Lindbergh asked for permission to examine the device, which was made of an oxygen tank, a pressure regulator, and several feet of rubber tubing.

"Would you show me how it works?" he asked.

"Of course," Flagg said.

After the demonstration, which filled the room with the sound of rushing gas, Lindbergh thought he'd finally found a doctor who would take his bioengineering ideas seriously. He was right about that, but Flagg didn't think his own specialized training in anesthesiology gave him the expertise to address the complex surgical issues raised by Lindbergh's ideas.

"But I know someone who could," he said: a Nobel Prize–winning surgeon Flagg once served as an intern in New York. The surgeon's name, he said, was Alexis Carrel.

When Dr. Carrel won the Nobel Prize for medicine in 1912 he was the first scientist in the United States to win a Nobel and, at thirty-nine, the youngest person yet chosen for any Nobel Prize. The prize honored his perfection of vascular anastomosis, the technique that enables a surgeon to reconnect a blood vessel after it has been cut, patch it if it has been punctured, or attach one vessel to another, without damaging the vessel being repaired or rerouted. This new ability to cut and sew arteries and veins—and keep them functioning—made Carrel the father of organ transplantation, which is unthinkable without vascular anastomosis. Likewise, open heart surgery, coronary artery bypass grafts, kidney dialysis, and countless other procedures that have saved millions would be impossible without Carrel's pioneering work. In the long history of cutting open the body to heal it, Carrel's achievement is perhaps second in importance only to the discovery of anesthesia.

Lindbergh, who did most of his reading in aviation journals, knew little of Carrel's achievements when he parked his car on November 28, 1930, near York Avenue and East Sixty-Sixth Street, a part of Manhattan where cows and goats had grazed on a dairy farm only thirty years earlier. Now this green campus was home to the nation's premier biological study center—the Rockefeller Institute for Medical Research (funded by John D. Rockefeller of Standard Oil)—and its most celebrated department head, fifty-seven-year-old Alexis Carrel, whose laboratories and surgery rooms occupied the entire top floor and attic of a five-story, brick and stone building constructed in the plain style favored by Mr. Rockefeller, a churchgoing Baptist.

Carrel, a bald barrel-chested man who viewed the world through pince-nez, was sitting at his desk when Lindbergh arrived for their meeting. Carrel's office, lined with glass-fronted cabinets filled with medical books and antique surgical instruments, was huge. His desk was situated at the far end, so any visitor had to walk some distance to meet him. Lindbergh, who'd been honored in government buildings all over the world, was used to large rooms, so he simply walked across the polished stone floor, indifferent to the echo of his footsteps. When he finally reached Carrel, Lindbergh announced himself and extended his right hand.

But instead of replying, Carrel raised himself up on his tiptoes—he was nearly a foot shorter than his six-foot-three-inch guest—and placed his nose almost flush against Lindbergh's. Once there, the surgeon began to examine the aviator's well-documented face, moving his own head back and forth, slowly and repeatedly, as if he were reading a contract. Lindbergh, his hand still in Carrel's, had no idea what was happening; nor did Carrel offer an explanation. If he had, Carrel would have said he was a believer in physiognomy, the now discredited "science" which holds that the qualities of a person's soul are imprinted on the individual's face by brain waves controlling facial muscles. Carrel was so convinced of this notion that before he would engage in a serious discussion with someone he was meeting for the first time he often insisted on scanning that person's face. Closely. If Carrel didn't like what he saw, he walked away— "discussion" over. Carrel's secretary, who escorted Lindbergh to Carrel's door, had seen it happen more than once.

Lindbergh's suspicion that he'd somehow entered Lewis Carroll's "Wonderland" became even more intense when he looked back at his examiner. Carrel's probing eyes, magnified by his pince-nez, did not match: one was brown, the other blue. The pilot had been stared at by millions, but never like this.

And so the silent scrutiny continued, for seconds that seemed an eternity, until Carrel decided he liked what he saw. "I am Alexis Carrel," he said in his French-inflected English. (Carrel had been born and raised in France.) "Would you do me the honor of signing my guest book?" That book, already autographed by Winston Churchill, Sarah Bernhardt, and hundreds of others, sat atop Carrel's mahogany desk. After Lindbergh signed, Carrel said he'd be happy to answer whatever questions his visitor had for him in the institute's lunchroom downstairs.

It was there that Lindbergh proposed his radical therapies for his sister-in-law to Carrel, ignoring the institute's junior staff members who gawked at the pilot from the lunchroom entrance. Lindbergh was not a garrulous man; he proved that in the nation's capital when, after receiving the Distinguished Flying Cross from President Coolidge, he expressed his gratitude in a speech lasting just over a hundred words.

But on this occasion Lindbergh was loquacious. He pushed aside his barely eaten sandwich and nearly full cup of tea and started talking. If there was one subject Lindbergh felt comfortable discussing besides aviation, it was engines. He'd been around them—in planes, in automobiles, and in the tractors and cow-milking machines on his family's Minnesota farm—for most of his life. He believed in the power of simple engineering to solve complex problems. Hadn't he bet his life on that assertion, and won, in the *Spirit of St. Louis*? Surely, Lindbergh said to Carrel, there's a mechanical solution for Elisabeth Morrow's heart problem. After all, isn't the human heart, for all its undeniable complexity, just a machine—a pump—that can be fixed or replaced like any other?

Unlike Lindbergh, Carrel was a garrulous man, someone far more likely to expound than listen, but on this occasion he listened attentively. He was impressed by the earnest, somewhat gangly person sitting across from him, who, to illustrate the treatment he was proposing with such conviction, took a notebook from his coat pocket and made a simple drawing of the artificial circulation system he envisioned. Carrel was struck by Lindbergh's commitment to his own ideas—self-confidence was near the top of the attributes Carrel believed necessary for greatness—but he was even more impressed with Lindbergh's hands-on approach.

Carrel was a hands-on guy, too. His specialty, cleaving and sewing blood vessels, some only a fraction of a millimeter wide, demanded exceptional dexterity. As a medical student in France, Carrel astounded his professors (and annoyed his fellow students, who thought him a show-off) by sewing 500 stitches in a single piece of cigarette paper. He'd learned to sew from his widowed mother and later took private lessons from Lyon's most famous embroiderer.

Moreover, Carrel was an *experimental* surgeon. His mandate at the Rockefeller Institute was to devise new surgical techniques for humans by testing them on laboratory animals. He'd performed the world's first coronary artery bypass graft there—on a dog—in 1910. If there was one doctor in the world ready to test a wholly new surgical approach to heart valve disease, it was Alexis Carrel.

But not the one suggested by Lindbergh. After praising the pilot for the boldness of his ideas, Carrel explained why his treatments were untenable, at least for now. For starters, he said, surgeons were as yet unable to stop the heart long enough to operate on it—or, even more important, *in* it—and keep the patient alive. And, even if they could, the hammering of pistons inside any mechanical replacement pump would surely damage the delicate red corpuscles of the blood coursing through it. Equally worrying, Carrel said, were the issues of clotting and infection raised by placing such a machine inside the body.

Lindbergh's disappointment was soothed by the respect with which Carrel responded to his ideas. He didn't feel patronized; he felt honored that Carrel took the time to explain the shortcomings of his theories. To be treated so courteously—and seriously—by a Nobel Prize winner, a man whose mind, Lindbergh later wrote, "flashed with the speed of light," was a heady experience for someone who'd flunked out of college and barely made it through high school.

What Lindbergh didn't know was that Carrel was similarly intoxicated to be talking to him. The pilot was a hero; and in Carrel's interpretation of Darwin this meant that Lindbergh was one of the elite selected by nature to play the most important role of all in evolution: promoting—by reproduction and example—the optimum growth of the fit. As Carrel saw it, Lindbergh's famous flight showed as much mental courage as physical: the pilot risked his life to challenge the experts who dismissed that flight as suicidal; and now the same young man was trying to smash similar barriers in medicine, even after virtually every doctor he'd talked to had brushed him off. This resolve in the face of skepticism greatly impressed Carrel. The same trait fueled his own experiments, many of which, just like Lindbergh's bid to fly to Paris, were ridiculed before the fact by his surgical peers as "impossible."

Carrel was so taken with Lindbergh that the surgeon heard himself making an offer he hadn't anticipated. So what if the artificial heart wasn't yet feasible? There were other ways Lindbergh might help his sister-in-law. "If you like," Carrel said to his guest, as they rose from their seats in the Rockefeller Institute lunchroom, "I will show you what I'm doing here."

SO THEY RETURNED TO the fifth floor, where Carrel led Lindbergh behind a white screen in the corner of his office. Carrel pointed to a sink located there and told Lindbergh to wash with disinfectant soap. While the pilot did so, Carrel reached into a nearby cabinet, pulling out a surgical mask and a long surgical gown, which he handed to his guest once Lindbergh had dried his hands. Both items, Lindbergh noticed, were black. "Please follow me," said Carrel, after donning his own black robe.

They walked out of the office, down a wide hallway, and into a rectangular room filled with small cages stacked two-high on several rows of long tables. "These," Carrel said, gesturing at the cats and dogs inside the cages, "are my patients." Several of the animals had open wounds on their backs, which had been shaved to make the wounds more visible. Carrel had excised strips of skin from those animals, then applied different substances to the wounds to see which speeded and which retarded the healing process. This study built on work that Carrel, aided by the chemist Henry J. Dakin, performed at a battlefront hospital during World War I. While serving as a surgeon in the French army, Carrel identified the first antiseptic—sodium hypochlorite, buffered with sodium bicarbonate—powerful enough to kill germs in soldiers' wounds, but gentle enough not to destroy healthy tissue along with them. As infection was by far the leading cause of battlefield mortality, this was a massively significant breakthrough, one that saved hundreds of thousands, if not millions, of lives.

In other experiments performed on animals in this room, Carrel sliced open the hearts of anesthetized beasts to stop the flow of blood, then sutured the hearts closed and resuscitated the animals so he could study the effects of the arrested circulation. "After five minutes, most animals are resuscitated easily," he told Lindbergh. "After ten minutes, some appear to have 'lost their soul.' Once, after a longer interruption, I finally succeeded in resuscitating an animal that could breathe and swallow food, but was completely paralyzed and blind."

The next room the men entered contained a small refrigerator, which Carrel opened. Inside were translucent containers, each holding a small object Lindbergh couldn't identify. Some of those objects, he could see, were surrounded by a clear, thin liquid, others with a more viscous substance. Carrel reached into the refrigerator, removed one container, opened it, and examined its contents. "The thyroid of a cat," he said.

This thyroid was in a state Carrel called "latent life." After perfecting his anastomosis technique, which enabled him to graft parts from one animal into another, Carrel began to wonder if the donor animal even needed to be alive for such donations to take place. In 1907 Carrel found that a severed blood vessel stored just above the freezing point, while immersed in saline solution or coated in petroleum jelly, could be successfully transplanted after as long as two months of refrigeration.

He then began to investigate the viability of grafts done with preserved human parts. In one test Carrel removed flaps of skin from the washed cadaver of an infant who had died the same morning at a nearby New York hospital. The skin was refrigerated, then used weeks later to patch four ulcers on the legs of two adults. "Most of the grafts took," Carrel reported to the *Journal of the American Medical Association* in 1912. Carrel also experimented with refrigerating animal semen. These studies anticipated the modern tissue bank and sperm bank by half a century.

Across the room from the refrigerator was a large glass-walled structure with its own door. "My 'mousery,'" Carrel said, inviting Lindbergh in. Inside were thousands of rodents in bins within interconnected cages. The sound of mice scratching through soil, gnawing on cage bars, and vying for alpha-male status was impossible to ignore. So was the smell.

For several years now Carrel had been doing a multigenerational breeding study. "Some mice are fed enriched diets, while others are given less," he said. Some mice, in fact, were fed carcinogens or alcohol; others were exposed to infectious diseases. Some were caged alone, while others were allowed to roam and fight—often to the death. The winners were given females to impregnate; the losers were given autopsies.

The purpose of the study was to see how physical and sexual competition, the resulting breeding practices, and the variations in diet and environment contributed to the creation of "heroic" mice—i.e., mice resistant to disease and endowed with greater strength and longevity. "If I could do the same tests on humans," Carrel said, "I might produce a man who could jump twenty feet in the air and live to be two hundred."

Carrel's research made him a target of animal-rights activists who urged the New York state legislature to pass laws banning his work as excessively cruel. But Carrel didn't experiment on animals because he was a sadist; he did so because he was a disciple of Dr. Claude Bernard, the great nineteenth-century French physiologist who created the field of experimental medicine. Whereas anatomists of Bernard's era studied the body's interior by dissecting cadavers, which made them experts on the internal organs' form, physiologists like Bernard investigated how those organs functioned in living bodies. The necessity of finding that latter truth, Bernard conceded, was matched by the unpleasantness encountered in searching for it. A physiologist "no longer hears the cry of animals," Bernard wrote in 1865, "he no longer sees their blood, he sees only his idea and perceives only organisms concealing problems that he intends to solve. . . . To learn how animals live, we cannot avoid seeing them die."

Besides, Carrel had used his animal research to make humans live. In a case that helped defeat an antivivisection bill then pending in the state legislature, Carrel saved the life of a five-day-old infant by using the vascular anastomosis technique he'd developed on dogs. This happened on March 8, 1908, when Carrel received an unexpected visit from Dr. Adrian V. S. Lambert, a professor at Columbia University's medical school. Lambert's wife had given birth to a daughter who exhibited symptoms of melena neonatorum, a rare disease marked by bleeding from the nose, mouth, and anus. When none of the known treatments proved effective, Lambert frantically searched the medical literature for alternatives until he found Carrel's paper on anastomosis. Perhaps a direct blood transfusion using Carrel's new technique would save his daughter's life. Lambert was desperate: he knew that no one in New York—not even Carrel—had ever attempted such a transfusion on a human patient.

Even so, Carrel agreed to do the procedure, rushing with Lambert to his apartment on West Thirty-Sixth Street, where Carrel found the baby's anxious mother, the father's two brothers, and the sick infant, unconscious and white as a sheet. After a brief discussion with his wife, Dr. Lambert said he would be his daughter's blood donor. (This turned out to be a lucky choice, as very little was then understood about blood groups.)

The baby was taped to an ironing board next to the Lamberts' dining room table, where Lambert lay down. Carrel would anastomose an artery in Lambert's wrist to a vein behind the child's knee, thus joining the healthy father to his sick daughter, so that his blood would replace that being lost by her. After Carrel began cutting, one of Lambert's brothers blotted up the blood with a dish towel. Once under the baby's skin, Carrel—working without the magnifying lenses today's surgeons take for granted—occluded the appropriate vessels, severed them, then used his revolutionary anastomosis method to join them end to end.

Carrel's Nobel-winning innovation involved placing stay sutures in the two vessels he planned to join; once gently tugged, those stays transformed two round vessels into two triangles, which Carrel sewed together, one straight line at a time. As every tailor knows, it's easier to sew a straight line than a curved one. What Carrel the surgeon knew was that sewing in a straight line prevented the tissue being linked from buckling, and so prevented damage to the vessel's inner lining. This in turn eliminated the possibility of clotting and infection, the problems that had defeated all previous attempts to devise a workable anastomosis method.

Soon the vascular bond was complete on the Lambert infant—no small feat, considering that her vein was thinner than copper wire and flimsier than wet tissue paper. Within moments the baby's color changed from white to pink, then red. "You'd better turn it off or she'll burst!" one of Lambert's brothers said, his joy tinged with anxiety. Two decades later Carrel attended his patient's twenty-first birthday party.

THAT OUTCOME GAVE CARREL great satisfaction, but, as he told Lindbergh while they walked toward another room at the institute, his research in recent years had moved beyond vascular surgery. These new experiments took place in Carrel's incubation chamber. This space was heated to thirty-seven degrees Celsius (98.6 Fahrenheit) and kept moist to mimic the interior of a mammal's body by steam jets in the walls. The experimental subjects in this room weren't likely to raise the ire of animal-rights activists; they weren't even fully formed animals. They were microscopic pieces of tissue—some from laboratory animals, others from humans—that Carrel kept alive in tissue culture flasks. More than a dozen such flasks, lined up on a workbench, were visible in the incubator room.

Carrel, a founding father of tissue culture, was the first scientist in the world to grow human tissue outside the body. He did that by adapting work done by the zoologist Ross G. Harrison. Harrison had made history himself in 1907 by keeping brain tissue from a frog embryo alive in a drop of frog lymph, sealed in a hollow glass slide, for several weeks. Carrel was impressed by Harrison's feat, but knew it wouldn't have any medical consequences until tissue from higher animals could be cultured as well.

This Carrel achieved in studies at the Rockefeller Institute that substituted mammalian and human tissues for amphibian, adult tissues for embryonic, blood plasma for lymph as the culture medium, and, in some tests, cancerous tissues for healthy ones. That last innovation, which enabled scientists to watch malignant cells as they grew, was the beginning of modern cancer research. In a similarly groundbreaking study, Carrel proved that cultured cells produce antibodies—just as cells inside the body do—after antigens are added to the culture medium. This was a hugely important step in the creation of the modern vaccine industry.

Careful maintenance was required to keep Carrel's tissue cultures nourished, infection-free, and rinsed of waste products. One of Carrel's technicians, dressed in a black robe like Lindbergh's but topped by a black hood covering his entire face, save for a slit revealing his eyes, was attending to those duties while peering through a microscope when Carrel and Lindbergh entered the room.

"Would you like to see what he's seeing?" Carrel asked.

"Yes, I would," Lindbergh said.

After the technician vacated his seat, Carrel took it, looked through the microscope, and then offered the same chair to Lindbergh, who lowered his eye to the lens. What the pilot saw was that the human tissue, which resembled a nearly transparent stone wall, was surrounded by a milky-bluish substance—Carrel's plasma medium—leaving a grainy substance at the shared border.

"Can you see the granulation?" Carrel asked.

When Lindbergh nodded, Carrel said the granulation was evidence that the human tissue was reaching out, taking in food from the medium, and proliferating. Carrel then invited Lindbergh to the corner of the incubation room where, on a table all by itself, lay Carrel's most famous tissue culture experiment—in fact, the most famous tissue culture experiment in the world.

This experiment began on January 17, 1912, when Carrel cracked open a nine-day-old fertilized chicken egg and excised the still-beating heart from the embryo inside. He then took a portion of this tissue, roughly the size of a match head, and placed it, surrounded by chick blood plasma, in a tissue culture flask. Carrel launched this study to silence the roar of skepticism that had emerged from his scientific peers a year earlier when he announced that he had been able to keep warm-blooded animal tissue alive in culture flasks indefinitely—with no signs of aging—by merely rinsing those tissues free of metabolites, the waste products left behind after the digestive system transforms food into energy. "My results," Carrel had declared in 1911 in the *Journal of the American Medical Association*, "demonstrate . . . that death is not a necessary, but merely a contingent, phenomenon."

It's no accident that Carrel chose heart tissue for his next experiment. What other tissue could answer his critics so dramatically? Only the heart illustrates the possibility of permanent life with its own distinctively visible liveliness: the ability to pulse. Carrel's goal was to keep his chick-heart cells alive and pulsing in his culture flask by feeding them and rinsing them free of metabolites until every doubter was overwhelmed by proof that his culture had grown and multiplied—and would continue to grow and multiply, forever, until stopped by outside means. The clincher to the argument, Carrel believed, would be to keep those cells flourishing beyond the typical life span of a flesh-and-blood chicken: approximately seven years.

And this he'd done. There, in the corner of the room where Lindbergh now stood, in a culture flask situated beneath a microscope, was the original "old strain," as Carrel called his chick-heart cells—still alive, nearly eighteen years after their "birth" in this same room. Carrel had announced his experiment in the *Journal of Experimental Medicine* in a report titled "On the Permanent Life of Tissues Outside the Organism."

The mainstream press was thrilled to spread the astonishing news: CARREL'S NEW MIRACLE POINTS WAY TO AVERT OLD AGE, trumpeted a headline in the *New York Times*. FLESH THAT IS IMMORTAL raved *The World's Work*. Many newspapers marked the birthday of the old strain each January, often with comical exaggerations. The *New York World* wrote in 1921 that had not Carrel's technicians been trimming back the culture's growth the cells would by now have formed "a rooster . . . so monstrous that if perched on this mundane sphere, the world, would look like a weather-cock."

Lindbergh had missed those press accounts and knew nothing of Carrel's chick heart. What he did know was that he'd never seen an eighteen-year-old chicken on his family farm in Little Falls, Minnesota, where he'd been the family's chief poultry farmer. When Carrel offered him the chance to look at the pulsing chick-heart cells under a microscope Lindbergh jumped at it. As a history-making pilot he'd already seen things few others ever had. But this left him awestruck.

Moments later Carrel invited Lindbergh to tour his operating rooms, located in the attic above his laboratory. As they headed up the staircase, Carrel explained why his chick-heart cells did not signify that immortality was easily attainable for humans—at least, not yet. He said it was easy to rinse cultured chick-heart cells free of metabolites on a nearly continuous basis, but it would be impossible to do the same procedure in a live human. So Carrel had decided to challenge mortality in a different way. There was a real possibility, he said, that his new experiments would have beneficial implications for Lindbergh's sister-in-law.

THE INITIAL STAGES OF these studies were performed in Carrel's operating suite, which the two men now entered. Lindbergh had never been in an operating room before, and this one defied his expectations. The floor, walls, and ceiling were painted black. The only source of illumination was a large skylight situated directly above the operating table, which was black as well, as were all the storage cases and cabinets in the room.

"Too much light inhibits the activity of the brain," Carrel said, anticipating Lindbergh's question. "Surely you've noticed that the world's great civilizations have formed far above the equator, where there is much less direct sunlight than in tropical regions."

Carrel told Lindbergh that black walls cut down on glare—no small worry when one is operating on tiny blood vessels. He also said that black surgical gowns were better than traditional white ones at illuminating dust, the elimination of which was an obsession for Carrel, who insisted on the highest standards of sterility and cleanliness in his operating rooms.

The experiments Carrel had recently started in this dark space grew out of the anastomosis technique that had made his reputation. Once he achieved that feat, Carrel immediately recognized its potential: if a surgeon could sever and reattach blood vessels, then a surgeon could sever, remove, and reattach just about anything. To demonstrate these possibilities, Carrel removed the kidney of a dog and grafted it to a new position in the same dog's neck. He wasn't trying to create a monster; Carrel chose the neck because it's packed with large blood vessels, thus making the anastomosis that much simpler. Once grafted there, the kidney continued to produce urine.

Carrel then took the next logical step: he transplanted kidneys from one set of dogs into another. (This was more than fifty years before a kidney transplant was achieved in a human.) At first Carrel's experiments seemed successful. The recipients recovered with normal kidney functioning and appeared to be in good health. But only for a while. Death ensued within days or weeks—caused, as we now know, by the body's immune system, which mobilized to reject the new organ. When doing autopsies on the animals Carrel saw that his transplant sutures had healed without infection. The problem, he realized, lay elsewhere. Virtually nothing was known about the immune response a century ago; Carrel guessed that one animal's blood serum was toxic to another's. He left it to other scientists to solve that problem. He knew that once they did, surgeons would use the technique he'd perfected to make the organ grafts.

But nearly three decades had passed without a solution to the rejection problem, so Carrel set out in a new direction: instead of trying to remove a diseased organ and replace it with an organ from a healthy donor, he would try to remove the diseased organ, keep it alive outside the body in a sterile machine, fix it there, then put it back in its original place, thus rendering the rejection issue irrelevant. In trying this, Carrel was building on the ideas of another forward-looking Frenchman, the physiologist Julien-Jean-César Le Gallois, who predicted in 1812 that "if one could substitute for the heart an injection of arterial blood, either natural or artificial, one would succeed easily in maintaining alive indefinitely any part of the body." The medical term for such an injection is "perfusion."

Carrel believed Le Gallois was right: an entire organ—even a human head—could be kept alive indefinitely outside the body. The challenge was to find the proper perfusion method. He knew this wouldn't be easy. A particle in the perfusion fluid as small as 1/400 inch was large enough to clog a capillary. A perfusionist would also have to control temperature, pH balance, and oxygen content; regulate the perfusion fluid's pulse and pressure; and, above all, prevent infection.

Carrel told Lindbergh that he'd already done perfusion experiments on the thyroids of cats, dogs, and chickens. The surgeon's long-range plan was this: Once he'd successfully perfused organs from those animals, he'd move to the next level—experiments on organs from primates—and then to the ultimate test: malfunctioning human organs would be kept alive by perfusion. Carrel said a temporary artificial heart pump, inside the body or out, was likely to be perfected in the future; once that happened, a diseased heart could conceivably be removed, perfused, repaired, and then reinserted.

This idea, Lindbergh realized, was nearly identical to one of the solutions he'd conceived of for his sister-in-law. It was even possible, Carrel said, that replacement parts for humans would be grown in vitro in the future, perfused from microscopic cell clusters into fully formed organs. (In saying this, Carrel anticipated the promise of stem cell research by more than half a century.) The potential, Carrel said, was limitless: the human body, revealed to be a living machine made of constantly reparable or replaceable parts, could be perpetually renewed. This raised a possibility almost too daring to be said out loud: the most intimidating of all human inevitabilities—death—wasn't inevitable.

Carrel explained these concepts to his famous guest on November 28, 1930, with an enthusiasm Lindbergh found thrilling. The pilot listened eagerly, flattered to be taken into the great scientist's confidence. He also felt validated in a way Carrel had no way of knowing.

CHAPTER 2

IF MAN COULD LEARN TO FLY

T WO YEARS EARLIER, IN THE UTAH DESERT, LINDBERGH HAD an epiphany. He hadn't traveled there on a spiritual or soul-cleansing mission. He'd been hired by Transcontinental Air Transport, an airline later to become TWA, to survey air routes for coast-to-coast passenger service. One cloudless fall afternoon, Lindbergh decided to land in the arid wilderness near the state's southern border. There was no engine trouble. He merely wanted to avoid the reporters he knew would find him if he checked into a hotel in the closest town. Several hours later, while lying on the sandy ground under a thin blanket and a moonlit sky, Lindbergh had a piercing moment of clarity. He looked at the unspoiled beauty around him, and the elegant flying machine that brought him there, a Ford Trimotor 4-AT-10, and realized he'd lost control of his life.

Lindbergh reminded himself that he'd become a pilot because of his love of science. (Lindbergh always referred to his transatlantic flight—indeed, all his record-breaking flights—as "scientific experiments.") But now the "hero business" had pushed him off course. Instead of pursuing science, Lindbergh was hiding from reporters and autograph hounds pursuing *him*.

In truth, Lindbergh never understood why he'd become the world's hero in the first place. He grasped the meaning of his Paris flight to commercial aviation—he'd demonstrated the ability of air-cooled engines to endure prolonged and stressful use—but he was bewildered by the impact of that achievement on everyone else.

"I wonder if I really deserve all this," he said on June 10, 1927, to Admiral G. H. Burrage, while standing on the deck of *Memphis*, the huge cruiser President Coolidge had dispatched to Europe to bring America's newest favorite son back home. He was referring to the four U.S. Navy destroyers, two U.S. Army blimps, and forty U.S. Army Air Corps fighter planes that were escorting *Memphis*—and Lindbergh—through Chesapeake Bay en route to Washington, where President and Mrs. Coolidge, leading members of Congress, and 300,000 other Americans waited to honor him in person, as millions more listened on radio.

But if Lindbergh didn't understand what the fuss was about, others did. "A young Minnesotan did a heroic thing and for a moment people set down their glasses in country clubs and speakeasies and thought of their old best dreams," wrote another young Minnesotan, F. Scott Fitzgerald. The *Washington Post* went even further: "Women see in Lindbergh the perfection of man, what they had conceived their husbands to be and what still constitutes their dream. . . . Men think wistfully of things they might have done, things they would like to do, things they wish they had the nerve to do." In a nihilistic and often hedonistic era still reeling from the carnage of World War I, Lindbergh's flight—achieved by a handsome, virginal teetotaler who never bragged about his accomplishment—seemed a glorious reminder of human potential and the power of moral rectitude.

Lindbergh's feat was all the more inspiring because he did it by himself. (Everyone else competing for the $25,000 Ortieg Prize, offered by a French businessman to the first pilot or pilots to fly nonstop in a heavier-than-air machine from New York to Paris, or vice versa, worked in crews of two or more.) Lindbergh's flight was "the greatest feat of a solitary man in the records of the human race," said the *New York World*. A plaque presented to Lindbergh by the National Geographic Society declared: "Courage, when it goes alone, has ever caught men's imaginations."

This is especially true of courage in the face of death. Six men died trying, or training, to cross the Atlantic before Lindbergh. Two of them, Charles Nungesser and François Coli, who flew west from Paris in their single-engine biplane, *L'Oiseau Blanc*, twelve days before Lindbergh, were still missing when he took off from Long Island in *Spirit of St. Louis*. They were never found. The public revered the lost Frenchmen as martyrs to the new religion of flight, as they did the others who perished in pursuit of the Ortieg Prize. It was a list that included the Americans Noel Davis and Stanton Wooster, who lost power on their final test flight in their trimotor plane, *American Legion,* on April 26, 1927, and dived nose-first into a Virginia swamp.

The public that listened to eulogies for those brave men worshipped Lindbergh for succeeding where the others failed. As genuine as this awe was, it was stoked into a frenzy by mass media becoming "mass" as never before. This point was made, as the process was happening, by the *New York World*, which compared Lindbergh's ticker-tape parade with one given twenty-nine years earlier for the first person to be so honored: Commodore George Dewey, the hero of the Spanish-American War.

"Dewey's world—1898—was a provincial place compared with Lindbergh's," the *World* wrote in June 1927:

> It had none of our modern machinery for following in great detail the exploits of a hero. . . . The movie was a toy; there were no newsreels. The wireless telegraph had not been perfected by Marconi; there were no pictures sent by radio, no 3,000-word cablegrams from Europe. In Lindbergh's case, more than in any other in our times, the full blare of the modern publicity machine, with all its blinding light, has been suddenly thrown on a single individual.

Despite Dr. Freud's efforts to foster introspection as a life strategy, most people in the 1920s and 1930s were not particularly eager to examine themselves and chose instead to look up to others who accomplished something physical and, better yet, something dangerous. As the *New Yorker*'s writer John Lardner, who lived through that era, later wrote, "The farther the hero went—whether upward, downward, sideways, through air, land, or water, or hand over hand on a flagpole—the better." And who'd gone farther than Lindbergh?

But Lindbergh wondered in the Utah desert if he hadn't gone too far. Sure, his celebrity had opened doors and made him a wealthy man: along with his contract with the future TWA, Lindbergh had consulting deals with Pan American Airways, the Pennsylvania Railroad, and the Daniel Guggenheim Fund to Promote Aeronautics. He'd also earned $250,000 in royalties—roughly equivalent to $3.5 million today—for *We*, his first book-length account of his transoceanic flight.

But Lindbergh wasn't that interested in money, and he was definitely tired of fame, especially if it meant being asked foolish questions, over and over again, by the press. (When Lindbergh arrived on Long Island in May 1927 to prepare for his flight to Paris, after setting a new cross-country speed record, he was not amused when reporters asked him: "Do you prefer blonds or brunettes?" "What's your favorite pie?" and "Do you carry a rabbit's foot?") In Utah Lindbergh reminded himself that life was too short to deal with anything but the big questions, which to him meant biological questions. As a child he'd dreamed of becoming a physician, but his grades weren't very good, and when someone told him, incorrectly, that he'd have to master Latin to qualify for medical school, he abandoned the idea as a vocational goal.

But not the idea behind it. As an adult Lindbergh was still curious about the mysterious inner workings of the body. After his ocean-conquering flight that interest grew dramatically. Having done something no one before him had ever done, something many people had thought could *never* be done, gave Lindbergh a nearly limitless belief in his own abilities. This was a self-confidence he'd never experienced as a shy, virtually friendless child, or as a money-losing farmer, or as a student who'd been asked to leave the University of Wisconsin after three semesters because of bad grades and an attitude his academic adviser described as "immature."

After all those failures, Lindbergh's early successes in the air—first as a gypsy flier selling fifteen-minute rides to country bumpkins for five bucks each, then as an army flying cadet, and then as one of the country's first mail pilots—had led him, even before his flight to Paris, to think of himself as a god. Lindbergh would write these words about his introduction to open-cockpit flying in *Autobiography of Values*:

> There was the earth spreading out below me, a planet where I had lived but from which I had astonishingly risen . . . Mine was a god's-eye view.

And these in *The Spirit of St. Louis*:

> I lived on a higher plane than the skeptics on the ground; one that was richer because of its very association with the element of danger they dreaded. . . . In flying, I tasted a wine of the gods of which they could know nothing.

Now, in 1928 in the southern Utah desert, Lindbergh decided to refocus his godlike powers on a subject he'd walked away from as a teenager. "The great and rapid success of aviation," the world's greatest aviator wrote in his journal, "rearrous[ed] my interest in biology." Actually, Lindbergh meant his interest in transforming biology.

That desire was born inside the cabin of the *Spirit of St. Louis*. While making his flight from New York to Paris, Lindbergh had an experience he later described as "penetrat[ing] beyond mortality." He felt his body disintegrate and then re-form—a "reincarnation," he called it. He saw "spirits" and "inhabitants of a universe closed to mortal men" moving about his plane, some of which spoke to him. Lindbergh was a different man after those otherworldly experiences. In Utah he changed yet again. In the desert Lindbergh came to believe that his next journey, even more than his flight over the Atlantic, would require him to stretch his wings beyond mortal realms. Lindbergh decided that his new mission was to reencounter those inhabitants of a world "closed to mortal men" that he'd met inside the *Spirit of St. Louis*—not merely to find them but to become one of them, and not as a ghostly spirit but as a flesh-and-blood human.

"If man could learn to fly," Lindbergh asked himself in the Utah desert, "why could he not learn how to live forever?"

THAT QUESTION HAD BEEN an obsession for Lindbergh ever since he was a boy spending summers on his family's 100-acre farm in Little Falls, Minnesota, a mill town settled mostly by Swedes (such as the Lindberghs) and other northern Europeans. On one unusually steamy afternoon there, when Lindbergh was about eight, he was searching for arrowheads near a dirt trail that cut through the pine forest behind the Lindberghs' home, a two-story wood structure, painted white, sitting on a bluff overlooking the

Mississippi River. Charles, shirtless and shoeless, wasn't alone on this trek; he was with his inseparable companion Dingo, a red-haired mutt who'd shown up begging for food at the Lindbergh place roughly a year earlier. Charles and Dingo were about a mile deep into the woods on this treasure hunt when the boy stumbled on something he never expected to find at all.

It was the swollen carcass of a horse. The massive body was mostly intact, but there were insects on its flesh and the decomposition process was under way, so this was one part of the pine forest that no longer smelled like pine. Though he was frightened and more than a little nauseated, Charles was mesmerized by this encounter. He shooed Dingo away and edged closer to the beast, pinching his nose shut with his thumb and forefinger. He scanned the carcass for wounds, or an obvious sign of disease, but couldn't find any. Why then, he wondered, was the horse dead? Who or what had caused its life to end? Why did life have to end for *anyone*? The inexplicability of it all troubled young Lindbergh with an intensity he couldn't control.

"The difference between life and death was so apparent in that rotting hulk, and yet it was not understandable!" he later wrote. "What stopped life from living?"

Charles wasn't the first child in history to ask his parents that question, but he was definitely one of the most persistent. C. A. Lindbergh, a lawyer who'd recently been elected as a Republican to the U.S. House of Representatives, and his wife, Evangeline, who'd come to Little Falls from Detroit to teach high school science, didn't agree about much—in fact, they spent nearly all their marriage living apart—but they were puzzled in identical measure by their son's puzzlement. After all, the Lindberghs raised chickens on their farm and Charles knew what happened to *them*. Years later he wrote about it: "I [saw] chickens whose heads were chopped off, the spurting blood, the body's violent kicking and flapping."

Besides, Charles was given a twenty-two-caliber rifle by his father when he was six and was an experienced hunter by his next birthday. "I crept up on some ducks near a lake shore," he wrote of a hunting trip he took when he was seven. "I aimed at one of them . . . and hit it in the head." Obviously, Lindbergh knew at least two ways to stop life from living.

But it was natural death that young Lindbergh couldn't understand. Like most children, he'd been taught by his mother that after life ended you met God, who took care of the dead in heaven, if they had worshipped him properly. "But I wondered," Lindbergh wrote in *Autobiography of Values*, repeating the questions he'd asked his parents as a boy,

> if God is so great, why did he make you die? . . . Was it not an inexplicable defect in his character? Even for the good people he would take to heaven, death seemed a horrible entrance. . . . Why should he not let you live forever?

Now, some twenty years later, Lindbergh was still troubled by the same questions. What would be the result, he wondered in the Utah desert, if the arteries of an old man and a youth were connected? Would the older man be rejuvenated? Suppose an old head were grafted onto a young body; could wisdom and knowledge be combined with eternal youth? Was aging inevitable? Or could man become immortal?

The pilot knew, of course, that questions like those could never be answered in an airplane. They could be addressed only in a well-equipped biology laboratory, which is where he now hoped to spend the rest of his life, even if it meant building his own. "With science at man's disposal," Lindbergh wrote of his desert epiphany, "nothing seems beyond his grasp." Lindbergh wouldn't stop flying; he could never give up the exhilaration he felt in the air, using his hard-won skills to experience things only a brave few had ever known. But from this point on, Charles Lindbergh decided that his major explorations would be in inner space. The man who solved the mystery of solo transatlantic flight would use science to try to solve the mystery of eternal life.

ONE CAN IMAGINE, THEN, how exciting it was on November 28, 1930, for Lindbergh to hear one of the most heralded scientists in the world raise identical questions about life, death, and ending the latter's dominion over the former: a Nobel Prize–winning scientist already researching those very questions with a method—organ perfusion—strikingly similar to an idea that had occurred to Lindbergh himself.

But Carrel's initial round of perfusion experiments at the Rockefeller Institute, studies in which he tried to keep whole organs alive outside the bodies that created them (not in a state of suspended animation while packed in ice, but fully functioning, with artificial blood coursing through them), had failed. These organs, after being removed from anesthetized animals, were placed inside a small glass machine designed by Heinz Rosenberger, a technician on Carrel's staff who'd recently emigrated from Germany. This machine perfused the organ within through motion from an interior metal piston that was oscillated from outside the glass by electromagnets. To Carrel's great frustration, the organs placed inside Rosenberger's perfusion machine became contaminated by bacteria in every experiment. "Infection," Carrel said on that autumn afternoon in New York, as he showed Lindbergh the device he'd been using, "always infection."

Lindbergh examined the machine after Carrel handed it to him in the operating room. "As impressed as I was by the perfection of Carrel's biological techniques," Lindbergh wrote in *Autobiography of Values*, "I was astounded by the crudeness of his device." The pilot told his frustrated host that he thought he could build a better one. He also grasped the implications of that claim. "If I could design a better perfusion pump," Lindbergh wrote, "I could keep those organs alive long after the body they supported had entered the state called 'death.'" Achieving that, the pilot understood, would raise even bigger questions: "Is death an inevitable portion of life's cycle or might physical immortality be achieved through scientific methods? . . . Suppose we could install artificial hearts and transplant [organs] at will. . . . How much closer we would [be] to solving life's most basic mystery"—the mystery of death.

So Lindbergh drove back to his home in Princeton and worked through the night on several designs for a perfusion pump. Two weeks later he was back in Carrel's New York office, spreading sketches for a new perfusion device on the surgeon's desk. Carrel was impressed. He promised to have a prototype made to Lindbergh's specifications and to use it in his future organ perfusion experiments. Even more to Lindbergh's liking, Carrel invited the aviator to join him in that research as a partner.

CHAPTER 3

A STUDENT WHO MAY AMOUNT TO SOMETHING

LINDBERGH WASN'T JUST SEEN ON THE FIFTH FLOOR OF THE Rockefeller Institute when he began working there in early 1931. He was heard. More often than not, he arrived from New Jersey carrying a metal toolbox which clanked with hammers, handsaws, T squares, screwdrivers, wrenches, wire strippers, nuts, bolts, and similar instruments. That Lindbergh would bring such tools, and that they would prove useful in America's leading medical research institution, tells us a little about Lindbergh but much more about the state of medical technology seventy-five years ago.

Carrel's laboratory, even with Mr. Rockefeller's generous support, lacked the resources today's medical researchers take for granted. It had no computers, no tissue-slicing machines, and certainly no lasers, scanning electron microscopes, or magnetic resonance imaging capabilities. Most of the machines at Carrel's disposal did little more than regulate room temperature and humidity, cool objects (such as body parts) or keep them warm, and take the most basic biological readings, using simple electric motors. The mortar and pestle, technology dating back to the Stone Age, was still used daily in Carrel's lab. This was acceptable to Carrel because, like most surgeons in the 1930s, he had little aptitude for, or interest in, mechanical engineering.

Lindbergh, who did, showed an uncanny facility with that subject at a young age. Probably the first adult to take serious notice was Martin Engstrom, an easygoing Swede who owned a hardware store in Little Falls in the same building that housed the local congressional office of Charles's father. Engstrom was often called on to service the products he sold in his store. Once, when summoned to the Lindbergh farm by Mrs. Lindbergh, Engstrom saw that Charles had designed a system for moving heavy blocks of ice from the family's icehouse to the kitchen, a distance of more than ten yards.

"What have you built there, son?" Engstrom asked.

Lindbergh, a shy child who rarely spoke to strangers—or anyone else—explained that he'd built a slide made of two-by-six-inch planks leading out of the icehouse. Using tongs attached to a rope, Lindbergh would tug the ice out of the shed on the slide, then push the block into his wagon, which he'd pull to the kitchen using a wire-and-pulley system rooted in a ring screw he embedded in the house wall above the kitchen porch. Engstrom could only marvel that the inventor of this efficient laborsaving system had yet to celebrate his tenth birthday.

Shortly after that birthday Lindbergh expanded his expertise to include the internal combustion engine. This happened when C.A. arrived at his farm one summer day driving a new Model T Ford. The automobile was the most beautiful machine his son had ever seen: shiny and black, with gleaming headlights, clincher-rim tires, and a noisy engine hiding under its vented hood, spitting vapors and intoxicating aromas from its tailpipe.

It wasn't long before "Maria," as Evangeline named the car, was put under Charles's personal supervision. "I took care of the car myself, cleaning spark plugs, adjusting coil points, filling grease cups and screwing them down after every long drive," Lindbergh wrote in *Autobiography of Values.* Many of those drives took place in 1914, when Lindbergh, then twelve, drove his father on campaign swings through central Minnesota. (Obviously, this was years before the state had an age requirement for drivers, or even driver's licenses.)

C.A. eventually traded "Maria" in for a Saxon Six, which Lindbergh took on his longest drive yet, from Little Falls to Los Angeles, chauffeuring his mother and uncle on a five-week, 2,000-mile trip, much of it on dirt roads. Lindbergh fixed a timer trigger that gave out in Iowa, a spring bolt that broke in Kansas, and a wheel shimmy that developed in the mountains of New Mexico.

That there were medical applications for such expertise was a lesson Lindbergh learned from his maternal grandfather, the dentist Charles Land. Lindbergh spent at least a month of every year at the Lands' home in Detroit, a place that for a science-minded boy like himself combined the best aspects of a world's fair and an amusement park. He'd peer at the prehistoric mammoth's tooth, nestled against an ancient human skull, in his granddad's display case. He'd ride up and down on Dr. Land's dental chairs, powered by hydraulic pedals. He'd pass hours in the shooting gallery Land built in his basement where, if you hit the bull's-eye—and Lindbergh often did—an iron bird popped up.

But over time, Lindbergh began to spend fewer hours in the shooting gallery and more in the room next to it. This was Land's laboratory, a room filled with microscopes, generators, ceramic ovens, Bunsen burners, anatomical charts, bottles of spirits and tinctures, and large cabinets filled with small hand instruments. It was here that Land did the work that made him one of the most innovative dental surgeons of his era. It was also the place where he became a person whose influence on Lindbergh's future would be almost incalculable.

That second feat was accomplished as a teacher. Land familiarized his grandson with nearly every tool in his lab. This gave Lindbergh a precocious mastery of chemical and electrical laws, but Land was actually teaching his protégé a larger lesson: "the key to all mystery is science," a key that if used properly, Land said, enables man to become "like a god."

What sounds grandiose to us today sounded just right to young Lindbergh. The "clear-cut language of science didn't hum in my ears like a church sermon or a political speech," he later wrote. People have argued about God and government for centuries, and still they don't agree. But "science," Lindbergh learned from his grandfather, "confronts opinion with facts." Giving Dr. Land's words even more authority was the fact that his most important scientific experiment had worked.

That happened in 1899. In a case published in the journal *Dental Cosmos*, Land recounted his treatment of a man who'd lost all his lower teeth and a large section of his lower lip and chin to cancer. Land, who'd already made dental history by inventing the porcelain crown, started by building his patient porcelain dentures. But this time he did something he'd never done before: he made a new lower lip and artificial skin to cover the gap at the bottom of his patient's face. The skin was made of gutta-percha, a rubberlike substance extracted from trees in southeast Asia, to which Land glued hair from his patient's beard. Land's report on this case, one of the pioneering efforts in modern reconstructive surgery, was illustrated with numerous photographs of what a medical historian later called Land's "ingenious maxillofacial prosthetic appliance."

That appliance had a huge impact not only on Land's life and career, but on Lindbergh's. Dinner conversations at the Land home had always been philosophical, with topics ranging from the future of dentistry to speculations on religion, evolution, and maybe the biggest question of all, "What is death?" Land's success at facial surgery made him wonder if perhaps the human body, as complex as it is, is really just a living machine: an aggregation of constantly reparable or replaceable parts. If scientific advances enabled doctors to repair or replace teeth, skin, and lips, Land wondered, why not internal organs as well? And once science made organ transplants feasible— and Lindbergh's grandfather had no doubt that it would—didn't that mean that human life could be extended indefinitely, maybe even forever?

NOW, JUST OVER A decade later, Lindbergh was using his engineering expertise to explore those questions in partnership with a Nobel laureate, the very man who made organ transplants possible, in a laboratory at the most important medical research center in the United States. But before Lindbergh could put his skills to full use in the quest he'd just joined with Carrel, he knew he'd have to observe virtually every procedure done in Carrel's laboratory. This was not a task for the impatient or the squeamish. Lindbergh spent dozens of hours, often after midnight, looking through a microscope in Carrel's incubator room, watching cells no longer in the bodies that created them—cells that his new mentor, with his famous chick-heart experiment, had "proved" were immortal.

Actually, Carrel had proved no such thing, but this wouldn't be known until 1961, when the biologists Leonard Hayflick and Paul Moorhead showed that normal human cells have a life span in culture limited to about fifty cell divisions. Robert Hay and Bernard Strehler later showed that chick cells have an even shorter life span in vitro: about twenty-five divisions. What both studies proved is that, contrary to Carrel, death is inevitable for normal cells, whether rinsed of metabolites or not.

So how did Carrel's "immortal" chick-heart cells live, without aging, for thirty-four years, until they were finally discarded, in 1946, by one of Carrel's former assistants, then working for a pharmaceutical company in New Jersey? The biologist Jan A. Witkowski, who has investigated this controversy, believes Carrel's chick-heart culture was regularly rejuvenated by live cells introduced to the culture as an ingredient Carrel began adding to his nutrient medium in late 1912. Called "embryo juice," this ingredient was composed of ground-up chick tissue and saline solution, which was cooled and filtered, then centrifuged to remove any lingering live cells. Witkowski's guess is that Carrel's centrifuge failed to remove all those live cells, and that survivors were inadvertently introduced into the heart culture at each feeding.

But in the winter of 1930–1931, Lindbergh's interest in cellular immortality thrived—untainted by any thought of the future discoveries of Hayflick, Moorhead, and the others. One day in Carrel's incubator room, Lindbergh examined his own semen, seeing thousands of "wigglers," as he called them, with "huge oval heads and thin lashing tails, . . . each one of them myself, my life stream capable of spreading my existence throughout the human race, of reincarnating me in all eternity." An hour later he looked through the microscope again. The dried semen he saw on the slide was "as desolate and lifeless as a plain upon the moon." It was a reminder—not that Lindbergh needed one—of why he'd come to Carrel's laboratory in the first place.

Lindbergh, who had his own key, came and went as he pleased in Carrel's laboratory, usually arriving late at night or early in the morning. Carrel, like most surgeons, was an early riser. Weather permitting, he began each day with a two-mile walk at dawn around the Central Park reservoir, followed by breakfast in his spartan flat on East Eighty-Ninth Street (where he slept in a bed like a monk's cot in a room without even one picture on the wall), followed by a brisk walk to the Rockefeller Institute. Carrel was married to the widow of a French marquis, a nurse named Anne de la Motte de la Mairie. Madame Carrel came to live in New York after World War I, but soon decided that the hectic pace of Manhattan did not suit her. She eventually chose to remain in Paris while her husband fulfilled his duties at the institute. They lived together only in the summer months, when Carrel joined her at their vacation home on the tiny island of Saint-Gildas, off the coast of Brittany.

When Lindbergh came early to the institute, Carrel often invited him into his office for a chat. The medical significance of organ perfusion, not surprisingly, was a regular topic. What did surprise Lindbergh was the frequency with which Carrel wanted to discuss politics.

That subject had become of great interest to Carrel after he'd joined a dining group hosted twice a month by Frederic Coudert, a prominent New York attorney of French descent. These gatherings, held at Coudert's lavish Upper East Side apartment or at the equally posh Century Association on West Forty-Fourth Street, were called the "Philosophers Club" by the members, a reference to the weighty subjects they enjoyed discussing while drinking fine Bordeaux wines and eating wild duck shot on Coudert's Long Island estate: Is democracy a workable form of government? What is the true relationship between faith and reason? How significant is the threat to western civilization posed by "the masses?" Is there really a white man's burden? and so forth. Other "philosophers" in the group included Boris Bakhmeteff, the engineer who was the last Russian ambassador to the United States before the Bolshevik revolution; Benjamin J. Cardozo, soon to be appointed by President Hoover to sit on the United States Supreme Court; and Frederick Woodbridge, a dean at Columbia University and one of America's leading experts on Aristotle.

Carrel would soon invite Lindbergh to the Philosophers Club, invitations the pilot accepted. In the meantime Carrel turned their early-morning talks at the institute into a more intimate version of those gatherings, one in which Carrel did all the philosophizing. Lindbergh, who valued straight talk above almost all other virtues, was awed by Carrel's willingness to speak his mind.

Americans have a "herd mentality," the Frenchman told the American. The white race is drowning in a sea of "inferiors." Democracy is an "error of the brain." Socialist governments "interfere with natural selection" by implementing programs that coddle the unfit. (Carrel often illustrated this last point by taking Lindbergh to his mousery, where strong rodents were given free rein to dominate, and even kill, their weaker brethren, a process that, according to Carrel, benefited the species as a whole.) The susceptibility of artists to left-wing movements filled Carrel with scorn. Not even Carrel's peers were safe from his criticism. Most surgeons, the prize-winning surgeon told Lindbergh, "are butchers."

And fools—especially in France, where, three decades earlier, Carrel had been told he had no future in surgery, even after he'd devised his history-changing anastomosis technique. What happened was this: In May 1902, when Carrel was a surgical resident at the University of Lyon medical college, he served as a physician on a train carrying sick religious pilgrims from Lyon to the shrine at Lourdes, where a holy spring was said to be responsible for spontaneous cures.

While still on the pilgrim train Carrel met a passenger named Marie. It was hard to miss her: she was stretched across two seats on a damp mattress, gasping for breath. Her face had the pallor of a corpse; her lips were purple. When Carrel examined her, he saw that her abdomen was distended with fluid and solid masses, her heart was beating irregularly, her legs were badly swollen, and she had a high fever. To him it was a clear case of tuberculosis peritonitis. Marie was in such pain he gave her a shot of morphine. He did not expect her to survive the two-day rail journey.

But she did and, once at Lourdes, asked to be carried to the grotto, so that she could be sprinkled with the healing waters. Carrel was there when Marie got her wish, then left to tend to other pilgrims. When he returned, Carrel was shocked. Marie's stomach had flattened, her legs had returned to normal size, her fever was gone, and her pulse—erratic before—had steadied at eighty beats per minute. All Carrel said about this incident after he returned to Lyon was that he saw something in Lourdes he could not explain medically, and that it should be investigated scientifically. He did not call it a miracle.

Even so, pro- and anticlerical newspapers in Lyon made a cause célèbre of the mysterious healing, and Carrel found himself in the middle of a war between two ideological enemies—the scientific community and the church—who shared only one idea: contempt for Alexis Carrel. His professors at medical school mocked him for abandoning science for mysticism. The church called him a coward for not declaring unequivocally that he'd seen the healing grace of the Virgin at work.

It was the disdain of Carrel's medical superiors that had the greatest impact. A month after his paper on vascular surgery was published in *Lyon Médical*, Carrel failed a test for a staff position at the medical college. Months later he failed another, and then another. Finally, a member of the surgical faculty took pity on him. It would be pointless for Carrel to continue, the professor said. Carrel had been blackballed—permanently—because of the controversy over Lourdes. Carrel left France, going first to Montreal and then to the University of Chicago, before joining the Rockefeller Institute in New York in 1906.

LINDBERGH'S EARLY-MORNING MEETINGS WITH Carrel weren't conversations as much as they were monologues, but Lindbergh didn't mind. The fast-talking, barrel-chested man with one blue and one brown eye embodied the best qualities of Lindbergh's earlier, now deceased, hero: Lindbergh's father, the congressman who sacrificed his political career rather than modify his opposition to America's entry into World War I. C. A. Lindbergh died in 1924.

After many of these conversations Lindbergh would don a black robe and hood, then follow Carrel up the staircase leading to the operating suite above. There Lindbergh witnessed the meticulous preoperative procedures demanded by the surgeon: how his surgical instruments were sterilized by steam for two hours, then covered with a black rubber sheet, after which the operating room was sprayed with disinfectant. Lindbergh looked on as anesthetized cats and dogs, their bodies cleaned and shaved, were bled to death by technicians. He watched Carrel slice open the animal's body, gently peeling back layers of skin and muscle, then covering the outlying areas of the exposed viscera with pads soaked in Carrel-Dakin solution—the antiseptic Carrel had used on wounds during World War I—to prevent infection, in a room lined with silent technicians who, as Lindbergh later described it, "glided about spectrally," silently anticipating Carrel's every need.

The aviator marveled at Carrel's concentration, how he lowered his round, closely shaven face just inches above the animal's abdominal cavity, his eyes so focused they seemed to stop blinking for minutes on end. He saw Carrel's nimble fingers cleave tiny blood vessels with special scissors, then close them with elegant silk stitches, sometimes using only one hand.

Then came the quick, confident motions of Carrel's scalpel, removing the organ (along with its surrounding tissues, arteries, veins, nerves, and lymph vessels) with a precision and economy of effort that left Lindbergh awestruck.

The color black had always symbolized death in Lindbergh's mind, but not anymore. The experiments he witnessed in Carrel's black operating suite made him focus on life—the animal's pumping heart and expanding lungs, which struggled to survive as technicians drained them of blood so that Carrel might, in Lindbergh's words, "see across the border separating life from death." The boldness of these biological inquiries shook the pilot to his core. In front of him "mortality was analyzed in its ultimate physical form. Life merged with death so closely I sometimes could not tell them apart." To wear his own black robe in that black room, Lindbergh wrote, was "a supernatural experience" unlike any he'd ever known.

But as yet it was a frustrating one. The organs removed in the operations Lindbergh witnessed in early 1931 were, at his request, placed in the same perfusion device Carrel showed him the first time Lindbergh came to the institute. Lindbergh's purpose was to identify the cause of the machine's infection problem. After observing it in operation several times, Lindbergh concluded that the contamination was the result of a design flaw that placed the machine's moving parts—specifically, the interior piston—in contact with the nutrient medium inside the pump. The device Lindbergh designed to replace it had no internal moving parts at all. It resembled a Roman candle, with twin glass spirals: one for the liquid nutrient, the other for oxygen. They would be mixed inside the device by gravity and a rocking motion created by a simple electric motor inside the base on which Lindbergh would set his machine.

That was the plan, anyway. But there were moments when Lindbergh, sitting alone at his workbench on the institute's fifth floor, looked at his perfusion-pump design, drawn on paper in his own hand, and wondered if he hadn't overstepped his capabilities. The logic of the design was clear enough: the nutrient medium would only come into contact with sterilized glass before perfusing the body part in the experiment. But it was one thing for a machine to exist on paper, and quite another to exist in reality.

The man responsible for making that transition was the institute's resident glassblower, Otto Hopf. Lindbergh, who was coming into Manhattan three or four times a week, spent dozens of hours in Hopf's basement glass-blowing station, where the pilot donned goggles and bulky fire-retardant clothing. The heat there was often oppressive, but Lindbergh never complained. He had enormous respect for Hopf, as he did for all engineers, mechanics, and other men who worked with their hands.

Each and every prototype Hopf fashioned from Lindbergh's blueprint was handblown in Pyrex glass. The demands made on Hopf's skills by Lindbergh's double-spiral design were significant. But by the spring of 1931, after several attempts that failed because of an inability to seal properly, Hopf had met the challenge. Carrel immediately scheduled an operation to put Lindbergh's perfusion device to the test.

Carrel removed a cat's carotid artery (the vessel that carries blood from the heart to the brain) from a refrigerator where it had been stored for weeks, just above the freezing point, while immersed in saline solution. As Lindbergh watched, Carrel cut out a segment of the artery and placed it in a petri dish. He then took two glass cannulas (a tool resembling a hollow fountain pen with a sharp metal point) and inserted one, point-first, at each end of the artery—no easy feat, because the vessel was narrower than a cocktail straw. Then he closed the dish, the cover of which had two tiny semicircles cut into it to make room for the cannulas. Each cannula was attached to a rubber tube: one tube led to a small waste collection tank, the other to Lindbergh's machine.

Moments later Lindbergh switched on the motor-driven base beneath his device, which began to rock gently back and forth. This motion sent the nutrient inside—blood serum—flowing up one of the spiral tubes, where it mixed in a reservoir with oxygen flowing up the other spiral tube, pumped in from a nearby tank. Once combined in the reservoir, the oxygenated medium, pulled by gravity, flowed down the device into a rubber tube, then into a cannula, and then into the artery.

The machine appeared to be working: the medium, tinted red with vegetable dye to make it more visible, slowly moved up one spiral into the reservoir, and then down into the artery, which expanded and contracted in rhythm with the flow of the medium. Lindbergh was visibly elated to see this. Might this small experiment be the first step in developing a treatment for his sister-in-law? Might it lead to a device that could temporarily perfuse, if not an entire human heart, a mitral valve from such a heart long enough to repair it and reinsert it in its proper place? They were a long way away from that reality, Lindbergh knew, if it was attainable at all. Besides, Carrel quickly reminded him that the question at issue in this particular experiment was whether or not the carotid would remain infection-free inside Lindbergh's machine. That question wouldn't be answered for several days.

Just under a week later, Lindbergh, Carrel, and several technicians met in the incubator room to assess the results. Carrel, showing signs of tension he rarely exhibited in public, carefully opened the petri dish with his gloved hands and slowly examined the perfused artery inside, looking for signs of infection. He was delighted to see that it was not only still alive, but unharmed: it retained its functionality and elasticity, without any evidence of tissue death.

Lindbergh was equally pleased. "We were, for the first time in the history of experimental perfusion, able to avoid infection," he wrote a friend. Even so, Carrel was cautious. To confirm this result he ordered more tests; in one subsequent experiment a carotid stayed alive in Lindbergh's machine without contamination for more than a month.

This was a significant triumph, but an incomplete one. It was clear to both men that the pressure generated by the rocking-coil pump, while sufficient to perfuse a blood vessel, was insufficient to perfuse a whole mammal organ, let alone one from a human. A more powerful pumping mechanism, and a larger organ chamber, would have to be designed for those studies. Even so, Carrel thought enough of Lindbergh's design that he submitted it to *Science*, which published it as "Apparatus to Circulate Liquid under Constant Pressure in a Closed System." At Lindbergh's request, however, the submission carried no personal credit, just "Division of Experimental Surgery, Rockefeller Institute for Medical Research."

Lindbergh was thrilled to be published, even anonymously, in such a prestigious journal. For him it was a marker—as visible as the famous dimple on his chin—showing that he'd begun to achieve the goal he set for himself in 1928 in the Utah desert. He was using his engineering skills to transform himself from an aviator into a laboratory scientist studying life's greatest mystery: death.

But this quiet, and very personal, sense of elation was short-lived. Whispers soon reached the mainstream press that the unnamed author of "Apparatus to Circulate Liquid" was none other than the pilot of the *Spirit of St.*

Louis. The editor of the *New York Herald Tribune* wrote to Carrel asking him to confirm or deny those whispers. "I have received your message about the rumor that Colonel Lindbergh* has been interested in some work in my laboratories," Carrel wrote back:

> I cannot make any comment on this. Men of science should not attract the curiosity of the public. They should be left at their meditations like monks at their prayers. Should any celebrated man choose to bring to scientific research the help of his imagination or wisdom, he must feel free to do so without being disturbed by any publicity.

Carrel had given many press interviews himself in the years after his Nobel Prize, though none on his secret work with Lindbergh. Even so, he knew Lindbergh's wish to avoid any and all press attention for himself was genuine. In July 1930, just after the birth of his first child, Charles Jr., Lindbergh announced he was no longer cooperating with five New York newspapers—the *American, Daily Mirror, Daily News, Evening Journal,* and *Post.* He said these publications had violated his privacy during his wife's pregnancy. When Lindbergh's silence continued after Anne delivered, other stories alleged that his son was born deaf (from the din of airplane engines heard inside Anne's womb when she was flying with her husband), deformed, or even dead, a situation the Lindberghs supposedly planned to rectify by purchasing a healthy infant to pass off as their own.

The public, which knew nothing of Lindbergh's work with Carrel, still saw the pilot as a hero. Three years after landing in Paris he received 3,000 fan letters a month, and his autograph fetched fifty dollars, more than that of any other living American. The press, however, took Lindbergh's rejection as an insult and began to sully his image. A young journal that took pride in not being written or edited for "the old lady in Dubuque" published an especially damning profile. The real Lindbergh, the *New Yorker*'s staff writer Morris Markey wrote, "is astonishingly uninteresting" and "grim with a Scandinavian grimness":

---

* Lindbergh was a colonel in the Army Air Corps Reserves. He'd graduated from the Air Corps' Advanced Flying School in 1925. He was not on active duty.

He spends his life creating sensations, and protests bitterly that the sensations he provokes make life miserable for him. . . . He is [like] a man who periodically dives off the Chrysler tower into a net after fitting announcements, expresses astonishment and anger that people should annoy him by gathering to watch, and contends indignantly that he does it simply for his own amusement.

It's not surprising that the story above, along with the inquiry made by the *Herald Tribune,* led Lindbergh to consider doing something he really didn't want to do: abandon his work with Carrel. The Frenchman summoned all his powers of persuasion to prevail on Lindbergh to stay. The surgeon was very impressed with his young protégé. While dining one night with Dr. Louis Gallavardin, perhaps the most revered cardiologist in France (and a man Carrel trusted to keep his secrets), Carrel turned to his colleague and said, "I have a student who may amount to something." "Who?" Gallavardin asked. Carrel grinned like the Cheshire cat, then whispered: "Lindbergh."

WHAT MADE CARREL SO excited was knowing that Lindbergh's first attempt at perfusion, with his rocking-coil pump, nearly solved a problem that had baffled the world's leading physiologists for more than a century. The surgeon also knew that Lindbergh's "failure" with this device—the fact that it didn't produce enough pressure to perfuse a whole organ—was nonetheless the most successful perfusion at the Rockefeller Institute since Carrel created a "visceral organism" there several years earlier. Of all the macabre experiments Carrel performed in his career, and there were many, his attempt to "cause a system of organs to live en masse outside that organism, in vitro," was without doubt the most macabre of them all.

Two cats were required for the procedure. After the cat "giving birth" to the visceral organism was anesthetized, cleaned, and shaved, Carrel cut into its windpipe. A catheter linked to a bellows was inserted into that hole, enabling artificial respiration to take place. Carrel then sliced open the cat's thorax and abdomen, cut and tied its aorta and vena cava, severed the small intestine and ureters, isolated the aorta and vena cava from the posterior walls, and tied off their posterior branches. The bellows was switched on, after which the outlying vessels of the thoracic aorta were cut. Even with artificial respiration, the cat was dead.

But its parts were not. Wearing latex gloves, Carrel lifted the cat's viscera—the heart, lungs, liver, stomach, intestine, and kidneys—and placed them in a tray of Ringer's solution (sodium chloride, potassium chloride, and calcium chloride, mixed in distilled water) warmed to thirty-seven degrees Celsius (98.6 Fahrenheit). The organs, connected by their blood vessels, floated on the surface. "The heart still pulsated," Carrel wrote in the *Journal of the American Medical Association*, "but slowly."

Then, using his anastomosis technique, Carrel linked the carotid artery of the second cat—alive, but sedated—to the inferior vena cava of the visceral organism. Seconds after he made that vascular bond, the lungs floating inside the tray, Carrel wrote, "became pink, the heart was beating strongly from 120–150 beats per minute, the abdominal aorta pulsated violently, and pulsations could be seen in the arteries of the stomach, liver, and kidneys."

The metal tray holding this organism was topped with a glass cover to facilitate observation. Carrel fastened the tracheal catheter to a small opening in the tray so that artificial respiration could continue. An esophageal tube was similarly connected, so that the organism could be fed from outside by injection. Carrel pulled the cat's intestine through a rubber tube inserted through the box's wall, then sewed the end of the intestine to the edge of the tube. The result was an artificial anus. "The pulsations of the heart and the circulation of the organs were normal," Carrel wrote:

> The intestine emptied itself through the artificial anus by regular peristaltic contractions. . . . In an experiment in which the stomach was full of meat at the time of death, normal digestion occurred.

Peritonitis eventually ensued in all of Carrel's visceral organism experiments; the organism that survived the longest lived for just over thirteen hours. Carrel was disappointed, but satisfied that he'd proved his point: by reducing the whole, even a dead whole, to its parts, then attaching those parts to a life-support system, he could negate death. The donor wouldn't be alive, but its organs would—organs that could be transplanted into others, extending their lives.

Surgeons, Carrel decided ninety years ago, should be thinking of the body not merely as a clinical workspace, but as a spare-parts warehouse. He was so convinced of this that he met with an attorney to get an opinion on his liability, should he make a visceral organism out of a human. An account of that meeting, written by the attorney's son, Arthur Train Jr., survives.

"What would be my responsibility if I bring people back to life?" Carrel asked.

"*What?*" said Arthur Train Sr., who had been expecting a question about a lease or a will.

Carrel repeated his question.

"Responsibilities for what?" the lawyer asked.

"For those I bring back," said Carrel. "Food and lodging and all that. If I bring back an old man too old to work. Or, in the case of a young man, suppose something happens and he isn't able to do anything for himself. Am I liable for his support?"

Train couldn't believe what he was hearing.

"Your point is that if a man is dead, and you resurrect him with, say, a transplanted organ from another man, you think the law might hold you responsible for the recipient's debts—regarding you as standing in loco parentis? Surely you don't suggest that's a practical question."

"Of course, I do," Carrel said. "I've done it with animals; humans will follow. Death takes place only at that mysterious and conjectural moment when life can no longer be re-instilled into the body. And that depends upon nothing except the technical skill and mechanical ingenuity of man."

Carrel now would have changed the end of that last sentence to "of two men": the technical skill of Carrel and the mechanical ingenuity of Lindbergh. Carrel's rhetoric may not have been effective with the dumbfounded attorney, but it did work with his new protégé: convinced that the press wouldn't disturb him at the Rockefeller Institute, Lindbergh returned there shortly after the inquiry from the *Herald Tribune,* to think about a design for an improved perfusion pump. Carrel was ecstatic.

A more powerful version of Lindbergh's pump, the surgeon believed, would eventually make it possible for him to move from simple experiments on laboratory animals to complex experiments on humans—perhaps even to remove a diseased organ from a living patient, treat it, and then put it back. Doing that would move Carrel a significant step closer to his ultimate goal: redefining the human body as a living, flesh-and-blood machine with perpetually reparable or replaceable parts: a body liberated from the tyranny of death.

Carrel was now continuing that quest atop the Rockefeller Institute with a man who understood bioengineering far better than he did—certainly the engineering part of bioengineering—a man who on one of his first attempts kept a carotid artery functioning outside its body without infection for more than a month, longer than anyone else in history. The future of their still secret enterprise, Carrel believed, was very bright indeed.

Lindbergh felt the same way, though he was extremely reluctant to talk about the venture in public. In the spring of 1931, however, Lindbergh allowed an old aviation buddy, the pilot Donald E. Keyhoe, to interview him at his home in Princeton for an article in the *Saturday Evening Post* marking the fourth anniversary of his flight to Paris. Keyhoe and Lindbergh had a grand old time reminiscing about the good old days, and catching each other up, so it's hardly surprising that Keyhoe's finished piece described Lindbergh as a happily married new father (Charles Jr., called Charlie, was eleven months old when the article was published) and a highly motivated "businessman" who drove fifty-five miles each way from Princeton to Manhattan, as many as four times a week. The nature of that "business" wasn't specified, surely because Lindbergh asked his pal to keep it secret.

But something else happened when Keyhoe was visiting the Lindberghs for his story—something that also never made it into the pages of the *Saturday Evening Post*. Early in the evening, when Charles had left the living room for a moment, and Anne was happily singing "You Must Have Been a Beautiful Baby" to her young son, who was loving every note of it, Keyhoe looked away from that charming scene for an instant and, much to his shock, saw a disheveled, wild-eyed man peering in through a window. As soon as the Peeping Tom realized he'd been spotted, he turned and dashed away into the night.

When Keyhoe told his hosts what had just happened, and suggested they might hire a security guard, Lindbergh, now back in the room, said, "We moved to the country so we wouldn't need a security guard." The man Keyhoe saw, Lindbergh guessed, had probably wandered off the grounds of a home for "epileptics" and mentally retarded adults, located a mile or so down the road. "I'm not worried about intruders," Lindbergh said.

CHAPTER 4

ISN'T HE IN THE CRIB?

S HORTLY AFTER THE *SATURDAY EVENING POST* INTERVIEW, Lindbergh left Princeton, and his still secret work with Carrel at the Rockefeller Institute, to make a series of intercontinental survey flights with his wife, who'd just learned Morse code, serving as his radio operator. Carrel wasn't happy about this interruption in their perfusion research, but he knew better than to challenge Lindbergh's determination to set his own schedule. Anne was reluctant to leave her infant son behind—her mother would oversee his care at Next Day Hill—but the idea of participating with her husband in such an adventure as crew, rather than as a mere passenger, won out in the end. Besides, she could communicate with him about aviation in a way she couldn't about his biology experiments. Lindbergh's talk of pulsing chick hearts, refrigerated arteries, and blood-drained cats made Anne so uncomfortable that he stopped bringing the subject up in her presence, and this was more than fine with her.

The Lindberghs would be flying on the three-month mission in the Lockheed Sirius plane Charles had bought on their honeymoon. The purpose of the trip—almost 8,000 miles in length one way, most of it over the frozen wastes of Canada and Alaska—was to chart the unexplored northern air route between New York and what was then called the "Orient."

The first Oriental capital Charles and Anne landed in was Tokyo, where their motorcade down Hirohito Boulevard was greeted on August 26, 1931, by more than 100,000 Japanese citizens, most of them shouting *"Banzai! Banzai!"* This display of affection was not wholly appreciated by the Tokyo police, which found its protective cordon around the Lindberghs' car under constant threat from "air-minded" fans of the aviators, who'd flown into Tokyo from Siberia.

In the week that followed, the Lindberghs were feted by the premier of Japan and virtually every ministry of the national government, from the foreign office to the department of fisheries. There were tea ceremonies, performances by geishas, and banquets attended by thousands (LINDBERGHS PICK UP KNACK OF CHOPSTICK, said the *New York Times*), which barely left time for gift ceremonies where the visitors were presented with ancient kimonos, swords, and scrolls. Anne thought she knew how famous her husband was. Now she realized she barely knew the half of it. Charles wasn't just America's hero. He was the world's.

After the final ceremony the Lindberghs flew over the Yellow Sea to Nanking, then China's capital. Lindbergh had pontoons installed on his plane before he left New York, so he was prepared to land on water, but nothing prepared him for what he saw outside Nanking. The Yangtze River, the third-largest in the world, was in high flood and the ancient walled capital was about to become a tiny island in a huge lake surrounded by thousands of refugees, many of them stranded on dikes, nearly all of them dying of cholera or dysentery. "Luckier" refugees were merely starving to death in overcrowded sampans floating on water that used to be dry land.

The Lindberghs set down in their plane on a real lake—Lake Lotus—on high ground several miles outside the city. After officials from the U.S. embassy secured the plane, two other diplomats drove the couple in a limousine on one of the few remaining dry roads into Nanking, where the Lindberghs rested, and then were taken to meet the Chinese president, Generalissimo Chiang Kai-shek, and his American-educated wife. While drinking tea in the generalissimo's home, Lindbergh learned that his Sirius, with its 450-horsepower engine, was the only plane in all of China with enough range to survey the outer limits of the flood. He immediately volunteered his services to the Chinese National Flood Relief Commission.

On one of those survey flights Lindbergh landed in a camp for displaced persons. "Whole families squatted in dirt-floored pens less than eight feet square," he later wrote. "Cholera was rampant. The stench was awful":

> If there were latrines about, they were apparently unused. As we walked by a family "pen," one of the children, a half-naked boy of eight or ten, collapsed and lay quietly in the filth. No member of the family moved; none seemed to even notice him.

A few days later Lindbergh saw something even more disturbing. On September 21 he flew from Nanking to Hinghwa, a city at the center of the flooded region where the disease outbreak was at its destructive peak. Lindbergh and two passengers—Dr. J. Heng Liu, Nanking's hygiene commissioner; and Dr. J. B. Grant, from the Rockefeller Institute's office in Peking (now Beijing)—were transporting medical supplies to the disaster area. They thought of bringing food as well, but decided the best use of the Sirius's limited storage space was for vaccines, serum, and two physicians who spoke the local dialect.

After Lindbergh touched down in water outside the city, a sampan approached. Dr. Liu negotiated with its owner to take him and the packages of medicine to Hinghwa's city's gate. Meanwhile, dozens of other boats surrounded the sampan and Lindbergh's Sirius. Lindbergh could see that still more boats were on the way. Within minutes more than 100 sampans, covering perhaps an acre, had surrounded his plane, forming a floating mass of frenzied humanity. The starving people on these boats thought the packages being transferred contained food. A ragtag army of emaciated men began to jump from boat to boat, heading toward Lindbergh's plane, many of them fighting with each other, nearly all of them pushing and yelling.

Suddenly a sampan with an open fire burning on its deck moved under the left wing of Lindbergh's plane. Seconds later smoke was curling around its forward edge. Was the Sirius on fire? Lindbergh jumped out of his cockpit to check. As he did so he saw that the boat Liu was standing in was sinking from the weight of so many uninvited guests. On the flight over from Nanking Liu had confessed to Lindbergh and Grant that he couldn't swim. Lindbergh now watched as Liu, growing more anxious, dropped a package of medical supplies into the water and jumped into an adjacent boat, which soon began to sink as well, after which Liu jumped into a third. Dozens of Chinese men jumped off their boats to retrieve that package, certain it held food. Now they were shouting and fighting over it, splashing and creating waves that put Lindbergh's plane in even more jeopardy. Finally, Dr. Grant, still in the plane's auxiliary cockpit, had seen enough.

"Do you have a gun?" he screamed at Lindbergh.

"Yes, I do," Lindbergh yelled back, "but some of those men must have rifles."

"You can be sure they don't," said Grant. "Every man with a gun has gone to the hills with the bandits."

Grant had lived in China for years. Even so, Lindbergh was skeptical. He'd be an easy target, standing in his cockpit (where the gun was), silhouetted against the sky. He didn't want to show his weapon—unless he had no choice.

Desperate men were now climbing onto his pontoons; one pontoon was already slipping below the waterline. Lindbergh thought of the stories he'd heard as a boy from his father in Minnesota, horrific tales of white settlers being killed by tomahawk-carrying Indians storming over a stockade wall. When an intruder on the left pontoon climbed onto his wing, Lindbergh knew the time had come: he pulled out his pistol and pointed it at the man's chest. The intruder stopped in his tracks. Now Lindbergh wheeled and pointed at a man on the right pontoon, who threw his hands in the air. Lindbergh turned back to his left and fired a single shot toward the sky. A loud crack echoed over the water. The Chinese to his right thought he'd shot someone on the left; the Chinese on the left thought he'd shot someone on the right. Hands let go of the plane's tail and wings, as dozens of men jumped off nearby boats into the water. Sampans began to scurry away. Lindbergh and Dr. Grant pulled Dr. Liu on board, hoisted anchor, and took off.

THERE WERE NO MORE relief flights in China for Lindbergh. Instead, he meditated on what he'd seen. A sense of being overwhelmed took hold in his mind: the sheer number of Chinese people struck him as dangerous not only to themselves, but to all of civilization. The nonresponse of the Chinese parents to their own dying child in the refugee camp struck Lindbergh as less a symptom of a disconnecting disease than a window into the dark core of the Oriental soul, which seemed indifferent, maybe even immune, to the civilizing values of western culture.

"There was not enough food to go around in the best of times, yet the population [in China] keeps rising," Lindbergh wrote in *Autobiography of Values*. The chief victims were the Chinese themselves, but how long before that growth would be a problem for the West? Didn't those masses threaten modern progress? After all, Lindbergh wrote, wasn't it "Occidental aeronautical science" that gave the world "the magic power of flight" and "Occidental medical science [that] injected us with antigens to make us immune to plagues" and "Occidental economy, founded on incentive and technology, [that] supplied [the western world] with plentiful food" and other life-enhancing products too numerous to mention? What could the backward, cruel, and overcrowded Orient offer the modern civilized world?

Just before he left New York, Lindbergh had attended a Philosophers Club dinner where the precarious state of western civilization was discussed at great length and with great passion by Carrel, who noted the danger posed to that civilization by another flood—that of semiliterate, faster-breeding immigrants from the "wrong" countries, who threatened to drown their "superiors" in a demographic tidal wave. Carrel was talking about this because, after three decades in experimental science, he found his research interests expanding from medicine into metaphysics. His real patient, Carrel told the Philosophers, wasn't a laboratory animal, or even a human being. It was the human race, which to Carrel meant the white race, a collective organism in danger of being infected or engulfed by lesser organisms.

Carrel had held some of these views since childhood. As a precocious schoolboy in France, he launched his own medical education: he'd capture a small animal, usually a rodent, cat, or bird; and then, to his mother's horror, kill it and dissect it, often in her house. Looking inside those beasts and observing the anatomical differences gave Carrel an intuitive, if unsophisticated, appreciation of evolution. The natural order, he decided, is one of stratification: all creatures are made by God; but they aren't created equal. The same hierarchical structure, young Carrel was certain, applied to the races of *Homo sapiens*. For Carrel, to think otherwise was unscientific.

When this way of thinking about humanity was codified into something called eugenics—defined in 1883 by Francis Galton (Charles Darwin's cousin) as the science of improving human stock by giving "the more suitable races . . . a better chance of prevailing over the less suitable," by selective breeding practices and other measures—Carrel became an enthusiastic supporter. The carnage of World War I left him depressed for many reasons, not the least of which was the eugenic damage created by the conflict, which Carrel saw as fratricidal. The white western European nations that gave birth to western civilization—the preeminent civilization on earth, in his view—had fought among themselves, rather than uniting to beat back their common foe: the faster-breeding, less creative races of Africa and Asia, along with the backward races of eastern Europe.

"There is no escaping the fact that men are not created equal," Carrel told the *New York Times*. The fallacy of equality, he said, was one of many asserted by democracy, "which was invented in the eighteenth century, when there was no science to correct it."

In 1928 Carrel traveled to Michigan to support one of America's leading eugenicists, John Harvey Kellogg, with whom he'd been corresponding for more than twenty years. Dr. Kellogg, the popularizer of cornflakes and the founder of the famous Battle Creek Sanatorium, was the organizer of a recurring symposium he called the Race Betterment Conference, which endeavored to find ways to prevent the deterioration of the white race. Carrel presented a paper on "The Immortality of Animal Tissues and Its Significance" at the Third Race Betterment Conference, at Battle Creek. It was the closest he'd yet come to making a public link between his research at the Rockefeller Institute and eugenics.

Though the adoring public didn't yet know it, Lindbergh shared the eugenical views expressed by Carrel at the Philosophers Club dinner. He'd heard the same words uttered by his father when, as a member of Congress, C. A. Lindbergh opposed the United States' entry into World War I for the identical "fratricidal" reasons.

But Charles's favorable opinion of eugenics wasn't something he'd inherited, or even something he'd learned from Carrel. It was based on personal experience. When, after his flight to Paris, Lindbergh embarked on the "girl-meeting project"—his words—that led to his marriage to Anne Morrow, he described his selection criteria this way: "Most of the girls I met tried to bring out their good qualities and hide their defects [but] it seemed to me that one attempt was about as unsuccessful as the other":

You did not have to be a scientist to realize the overwhelming importance of genes. . . . My experience in breeding animals on the
farm taught me the [role] of good heredity. I knew that the qualities
of the father and the mother, and their ancestors before them, invariably came out in the offspring. . . . I heard people say that environment was the most important thing, but it seemed to me that they
had just closed their minds to what was obvious.

Now, after his terrifying experiences in China, Lindbergh found his intellectual bond with Carrel on the subject of eugenics taking on a new urgency. Lindbergh's scientific goals were evolving, too. Previously, he was
experimenting in organ perfusion to save his sister-in-law. Now his goal was
to use that same research to aid white civilization. Carrel was right: something had to be done to make that civilization prevail—if possible, forever.
If diplomacy and war couldn't do it, maybe experimental biology could. Perhaps, working together at the Rockefeller Institute, Carrel and Lindbergh
could find a technique (organ perfusion seemed highly promising) that
would be made available only to white people—actually, to only the right
white people.

Once perfected, that technique might enable the elite of western civilization to live beyond mortal limits and thus maintain their rightful superiority over those who wouldn't. This didn't strike Lindbergh as cruel. It
confirmed what he'd learned as a boy working on his family farm: some
animals are fitter than others. A smart farmer breeds from those animals
and eliminates the weaker stock.

Lindbergh's trip to China convinced him that the stakes inside Carrel's
laboratory were even higher than he originally thought. As for his views on
mortality, the frustrations of that "accidental" limitation were again made
clear to Lindbergh just before he left China. After the Lindberghs flew
from Nanking to Hankow, they received a telegram from Anne's sister Elisabeth, who was at Next Day Hill. Anne and Elisabeth's father, Senator
Dwight Morrow, had died in his sleep of a cerebral hemorrhage, on October
5, 1931. He was fifty-eight. Among those attending his funeral, which took
place before Charles and Anne could return, were former president Calvin Coolidge (Morrow's college classmate at Amherst); more than twenty
members of the United States Senate (Morrow had been elected to that

body in a special election held in March 1931); the renowned newspaper columnist Walter Lippmann; the publisher of the *New York Times*, Adolph Ochs; the financier Bernard Baruch; and Judge Learned Hand of the U.S. Court of Appeals for the Second Circuit.

Lindbergh hadn't always agreed on political matters with his father-in-law, a former partner at J. P. Morgan and Company who, by today's standards, would be called a moderate Republican. But Lindbergh definitely thought him a great man and a shining example of western civilization at its best. That such a productive life was cut short struck Lindbergh as a pitiful waste, and maybe—Lindbergh's and Carrel's experiments were just beginning, after all—a preventable one.

THE LINDBERGHS RETURNED FROM China by steamship, commercial airliner, and, finally, chauffeured limousine, to Englewood, on October 20. (The Sirius was sent back in crates to Los Angeles, for repairs.) Charles and Anne had been gone so long that little Charlie, now a handsome golden-haired toddler, didn't recognize them when they got back, a situation that caused Anne no small measure of guilt, as did missing her father's death and funeral. A wing of rooms at Next Day Hill would be the Lindberghs' home for the next several weeks as they waited for construction of their new house to finish. The house was situated in the woods near the small New Jersey town of Hopewell, roughly sixty miles to the southwest of the Morrow estate.

Lindbergh quickly returned to Carrel's laboratory at the Rockefeller Institute, a place he now considered his professional home. Compared with China, the institute was a Trappist monastery, its well-ordered silence broken only by the muffled scratching sounds coming from Carrel's mousery. Lindbergh was delighted to be back—and Carrel to have him. There were some mechanical issues Carrel wanted his protégé to address before returning to the perfusion pump, however. Within three months, Lindbergh designed two machines that made Carrel's ongoing tissue culture research significantly easier.

The first of these was a centrifuge that enabled Carrel to extract fibrin-free serum from coagulated chick-blood plasma, an important step in preparing the nutrient medium for cultured tissues and organs. Previously, that serum was removed in a laborious two-step process that involved grinding the plasma by hand in a mortar with sterile sand, then separating the serum from the sand by centrifugation. Unfortunately, the serum obtained this way often clotted because some fibrin remained behind.

Lindbergh solved that problem with a device made of two silver tubes, one fitting inside the other. The inner tube, about an inch shorter than the outer, had tiny holes drilled in its bottom, which were covered with a layer of quartz sand. A layer of glass beads went on top of the sand; these pulverized the plasma when the centrifuge was turned on. Once freed from the plasma, the serum dripped through the sand, then dripped through the holes, then collected at the bottom of the outer tube. This simple technique, Raymond C. Parker wrote in 1938 in *Methods of Tissue Culture*, rendered more than 95 percent of the chick plasma into fibrin-free serum, with virtually no danger of contamination.

Lindbergh then created a device that washed red blood cells in suspension, yet another procedure that previously had been done by hand. Lindbergh started by altering a centrifuge so that it could make 4,000 revolutions per minute, 1,000 more than before. The blood to be washed was placed in a conical chamber attached to the centrifuge. The genius of Lindbergh's design was that during centrifugation, the rate of flow at the narrow end of the conical chamber was too fast to permit the packing of the corpuscles, and too slow at the wide section to flush them out. The washed corpuscles remained suspended in the center.

Just as Lindbergh was completing this work, the *New York Times* published an editorial marking the twentieth anniversary of Carrel's chick-heart tissue. Once again Lindbergh was reminded of the historic task his mentor and he had undertaken.

"The piece of chicken heart which Dr. Carrel has kept alive for twenty years—longer than any chicken lives—is not quite a perpetual motion machine," the editorial began. "Like the lamp of the vestal virgins, it needs constant attention. It must be kept at the proper temperature. It must be fed with embryonic proteins. It must be washed to remove wastes. . . . In the next century, if infection, starvation, physical injury and poison are warded off, it may become as sacred as a venerated religious relic."

Three weeks later Lindbergh celebrated his thirtieth birthday, a milestone he noted with satisfaction. The "hero business" was slowly fading away. So were his name and photograph from the front page. He was still a public figure: 15,000 people a day admired his mementos at the Missouri Historical Society, a collection that now included items from his trip to the Orient. But the news about Lindbergh, the press reported on February 4, 1932, was that there *was* no news. "The world is left to guess whether there'll be frosted cake at the Lindberghs' New Jersey home for Charles A. Lindbergh Jr., now a toddling youngster, to admire," one columnist whined.

What really pleased Lindbergh was that he was living the life of science he'd dreamed of in the Utah desert in 1928. If he had to be admired by the public, he was happy that admiration came from afar. It was still difficult for the Lindberghs to attend a Broadway show or sit in a restaurant without causing an uproar. That hadn't changed. Even so, Lindbergh felt free, as he turned thirty, to be himself in a way he hadn't experienced since his flight to Paris. His personal life and working life were becoming more and more private.

Lindbergh was also happy to see that construction work on the house in Hopewell, which his wife and he decided to call High Fields, was finally finished. But Anne, then pregnant (a fact still unknown to the press), preferred to stay during the week at Next Day Hill, where her mother and a huge staff of servants doted on her and Charlie. For this reason the Lindberghs typically spent only weekends at their new home. Anne also found the solicitous environment at Next Day Hill conducive to her new career. She was writing a book about the Lindberghs' trip to Japan and China. The working title was *North to the Orient*.

Charles was willing to spend most of his domestic time at Next Day Hill, too. He could read his medical journals as easily there as he could anywhere else, and the commute to the Rockefeller Institute was an hour shorter than from Hopewell. Best of all, the public was still unaware of his destination when he drove into New York. At Carrel's urging, however, that anonymity was about to change: Carrel was so impressed with Lindbergh's corpuscle washing machine that he asked him to write it up for *Science*, and this time to sign the article. Lindbergh was nervous about the attention this would bring, not only to him but to their research, which the public still knew nothing about. Carrel told Lindbergh he'd earned the honor and reminded him that the machine wasn't directly related to their secret work, so no one was likely to make a connection. The protégé acceded to his mentor's wishes.

AND SO IT WAS that Lindbergh spent the final week of February 1932, and the opening days of March, lost in concentration at the institute, working on his paper for *Science,* which gave him enormous personal satisfaction. It had always annoyed Lindbergh that the tabloid press considered him little more than a successful stunt pilot or, even worse, "Lucky Lindy." The corpuscle washing machine he'd invented was neither a stunt nor a product of luck. Knowing this—and knowing that the scientific community would soon know it as well—filled Lindbergh with pride. He drove into New York on Saturday morning, February 27, to continue fact-checking his *Science* paper. Later that day a chauffeur drove Anne, Charlie, and a maid from Next Day Hill to their home in Hopewell. They were met there by Aloysius "Olly" Whately and his wife, Elsie, the Lindberghs' butler and cook, who had moved full-time onto the High Fields estate three months earlier.

Sitting on 400 acres, High Fields was assembled from thirteen farms quietly purchased by Lindbergh in 1930. The house, designed by the architect of Next Day Hill, Chester Aldrich, had two wings running perpendicular to, and projecting slightly in front of, the main central section. It was constructed of whitewashed fieldstone boulders, each more than two feet thick. Heavy-duty wiring and a seven-zone heating system were installed to serve the biology laboratory Lindbergh planned to put in the basement. After walking through the front hall, one entered a large living room, paneled in mahogany. To the left were the library and a guest bedroom; to the right, the dining room and kitchen, which led to the servants' quarters. Upstairs were the master bedroom, three more guest bedrooms, and, in the back corner, farthest from the main entrance, the nursery. The only access road to High Fields was unmarked. Even so, the press knew its location. A photograph of the estate—and the floor plans of the house—had been published in the *New York Sunday Mirror* under the headline THE LONE EAGLE BUILDS A NEST.

Anne spent the night of February 27 at High Fields with her twenty-month-old son, who was getting a cold. Charles returned to the house from Carrel's laboratory later that night. After dinner the Lindberghs went upstairs to the nursery around eleven p.m., where Anne medicated the baby's stuffed nose with nasal drops. Still worried about his condition, she decided to keep Charlie inside the house all day Sunday, hoping that this would ensure his recovery by the start of the new week.

But on Monday Charlie still had symptoms, which led to a change of plans. Instead of returning to Next Day Hill, Anne and the baby stayed in Hopewell. Lindbergh drove into Manhattan and returned to work at the institute. (He'd be so busy there, in fact, that he'd phone Anne later to tell her he was staying overnight.) Anne was hoping to get back to Next Day Hill on Tuesday, but that was not to be, either. Charlie was still ill, and now Anne was feverish. To get help for his wife, Lindbergh called Next Day Hill and asked that a chauffeur drive Charlie's young Scottish nursemaid, Betty Gow, out to High Fields. She arrived around 1:30 p.m.

Four hours later Gow took Charlie to his nursery for his supper. Not long afterward, Anne entered to help put her son to bed. After rubbing the boy's chest with Vicks VapoRub, the women decided he'd sleep better with an extra garment underneath his pajamas. Gow quickly made a little shirt from a flannel remnant. Underneath the shirt the boy was wearing two diapers and rubber underpants; on top of it, his gray pajamas. Gow hooked on metal thumb guards that prevented Charlie from sucking on either thumb while he slept, lowered the guardrail on his crib, and laid the boy on the mattress. She covered him with a blanket, which she secured to the mattress with two safety pins, then raised the guardrail. Anne went down to the living room to await her husband while Betty washed Charlie's clothes, then checked up on him, opening one of the room's windows slightly, before leaving. At around eight o'clock, she told Mrs. Lindbergh her son was sleeping peacefully.

A half hour later Lindbergh sounded the horn of his car before pulling into the garage. He entered the house through the kitchen, greeted the servants, washed up, and then joined Anne for dinner. After their meal the Lindberghs sat by a fire in the living room, chatting and perusing the day's mail. A little after nine Lindbergh heard what he later described as a sound like a wooden box being dropped nearby—he assumed in the kitchen. He didn't check on it, though, choosing instead to go upstairs to bathe, after which he returned to the den to do some biology reading. Anne took a bath as well, then prepared for bed. Just before ten, she rang a bell for Elsie

Whately and requested a hot lemonade. At almost this same moment, Betty Gow decided to look in on Charlie. She entered the nursery, closed the window she'd opened before, and was about to turn on a small space heater when she realized she couldn't hear the baby breathing.

Moving to the crib, Gow realized she couldn't see him, either. She ran her hand over the mattress, in case shadows were playing tricks on her. They weren't. She ran to the master bedroom. "Do you have the baby, Mrs. Lindbergh?" Startled, Anne said no. "Perhaps Colonel Lindbergh has him, where is he?" Gow asked. "Downstairs in the den," said Anne, now growing anxious. Gow ran all the way there. "Colonel Lindbergh," she said, gasping for breath, "have you got the baby?"

"Isn't he in the crib?"

Without waiting for the answer, Lindbergh jumped out of his chair and ran up the stairs, brushing by his wife as if she weren't there. When he got to their bedroom, he yanked open a closet and pulled out a rifle. Then he ran back to the nursery, followed by his wife and Betty Gow. He saw that the crib was empty. He saw a white envelope on the window sill. Lindbergh looked directly into his trembling wife's face.

"Anne," he said, "they've stolen our baby."

CHAPTER 5

THE CHAMBER OF LIFE

LINDBERGH BABY KIDNAPPED FROM HOME OF PARENTS ON FARM NEAR PRINCETON/TAKEN FROM CRIB/WIDE SEARCH ON, said the *New York Times*, which reminded readers in the article below that the missing baby had received more worldwide attention at his birth twenty months earlier than newborns of royal blood. Virtually identical headlines appeared in Paris and wherever else Lindbergh had landed since his historic flight—in all forty-eight American states, London, Mexico City, Caracas, Nanking, and Tokyo—and even in places he hadn't.

The Hearst Corporation's International News Service sent its subscribers 50,000 words on the day the story broke, 30,000 the following day, and 10,000 per day for the next two weeks, "outdoing themselves in verbosity and vulgarity," wrote the media critic Silas Bent. One wire syndicate spent $3,000—twice the average American's yearly income in 1932—in just one day to transmit photos of the handmade wooden ladder used to gain entry into the sleeping baby's nursery.

The abduction of little Charlie Lindbergh triggered the largest manhunt in American history to date. The Federal Bureau of Investigation, the Secret Service, the Postal Inspection Service, and even the Internal Revenue Service were put on the case, joining the New Jersey state police and detectives from New York City, Newark, Jersey City, and Trenton. The Coast

Guard went on alert, the Department of Commerce monitored the nation's airports, and the War Department placed the pilots and planes of the Army Air Corps, Colonel Lindbergh's aviation alma mater, at the service of its most illustrious graduate. All told, more than 100,000 law enforcement officers, government agents, and military personnel took part in the dragnet.

Several of those officials—and even a few gangsters—moved into High Fields. Lindbergh was advised by his attorney, Henry Breckinridge, to let the criminals in because it was thought the kidnapping had been committed by a gang with whom those criminals could negotiate. Once inside, Morris "Mickey" Rosner, Salvatore "Salvy" Spitale, and Irving Bitz, each supposedly connected to the gang led by Jack "Legs" Diamond, answered the Lindberghs' phones, bossed around their servants, flicked cigar ash onto the carpets, and screened visitors. This bizarre situation forced Anne, pregnant with her next child, to retreat behind a locked door into the master bedroom upstairs, where she spent most of her time crying and having her head stroked, literally and figuratively, by her mother, Betty, and her three siblings: Elisabeth, Constance, and Dwight Jr. Charles spent nearly all his waking hours downstairs, tearlessly directing the investigation.

The tension increased after a $50,000 ransom was paid on April 2, 1932, by Lindbergh's intermediary, James F. "Jafsie" Condon, a mustachioed, seventy-one-year-old retired school principal who'd made contact with the kidnapper through a series of letters he'd published in the *Bronx Home News*. As Lindbergh watched from a few hundred feet away, the cash was accepted in a Bronx cemetery by a foreign-sounding man who, in return, passed this handwritten note:

> The boy is on Boad Nelly. It is a small Boad 28 feet long. Two person are on the Boad. The are innosent. you will find the Boad between Horseneck Beach and Gay Head near Elizabeth Island.

(By "Boad" the writer meant "boat." The capitalization of this word suggested to authorities that it was written by an immigrant thinking in German, a language in which nouns are capitalized.)

An aerial search led by Lindbergh himself proved the note to be a lie. Instead, Charlie's body was discovered on May 12, five weeks later, in a wooded area not even five miles from High Fields by a truck driver who'd stopped to answer a call of nature. Found facedown in the mud, the corpse was missing most of its left leg, all of its left hand, and its right arm below the elbow. Much of its viscera was gone too, eaten by wild nocturnal scavengers.

Soon human scavengers—peddlers selling snacks and postcards—set up shop near the spot where the body was found. The autopsy, during which the baby's brain spilled out of the cranium like putrefied pea soup, located a fracture behind the child's ear. The coroner ruled that death was caused by a blow to the head, intentional or not. The Lindberghs' grief turned to horror when pictures of the mutilated child taken at the mortuary by a press photographer turned up for sale at East Coast speakeasies for five bucks each. Anne's crying, already chronic, now became nearly incessant. "I'll never believe in anything again," she wrote in her diary. Even worse, she began to hear "demon voices" in her head telling her she'd neglected her son when he was alive.

The fact that no one had been arrested in the case led some people in the press, and even in law enforcement, to point accusatory fingers at members of the Lindbergh and Morrow families. In one theory Charlie was supposedly killed by his own father—either accidentally, when Charles Sr. went too far with a physical prank, or intentionally, because the baby had been born physically or mentally defective, a situation his "perfect" father found so abhorrent that he murdered him. In another theory, Charlie was said to have been thrown out the nursery window by Elisabeth Morrow, a woman so jealous that the aviator had chosen her sister Anne, and not herself, to be his bride that her rage became homicidal.

In truth, Elisabeth wasn't even at High Fields on the night of the kidnapping; she was in her own bed at Next Day Hill nursing an impacted wisdom tooth. Moreover, her heart condition had by this time left Elisabeth so weak that it was impossible for her to have committed such an act. In all these theories, the kidnapping of Charlie was said to be an elaborate cover-up executed at Lindbergh's command to obscure the involvement of either Elisabeth or himself.

These accusations, some made to Lindbergh's face, led him to have even more contempt for the press than he already had, and to loathe the "normal citizens" who bought newspapers. As time continued to pass without an arrest—months, then years—Lindbergh became more and more depressed. Finally, he sought refuge from the maelstrom swirling about him in a familiar place.

CARREL HAD SENT LINDBERGH a letter after his son's body was found: "There are no words that can properly express what I feel at the end of your great tragedy," he wrote. "Today, the final blow has come. But life continues. I wish for you what all the future holds in store for those who possess indomitable courage." In subsequent telephone conversations Carrel urged his protégé to show that courage by returning to his workbench at the Rockefeller Institute. To be distracted from his suffering by serious scientific research, in a setting where his privacy would be scrupulously respected, Carrel said, would surely speed Lindbergh's recovery. The hero took his hero's advice, arriving there at the end of May 1932.

Rather than working immediately on a new perfusion pump, Lindbergh began with a smaller task: redesigning the flasks Carrel was using in his ongoing tissue culture experiments. This was the kind of work Lindbergh loved. He could spend hours studying a piece of equipment, putting it through tests of his own creation, silently scribbling notes on a nearby pad—never feeling bored, or feeling the need to interact with another human. His patience in such projects was boundless: prototypes were drawn, built, and tested, and failures analyzed, with a wordless concentration that seemed almost otherworldly.

Carrel understood this aspect of Lindbergh's personality better than anyone except Lindbergh's mother, still teaching science at a high school in her hometown, Detroit. The surgeon gave his protégé a quiet place to work, then left him alone. Days would pass before Carrel would stop by Lindbergh's station. Usually, the Frenchman stood there without saying anything. Lindbergh was so focused on his work he often didn't know Carrel was there.

Lindbergh's final design for the new tissue flasks made significant changes to the original. Whereas the old flasks circulated the nutrient medium as needed, the new ones circulated it continuously, using upright inlet and outlet tubes on each side of the culture chamber. A difference in air pressure between those tubes propelled the nutrient through the chamber; special valves and a gas lift regulated by a capillary tube created the pressure differential. In one imaginative improvement, Lindbergh embedded the tissue in his new flasks in a layer of fine sand between thin, converging glass walls, above which he placed a microscope. The sand was only one grain thick at the center of that tiny glass box. There, individual cells formed group structures, which greatly facilitated microscopic observation. Lindbergh sat for hours at the microscope. Occasionally he saw cell division, the primal act of multicellular life. For the first time in months, Lindbergh felt at peace.

After the institute's summer break, Lindbergh refocused his attention on organ perfusion. The problem with his original rocking coil perfusion pump was pressure—actually, a lack of pressure. Although the device generated enough force to perfuse an artery, it wasn't forceful enough to perfuse a whole organ. Lindbergh thought capillary action, the physical process that enables plants to draw water up from the earth through long narrow roots and stems, might be the solution. So in early 1933 he designed a capillary lift pump, made of Pyrex glass.

This device created sufficient pressure, but it was extremely fragile and had to be nearly six feet tall to generate the necessary capillary force. This struck Carrel as impractical, especially after one of his technicians nearly destroyed the machine by merely brushing against it. After doing some simple tests Lindbergh began to have his own doubts as well: he became convinced that the nutrient medium should be perfused through the organ inside his device in pulses, rather than in a steady stream, as was the case in his latest design. The capillary lift pump, Lindbergh decided with regret, was a failure.

But he had no regrets about having returned to the Rockefeller Institute. Everything Carrel promised was true: the staff let Lindbergh work in peace; the in-house glassblower, Otto Hopf, was tireless in answering Lindbergh's questions about Pyrex; and whatever additional matériel Lindbergh needed was just a requisition form away. Soon tools from his own toolbox vied for space on his table with jars of sterilizing acid, tubs of quartz sand, platinum screens, and silvery gas tanks linked by yards of rubber tubing. Best of all, Lindbergh was treated as a peer by the institute's scientists. Dr. Merrill Chase, then a young immunologist there, told *American Heritage* in 1985 about a lunchroom conversation he had with Lindbergh more than fifty years earlier. "As soon as we sat down," Chase said, "he began talking about what was wrong with centrifuges. His most salient point was that they were all made of bronze, and at certain speeds bronze will crack. I was amazed at his knowledge."

Not long after this conversation, Lindbergh designed an even faster centrifuge made of a sturdier metal. Several scientists at Rockefeller, including Chase, were using that centrifuge in 1933, as well as the corpuscle washing machine Lindbergh had designed there a year earlier, in their own studies. In fact, the centrifuge made by Lindbergh helped Chase make one of the most important immunological breakthroughs of the twentieth century: the discovery of cell-mediated immunity. Chase demonstrated this in the early 1940s by transferring immunity against tuberculosis by transferring white blood cells from one rodent to another—cells he'd separated from the first rodent's blood by using Lindbergh's centrifuge.

To be treated so respectfully by men of such great achievement was a healing experience for Lindbergh. What he didn't realize was that those scientists were just as starstruck by him as he was by them. It was customary for workers at the institute to sign a chit for their lunch; they were then mailed a monthly bill. Many of Lindbergh's chits never reached the institute's accounting office. They were taken by other researchers, who gave them to their friends, spouses, or children as autographs.

The intellectual energy shown by those chit-pocketing scientists thrilled Lindbergh as only one thing in his life had done before: flying. He saw that courage, a quality he'd always defined in physical terms, was achievable at an equally heroic level by brainpower. And surely no one showed more courage than Carrel, who recognized no limits to his curiosity or expertise. One day at the institute Lindbergh heard Carrel lecturing an animal trainer on the impossibility of teaching a camel to walk backward. Another time he heard Carrel tell the institute's lunchroom chef that processed white bread lowered the intelligence of all who ate it.

The Frenchman's boundless confidence was actually his most American attribute. This observation was made in the French magazine *Gaulois* by the journalist Bessie Van Vorst, who visited Carrel at the institute to see his "immortal" chick heart. "The American is a dominator of obstacles," Van Vorst wrote. "He has, in his new country, mastered all the forces of nature."

There remains one sole enemy that he has not been able to conquer: death. It is this opponent that he is determined one day to vanquish. . . . So, to accomplish this indispensable work, [the American] summons the modern alchemists and furnishes them with incomparable facilities. . . . Why should not incurable maladies yield to him so that he shall attain immortality? Carrel has gone to live in the United States because [there] he is able to pursue that dream without hindrance.

Now Carrel was pursuing that seemingly impossible dream with his mechanically gifted collaborator, the most famous American of them all. Lindbergh, still dealing with his grief, was nonetheless ready to refocus his mind on their secret organ perfusion project. But he was momentarily at a loss for ideas on how to design a perfusion device powerful enough for their needs. Carrel, hoping to recharge his protégé's inspiration, sent him to the Rockefeller Institute library to read about the founding experiments of perfusion. Lindbergh entered the elegant basement room, wandered through the stacks, found the books and journals that his mentor recommended, and came away, weeks later, with a profound appreciation for the courage showed by his predecessors.

He began his reading at the obvious starting point: an English translation of J. J. C. Le Gallois's century-old perfusionist manifesto, *Expériences sur le principe de la vie*, in which the physiologist predicted, "If one could substitute for the heart an injection of arterial blood, either natural or artificial, one would succeed easily in maintaining alive indefinitely any part of the body." Then Lindbergh read the scientists who tried to prove Le Gallois was right. The boldest—and certainly the bloodiest—of these was Dr. Charles-Édouard Brown-Séquard, who taught experimental medicine at the Collège de France and Harvard in his long academic career.

Brown-Séquard began his perfusion research in Paris in the 1850s, reversing rigor mortis in rabbits by injecting the cold cadavers with warm blood. Then he tried the same experiment on a human. The corpse of a twenty-year-old murderer guillotined that morning was brought on a horse-drawn cart to the professor's private laboratory, where it was placed on an operating table and left untouched for thirteen hours. Then Brown-Séquard took half a liter (about one pint) of his own blood and injected it into the cadaver's radial artery. Within forty-five minutes, Brown-Séquard observed contractions in twelve distinct muscle groups on the headless body lying before him.

This got Brown-Séquard thinking about bodiless heads. He captured a stray dog from an alley outside his laboratory, anesthetized it, withdrew a quarter of a liter of the animal's blood, which he placed in a sterile container, and then decapitated the dog. After half an hour passed, Brown-Séquard injected the preserved dog's blood back into its severed head via the carotid artery. Minutes later Brown-Séquard saw facial muscles twitching and the unmistakable blinking of the dog's eyelids. Rumors soon spread that Brown-Séquard had tried the same experiment on a decapitated human head, taken from the guillotine. But if he did, he left no record.

It was left to other scientists to move perfusion research away from necromancy. These researchers began by building perfusion devices more complex than the one favored by Brown-Séquard—the lowly syringe. They also focused their research more specifically on perfusing whole organs. Working at Carl Ludwig's Physiological Institute in Leipzig, Germany, Élie de Cyon perfused a frog's heart in 1866. Three years later, Axel Schmidt invented the first machine to artificially oxygenate blood; he and other scientists at Ludwig's institute used it to perfuse a cat's kidney. Three decades later, Oscar Langendorff used a similar device at the University of Rostock in Germany to perfuse a cat's heart. In none of these pioneering studies, however, did the perfused organ function normally—or even survive—for more than a few hours.

LINDBERGH'S EFFORTS TO IMPROVE on those (and his) early results began in earnest in 1934. In January the Lindberghs rented a penthouse apartment on East Eighty-Sixth Street in Manhattan. They'd abandoned High Fields shortly after Charlie's body was found, moving back to Next Day Hill. That was just a temporary solution; Lindbergh was eager to live in his own home—as long as that home wasn't the crime scene—and it surely crossed his mind that the apartment on Eighty-Sixth Street was close to the Rockefeller Institute. But there was another reason he chose to live in Manhattan: Anne gave birth to their second son, Jon, on August 16, 1932. Though Lindbergh had few warm feelings about urban life, he told Anne that the only place he'd feel safe leaving a baby now was in a New York apartment with round-the-clock security.

Lindbergh's work at the institute in 1934 focused on finding a method to generate sufficient force to perfuse whole organs with pulsatile pressure. After a few false starts, he decided that compressed gas might be a solution. The gas pressure would have to be regulated; too much would damage the nutrient medium just as surely as any metal piston. Lindbergh would also have to be certain that the gas entering his device in those pulses wasn't carrying any bacteria along with it. He knew that unless he could prevent infection, it wouldn't matter how ingenious his as yet undesigned gas-powered pulse system was, so he began his work by addressing the sterility issue.

If Lindbergh had underestimated this problem before, he was soon disabused of his confidence. A series of tests in which he sent a steady flow of compressed air into a culture broth showed all too well that piped-in air could be a powerful infecting agent. Rather than abandon this approach, Lindbergh searched for a filtering agent—the simpler the better. After weeks of trial and error, he found one: a wad of sterile nonabsorbent cotton, packed tightly into a glass bulb. In an experiment lasting thirty days, compressed gas directed through such a cotton-packed bulb—and then into the culture broth—left the broth free of infection. Carrel was amazed to see that this simple "technology"—one available at any neighborhood pharmacy—really could maintain a state of asepsis, an absolute requirement for his organ perfusion research to proceed.

Now Lindbergh devoted his energy to designing a pulse-creating mechanism. His still undesigned organ perfusion machine, Lindbergh knew, would use "control gas"—a mixture of oxygen, nitrogen, and carbon dioxide—to perform two functions: to oxygenate the nutrient inside the perfusion device, and to propel that nutrient through the device in pulses resembling those created by the human heart. But something outside was required to propel that gas mixture into the perfusion machine. In 1934 Lindbergh decided it would be a liquid piston called an oil flask.

The idea was as simple as it was efficient. The oil flask was made of two glass chambers: a rounded outer chamber, which sat at the bottom; and a tall cylindrical inner chamber, which rose up from the outer one like a stalk from a germinated onion. The outer chamber was partially filled with liquid petrolatum (petroleum jelly); the inner one was filled with control gas, pumped in from an external tank. Compressed air from a second external tank was released by a rotating valve into the outer chamber of the flask, which forced the liquid petrolatum into the inner chamber.

Once there, the petrolatum became a liquid piston, moving up the inner chamber with enough force to drive the control gas out of that chamber and into whatever final destination Lindbergh devised. One turn of the valve activated the piston upward by sending compressed air into the outer chamber. Another turn allowed that compressed air to escape, thus sending the piston down, and the oil back into the outer flask. (The expended control gas was replenished from its external source after each upstroke by the piston.) This entire process—which could be repeated indefinitely—was powered by a small electric motor that regulated the air-pressure valve in an almost exact duplication of the systolic and diastolic pressure created by the heart.

Lindbergh started sketching designs for his perfusion pump in the spring. He knew that the exterior would be made of Pyrex glass and that there would be valves and screens inside the device to monitor the internal pressure and filter out impurities in the nutrient medium. The toughest hurdle would be to create a glass chamber in which to place the organ without bringing infection along with it.

Lindbergh spent hours at his workbench designing prototypes. Then he'd walk down to Otto Hopf's basement glassblowing station. The two men, now quite close, were convinced that they were on the brink of a major biological breakthrough. If successful, they would create a glass chamber of life—a place where whole organs could live, fully functioning, outside the body in which they'd been formed. These organs could be kept alive indefinitely and, if Carrel was right, could be used as replacement parts to keep a human being alive. Maybe even forever.

Hopf was unaware of that ultimate (and still secret) goal of Lindbergh and Carrel's collaboration. Several prototypes were manufactured before the institute closed for summer recess. Carrel put mammalian organs into these pumps after lifting them out of animals who gave their lives in the name of science. In all these tests, Lindbergh saw to his horror that the perfusion pump failed almost immediately. "Necrosis," Carrel would say, pointing to discoloration on the organ inside Lindbergh's machine, which indicated that the tissue was dying. On other occasions he'd point to signs of bacterial infection.

The technicians who assisted in those unsuccessful experiments were shocked by Carrel's patience with Lindbergh. He was rarely this patient with *them*. Lindbergh's frustration with himself was showing, however. There were no tantrums, but there was a visible, if not verbalized, disappointment that would later evolve into a renewed sense of determination. Lindbergh's single-mindedness served him well in Carrel's lab; when Lindbergh set a goal for himself, he rarely walked away from it. Besides, significant progress had been made. The idea of finishing the work in the fall—and solving the problem of infection permanently—filled Lindbergh with the same pleasurable anticipation he felt in planning a long air journey.

It had been a good six months for Lindbergh, even if he hadn't yet perfected his perfusion pump. Carrel was right: Lindbergh's healing process had quickened at the institute. Lindbergh decided to make one change in his life, however. In June 1934 he terminated the lease on his New York apartment and moved to Next Day Hill. The noise and crowds of the city had become more and more of a distraction for him as the year progressed. Things weren't perfect at the Morrow mansion—Lindbergh never got used to the social events his mother-in-law was so fond of hosting there—but it was a large house, well-guarded, with servants who doted on his son. Better yet, it was situated on a rolling fifty-acre plot of land where Lindbergh could smell the grass and see the stars. It was a trade-off, living there, but one he was willing to make.

Shortly after the Lindberghs reestablished themselves at Next Day Hill, Charles and Anne left their son there with servants and flew in their own plane to California. They hoped to spend a quiet end-of-summer vacation with Anne's sister Elisabeth and her husband, the Welsh businessman Aubrey Morgan, then staying at Will Rogers's ranch in Pacific Palisades. Rogers, whose hugely popular radio show and newspaper column had been instrumental in making Lindbergh the world's most famous man, was away at the time.

This was the second trip the Lindberghs had made to California to visit the Morgans, who'd moved to Pasadena, from Cardiff, Wales, a year earlier. The first visit, made over the previous Christmas, showed Anne and Charles—to their identical dismay—how weak Elisabeth's heart condition had left her. Because she had trouble walking, Elisabeth had to be wheeled around her garden in a custom-made bed. Elisabeth joked that her strange mode of transport was surely encouraging her neighbors to think lascivious thoughts about her married life. In reality, Elisabeth's health was so precarious that her husband had to get permission from her physician each and every time they wanted to make love.

The Lindberghs were with the Morgans at Rogers's ranch for only three days in September 1934, when they got some unexpected news. Colonel H. Norman Schwarzkopf, the superintendent of the New Jersey state police, announced in a long-distance telephone call that a suspect was about to be arrested for the kidnapping and murder of Charles Jr. "Oh, God," said Anne, "it's starting all over again." "Yes," Charles said, after he hung up the phone, "but they've got him at last." The Lindberghs ended their vacation and flew back east.

Bruno Richard Hauptmann, a thirty-four-year-old German immigrant living in the Bronx, was arrested for the murder of Charles Lindbergh Jr., on September 19, 1934. At the time of the arrest he was driving his blue 1930 Dodge sedan on Tremont Avenue, only minutes away from St. Raymond's cemetery, where Lindbergh's intermediary, "Jafsie" Condon, handed $50,000 to a foreign-sounding man two years earlier. Within two weeks of that payment, bills known to be part of the ransom began surfacing in the Bronx and upper Manhattan. After some good legwork by the New York police, the authorities heard, on September 18, from a teller at a Bronx branch of the Corn Exchange Bank, who had checked a $10 gold certificate in his till against the list of serial numbers on the ransom money provided to the bank by authorities and found a match.

All the ransom bills were gold certificates. In April 1933, President Roosevelt had ordered that gold certificates valued above $100 be deposited or exchanged for non-gold certificates at Federal Reserve banks, to end the Depression practice of hoarding gold. This made the remaining gold certificates—those worth less than $100 were still legal to possess—much easier to spot.

The gold certificate the bank teller called the police about had a license-plate number written on it: 4U-13-14 (N.Y.). A New York police detective guessed the bill might have come from a cash drop made by a local gas station. He was right. The manager of a service station in Harlem had written down that plate number when he accepted the gold certificate from a customer who bragged that he had plenty more at home. The New York motor vehicle bureau gave the police the name and address of the car's owner: Bruno Richard Hauptmann, of 1279 East Two Hundred Twenty-Second Street, in the Bronx. The officers set up a stakeout on September 19, and arrested Hauptmann shortly after he drove away from his home.

The suspect had a $20 gold certificate from the ransom payment in his wallet when he was arrested; the police soon discovered nearly $15,000 in ransom bills in his garage. Later, a forensics expert established that portions of the ladder found at the crime scene were made from wood planks taken from the attic of Hauptmann's home. Hauptmann, who'd served time in Germany for armed robbery and burglary, was a carpenter.

Shortly before New Jersey put Hauptmann on trial, the Lindberghs' stress level climbed even higher. In November 1934, Elisabeth Morrow Morgan underwent an emergency appendectomy in California. She seemed to be recovering well, and she sent Anne a number of cheery notes; but then an infection set in, followed by serious respiratory problems. On December 3, Elisabeth Morrow Morgan, just thirty years old, died in her hospital bed in Pasadena. The death certificate said pneumonia, but Lindbergh felt he knew better.

The real cause was her damaged heart—the same organ that had brought him to Carrel's laboratory at the Rockefeller Institute in the first place. Elisabeth's death did more than sadden Lindbergh, it reminded him of the capricious nature of death. Why did someone as intelligent and vivacious as Elisabeth have to die so young when other, less gifted, individuals lived so long? Why did she have to die at all?

Hauptmann's trial convened in Flemington, New Jersey, on January 2, 1935. The biggest names in American journalism arrived to cover it, using specially installed electronic equipment that gave reporters the collective capability to file more than a million words per day from the courthouse. The *New York Times* published a nearly complete transcript of each day's session.

When Lindbergh, who attended every one of those sessions—often with a pistol visible beneath his jacket—identified Hauptmann as the man who accepted the $50,000 ransom, the *Times* gave bigger play to his testimony than it did to President Roosevelt's state of the union address, delivered the same day. Fox Movietone News showed newsreels of Lindbergh's testimony and the prosecutor David Wilentz's cross-examination of Hauptmann in movie theaters while the trial was still in session. (Here is a typical exchange. Wilentz: "You think you're a big shot, don't you?" Hauptmann: "Should I cry?" Wilentz: "Bigger than anybody!" Hauptmann: "No, I know I'm innocent.") So great was the rage created by these films, and so real the possibility that a lynch mob might form outside the courthouse if more footage was exhibited, that Judge Thomas W. Trenchard ordered Fox's equipment removed from the proceedings. As a result, cameras were barred from virtually every trial in America for the next fifty years.

Jack Benny, Lynn Fontanne, and other celebrities pulled strings to get seats in the courthouse gallery. Roughly 60,000 tourists descended on tiny Flemington almost every weekend. Some locals sold these visitors necklaces with a replica of the kidnap ladder dangling from a cheap chain. The witness chair had to be nailed to the floor after one souvenir hunter tried to steal it.

Hauptmann, who loudly proclaimed his innocence in that chair, was convicted on February 13, 1936, after a summation from Wilentz that called him the "vilest snake that ever crept through the grass" and "Public Enemy Number One of this world." The verdict was met by cheers from a crowd of several thousand outside the courthouse who had been chanting *"Kill Hauptmann! Kill Hauptmann!"* for nearly an hour. They got what they screamed for. The defendant was sentenced to death.

LINDBERGH RETURNED TO CARREL'S laboratory almost immediately after the final gavel. Whether he was inspired by his sister-in-law's death, or maybe just the idea of moving his life forward, we know that he worked there at a feverish pace. Dozens of pencil drawings made by Lindbergh in early 1935 show that he was continually reworking his designs for the perfusion pump. His notes to Otto Hopf show a willingness to experiment with even the smallest degrees of difference. A page in Yale University's Lindbergh Collection contains these handwritten instructions:

> *Pump #1: Same [as before] except neck slightly more flared to permit cork to go farther in. Also: neck to be at slightly more angle—about 45 degrees.*
> *Pump #2: Same except neck even more flared.*

Pump #3: Exactly the same [as before].

*Pump #4: Same size but without lip. Leave at least 1 cm. clearance
between bottom of cover and top of inlet tube for rubber band seal.*

All these designs were made and tested. Index cards written by Lindbergh left terse instructions for Carrel's black-hooded technicians:

To remove pump from incubator—

1. *Clamp hose above red cork on side of incubator. (Marked with white thread.)*
2. *Remove ribbon tubes from the two cotton filter bulbs.*
3. *Remove pump from incubator.*
4. *Do not stop outside apparatus.*

To start pump again—

1. *Replace rubber tubes on filter bulbs (upper tube to lower bulb, lower tube to upper bulb).*
2. *Remove clamp from hose marked with white thread.*

Another index card has a small conversion chart:

1 Kidney man = 160 gr
1 Kidney cat = 15 gr
Secretion 1K man + 750 cc3 / day
275 liters thru 1 K (man) / day

(Clearly, Lindbergh was already thinking of using his pump on a human.)

A piece of loose-leaf paper gives directions for another experiment:

1. *Insert tissue in chamber unattached to very small cannula. (To be placed on filter paper.) Wet tissue with Tyrode* so that filter paper is*

---

* A solution made of sodium chloride, potassium chloride, calcium chloride, sodium bicarbonate, and sodium dihydrogen phosphate, mixed with glucose and distilled water.

saturated. Place pump in incubator for five days. (Pump to be already filled with Tyrode.)

2. *If no infection at end of five days then circulate for five days more.*

Then, below a hand-drawn line, there are instructions for a follow-up study (or perhaps notes to himself):

Place tissue in flask with Tyrode. (About 2–3 mm. on bottom. Leave in incubator for five days then if no infection pulsate for five days more.)

Then, below another hand-drawn line:

Place 30–40 fragments of rabbit spleen in sand in flask. Try to scatter thru sand as much as possible. Add 75% normal rabbit serum, 25% Tyrode. Change fluid after 48 hours. Save old fluid.

Then, on a separate page, a handwritten note from Carrel, marked 6:15 p.m.:

Dear Colonel Lindbergh,

The apparatus worked well until around 4:30—then something went wrong. We put the clamp on the rubber tubing.

The kidney is in excellent condition. We are adding 100 cc of Tyrode and hemoglobin. If you can restart the apparatus, I believe that the kidney will recover in spite of the interruption of the circulation.

Yours sincerely,
Alexis Carrel

(As close as their friendship was, Carrel and Lindbergh always addressed each other in writing as "Colonel" and "Doctor.")

Despite the setback noted in Carrel's letter, technicians in his laboratory saw a new optimism on Lindbergh's face as spring came to Manhattan. His latest design for the perfusion pump, Lindbergh believed, gave the best hope yet for solving the problems of infection and tissue death that had plagued the project so far. On April 2, 1935—not even two months after the end of Hauptmann's trial—Lindbergh put on his goggles and fire-retardant suit for what must have felt like the millionth time and took a seat next to Hopf in his glassblowing station. A few hours later they emerged from the institute's basement with the finished product.

A musician might have likened it to a small glass saxophone, roughly half normal size, with two bulbous mouthpieces, resting on a cylindrical glass stand. A naturalist or sculptor might have seen three glass herons—two parents and a child—somehow sharing the same torso. Whichever description one preferred, there was no doubt that this was an impressively designed machine. The exterior of the pump, made entirely of Pyrex glass, had three openings to the exterior, each protected by bulbs stuffed with nonabsorbent cotton. The device had three main chambers, one above another: the highest, the organ chamber, was slanted and curved; the lowest was the upright reservoir chamber; in between was a pressure equalization chamber.

The machine, Lindbergh told Carrel, would work this way: The nutrient medium—a mix of natural components (cat blood serum) and artificial components (sodium chloride, potassium chloride, calcium chloride, magnesium chloride, sodium acid phosphate, sodium bicarbonate, and glucose, in triple-distilled water)—would be poured through an inlet into the reservoir chamber. When the oil flask went into operation, it would send pulses of control gas into the reservoir chamber, where the resulting pressure would propel the medium into a feed tube. Once inside that tube, which ran up the device to the organ chamber, the nutrient medium would be filtered by two platinum screens before it entered a cannula attached to an artery in the organ. After the medium perfused that organ, it would exit through another cannula to be filtered through a layer of sand. Floating valves in the pressure equalization chamber would prevent the medium from flowing back into the organ chamber. Then gravity would send the medium back to the reservoir chamber, where the process could start again.

That was the idea; only a careful test would show if Lindbergh had gotten it right.

THE AIR WAS THICK with expectation on April 3. Just after daybreak Lindbergh connected his perfusion pump to his oil flask, and his oil flask to two gas tanks on a shelf inside Carrel's incubator room. He then donned a hooded black robe and climbed the staircase to Carrel's operating room, where other black-clad technicians already lined the walls. Over the years Lindbergh had learned to recognize most of those men and women by their eyes, the only parts of their faces visible through the slits in their hoods. Today he saw a pair of eyes he knew well, but hadn't seen before in this room: they were Hopf's. If history was going to be made, the glassblower wanted to see it.

Gleaming surgical tools, reflecting the early morning sunlight pouring through the skylight above, lay on black towels on black metal tables in the black operating room. The simple lighting system demanded by Carrel had the effect of making the sunlight appear almost solid as it entered through the glass ceiling—like a pillar of light, offering divine guidance from above. A cat lay motionless on the operating table under that illuminating pillar in the room's center. An hour earlier the cat had been anesthetized, shaved, and bled to death. It was now strapped to the table and draped in black linen that left only one small area exposed: its throat.

The silence was broken by the arrival of Carrel. The surgeon looked around the room to make sure his hooded technicians were in place. They were. Then he peered through his pince-nez at the skin he was about to cut and drew a line there with a crayon to mark the site of his incision. Carrel extended his gloved hand, palm up; a technician took away the crayon and replaced it with a scalpel. Carrel accepted the blade, clutched it between his thumb and fingers, then sliced open the animal's neck. It parted like a theater curtain, revealing the right carotid artery, vagus nerve, and thyroid glands. Carrel stepped back from the table, handing his scalpel to a nearby aide. Two other technicians carrying infection-fighting gauze pads soaked in Carrel-Dakin solution stepped forward. They placed the pads on the surrounding tissue exposed by Carrel's incision. The surgeon refocused his gaze on the cat's opened throat and, without looking up, again extended his palm. A technician dropped in a cannula, which Carrel inserted into the lower part of the thyroid artery like a pointy glass plug into a socket. Then, after being handed a syringe, he injected that artery with saline solution. When finished, Carrel tied off the carotid artery, then accepted the scalpel once again. He sliced through the carotid two centimeters (not quite one inch) above the thyroid artery. The glands were ready for removal.

It was now nearly twenty minutes into the procedure. Carrel reached into the cat's throat and with cupped hands carefully and slowly lifted out the thyroids, one at a time, along with the carotid artery, vagus nerve, and surrounding connective tissue. The left thyroid, which would serve as a control, was given to a technician who placed it in a small container of formalin. The right thyroid—which would be used in today's experiment—was given to another technician, who wrapped it in cellophane.

With another nod Carrel directed his assistants to begin cleaning up. He took the cellophane-wrapped thyroid himself and carried it downstairs in a small black tray to the incubator room, where Lindbergh's perfusion pump was already set up. Only Lindbergh followed. Once inside that room, Carrel removed the rubber cork from the inlet of the pump's organ chamber and, as the designer of that device looked on intently, slid the gland out of the cellophane and into that glass enclosure. Carrel then attached the gland to the cannula at the end of the feed tube, restoppered the inlet, and sealed the inlet shut with cellulose acetate cement.

Lindbergh switched on the electric motor that powered his apparatus. The valve began to rotate. There was a noticeable hiss as the control gas and compressed air entered the oil flask, followed by a subtle sloshing sound as the liquid piston began moving up and down. So far, so good. The nutrient medium, tinted with sterile red dye, surged in pulses—sixty per minute—up the reservoir chamber into the feed tube, through the filtering mechanisms, and then through the cannula into the thyroid itself. Lindbergh and Carrel could see it coursing through tiny arteries near the surface of the gland in the organ chamber.

But what were they seeing? This experiment, both men knew, was penetrating farther than any other in medical history into the unexplored territory dividing organic life from artificial life. Carrel had already proved, at least to his own satisfaction, that death was an "accident" for individual cells and small pieces of tissue: his world-famous chick-heart fragment was still pulsing outside its body in a nearby room, twenty-three years after Carrel "gave birth" to it.

But the cat thyroid inside Lindbergh's glass perfusion pump was neither a single cell nor a tissue fragment. It was an entire organ—a society of cells functioning as an organic whole. This was a leap. To keep some cells alive outside the body was—to borrow a metaphor from the construction trades— like making a few bricks. To keep a whole organ alive outside the body was like building an entire room. Lindbergh and Carrel knew they were trying something very difficult and brimming with almost unimaginable promise. What would they find inside their perfusion pump when they opened it weeks later?

The mechanical aspects of the pump were working, but what about the biological aspects? Would the thyroid, removed from the body that formed it and placed in a tiny "glass house," survive for a meaningful period? Would the nutrient medium give it the sustenance it required? Would the nutrient fool the organ into performing its normal functions? Would the pressure-regulating valves perform as planned? Would the filtering mechanisms work? Would the process remain free of infection?

Would Carrel and Lindbergh be able to show, in a way no scientists had shown before, that the body is a living machine made of infinitely removable, reparable, and replaceable parts? Could those parts—made, as Carrel believed, by God, and now removed by human hands from their God-given environment—be kept alive by science without God? Had two unlikely partners—the world's most famous man and the man some people thought to be the world's smartest—attained that godlike power? Here, in a secret experiment atop the Rockefeller Institute, had they taken the first scientific step in ending the tyranny of death?

Carrel stepped out of the incubator room and walked back to his office, unbuttoning his black robe. Lindbergh stayed behind to monitor his machine. Carrel had told him there was no point in doing so; for the experiment to have any validity, the perfusion device had to pass this test without interference. But Lindbergh was unmovable. When the sun went down nearly twelve hours later he was still there, his black robe thrown over a nearby chair. One of Carrel's assistants, about to leave for home, peered in through the glass window and returned a few moments later. Without saying a word, the technician left a bowl of rock candy—Lindbergh's favorite—outside the door of the incubator room, then walked down the hallway and into the night.

CHAPTER 6

EVERY ACT OF HIS IS NOT A FLUKE

N O ONE AT THE ROCKEFELLER INSTITUTE WAS SURPRISED TO see Lindbergh in the incubator room the next morning. For the whole week, in fact, the nervous inventor spent most of every working day and a good many more sleepless nights in that steamy space. This self-imposed schedule eased up a bit in the second week; and halfway through the third—eighteen days after the experiment began—Carrel decided it was time to analyze their results.

As Lindbergh watched, the surgeon unstoppered the glass organ chamber in the perfusion pump and carefully removed the thyroid, cupping it in his gloved hands as he carried it to a sterile glass dish. Already lying on the glass surface was the control gland Carrel had removed from the same donor animal when the experiment began, and had then stored, immersed in formalin, in a refrigerator. Carrel took a cloth measuring tape and compared the measurements of the two glands from top to bottom, and fully around: they were identical. He weighed them: identical again. He tested to see if the cultured gland was producing thyroxin, as a normal one would: it was. He looked for signs of infection on the cultured gland and found none. He excised fragments of the cultured gland, then embedded them in tissue flasks to see if they would grow there: they did. Each of these tests demonstrated the same thing: after eighteen days outside the body that created it, the cat thyroid just removed from Lindbergh's perfusion machine was still alive and functioning.

Carrel was ecstatic. He embraced his lanky protégé, who was visibly embarrassed by the show of affection. Carrel sent a technician to fetch Otto Hopf, whom he immediately ordered back to the basement to create more of Lindbergh's pumps. Additional tests were needed to confirm these results.

After Hopf completed his work, twenty-five more organs from cats and dogs—ovaries, spleens, kidneys, thyroid glands, and adrenal glands—were perfused (some for more than twenty days) in Lindbergh's machines. Organs cultured in diluted serum decreased in size; those perfused with a growth-promoting substance created by Carrel's chemist, Lillian E. Baker, grew rapidly. In one especially intriguing test the weight of a cat ovary increased from ninety milligrams to nearly 300 milligrams in just five days. Even more amazing, the ovary developed three corpora lutea—the yellow masses formed by an ovarian follicle that has matured and discharged its ovum—where none had been seen before. Lindbergh's pump wasn't just keeping the ovary alive, Carrel decided: it was creating real life out of artificial life.

This, he said, was news the world had to hear. Carrel and Lindbergh cowrote an article for *Science*, published on June 21, 1935. Compared with most articles in that journal, "The Culture of Whole Organs" was noticeably short on details. The study's results were summarized, with special attention given to the fast-growing cat ovary that had produced several corpora lutea. But anyone hoping to duplicate this experiment wasn't given enough information to do so: Lindbergh's perfusion apparatus wasn't described at all. That description would come later, the article in *Science* said, in a submission written by Lindbergh alone. ("An Apparatus for the Culture of Whole Organs," by Charles A. Lindbergh, was published in September in the *Journal of Experimental Medicine*.)

Most of the *Science* piece was instead devoted to a history of failed efforts to achieve organ perfusion, all of them precipitated by Le Gallois's declaration of its inevitability in 1812. Critics who saw Carrel as a glory hunter weren't surprised to see that this account gave most of the credit for the current successful perfusions to advances in vascular surgery, antiseptic agents, and transplantation techniques made by none other than Carrel himself. After those breakthroughs, the article declared, the only thing left to be done was to develop "an apparatus capable of playing the role of the heart and lungs and of keeping an organ free from infection indefinitely." In Carrel's laboratory, the *Science* article asserted, "a long search has been made for [that] apparatus. . . . The purpose of this article is to show that, after 123 years, the conception of Le Gallois has finally been realized."

Newspaper editors—men who'd probably never even heard of Le Gallois—covered the invention of the perfusion pump on their front pages. ONE STEP NEARER TO IMMORTALITY blared the headline in the *New York American*. The headline in the *New York Times*, CARREL, LINDBERGH DEVELOP DEVICE TO KEEP ORGANS ALIVE OUTSIDE BODY / INVENTION SEEN AS EPOCH-MAKING IN MEDICAL SCIENCE, didn't use the word "immortality." Even so, there was no doubting the awe the pump inspired in the *Times*'s science correspondent, William L. Laurence:

"The development of an 'artificial heart' and of a man-made 'bloodstream' enables science for the first time to keep the vital organs of man and animals alive and functioning indefinitely outside the body in a 'chamber of artificial life,'" Laurence wrote. "This announcement . . . opens up a vast untrodden field for peering into the mysteries of life and death."

Anne Lindbergh, who detested the press nearly as much as her husband did, was delighted by Laurence's piece. "Today it came out in the paper about C.'s pump," she wrote in her diary. "It is thrilling to me—the accurate, bare account in the *Times*. I am glad to have it known, and glad, in a way, for the publicity. [For once] the right values are being set, even for his other achievements. Every act of his is not a fluke, not chance, not charm and youth and simplicity and boyishness, but the expression of a great mind that can turn its searchlight in more than one direction."

Even Hollywood was starstruck. Shortly after Lindbergh's invention was announced to the public, a film called *The Walking Dead* went into production on the Warner Brothers lot in Burbank. The movie starred Boris Karloff, who'd recently become world-famous playing the monster in *Frankenstein*. In *The Walking Dead* Karloff played a musician who was framed for murder, convicted, and then electrocuted, only to be revived—after a supporter belatedly established his innocence—by a doctor using a miraculous new machine. That machine, called a "glass heart" in the film, was a nearly perfect replica of Lindbergh's pump.

Lindbergh was wary of such attention, but his anxiety had little effect on his mentor. Carrel was on vacation in France when the press began beating the drum—usually responsibly, sometimes not, but always loudly—about their device. Several tabloids predicted that Lindbergh would soon ask his surgeon friend to install the glass pump as an artificial heart in Lindbergh's own chest, so that the flier could live forever. Other stories talked of babies being created in the device or its being used to keep a severed brain alive and thinking. Carrel dismissed these reports when speaking to journalists in France; even so, he virtually exploded with praise for the man whose invention triggered the wild speculation in the first place.

"Lindbergh is considered in America and in France, too, exclusively as a flier, and particularly as the conqueror of the Atlantic," Carrel said. "Certainly that is a title of glory, but he is much more than that. He is a great *savant* . . . with one of the keenest and most intuitive minds possible to imagine." Instead of being surprised by that, Carrel said, the public should understand that "men who achieve great things in one area are capable of great accomplishments in all domains."

When Carrel spoke to a *New York Times* reporter on an ocean liner crossing the Atlantic back to America, the surgeon said that in the future his therapeutic focus would expand from the human body to include the human mind. He wasn't abandoning his work on organ perfusion; the quest for immortality, Carrel told the reporter, would always intrigue mankind and, of course, himself as well. If at present science is unable to find a technique for fully achieving that goal, he said, it means only that "science has yet to discover that technique"—which it inevitably will, just as Lindbergh and he solved the mystery of perfusing whole organs. Even so, said Carrel, the ultimate goal of science must transcend the modification of the human organism, no matter how permanent that transformation. It must also include the "modification of the human intellect."

One method to achieve that second goal was a new approach to breeding and child rearing. "I do not necessarily mean that we should pick out a few children and make supermen of them," Carrel said. But he left no doubt that he thought mankind would benefit if superior men were given leadership roles in shaping human consciousness—and everyone else either followed or got out of the way.

"There is no escaping the fact that men are not created equal, as democracy, invented in the eighteenth century—when there was no science to refute it—would have us believe," he told the reporter. The human race is moved forward, Carrel added, by "great men." Unfortunately, "we don't yet understand the genesis of great men. Perhaps it would be effective to kill off the worst [of us] and keep the best, as we do in the breeding of dogs."

IN MAKING THAT SUGGESTION, Carrel was saying in public something he'd often said in private at meetings of the Philosophers Club. For years, members of that group, which now included Lindbergh semi-regularly, urged Carrel to collect his thoughts on science, art, politics, family life, and religion, and so on, in a book. (Frederic Coudert, the founder of the club, thought so much of Carrel that he named his son Alexis Carrel Coudert.) At first, this suggestion provoked uncharacteristic modesty in Carrel: "Who on Earth would be interested in reading *that*?" he asked. Finally, he decided his friends were right: *Man, the Unknown*, most of it written in early 1934, had just been published in the United States by Harper, and in France by Plon. Its official release date, in fact, was just a day or two before Carrel granted his shipboard interview to the reporter for the *Times*, and this may explain why he gave the interview in the first place.

Whatever the ultimate reason for Carrel's cultivation of publicity, it bore fruit. The public's interest in his work with Lindbergh at the Rockefeller Institute—already high, now that they'd published their preliminary results—became even higher. One autumn morning at the institute, Lindbergh, who had several perfusion pumps in operation nearby, looked up from his workbench and saw Carrel pointing them out to a tall man with a sad face, a droopy mustache, and a shock of white hair: Albert Einstein. A few weeks later, members of the general public got a chance to express their fascination with Carrel and Lindbergh's death-defying research, a fascination so intense that the police had to be called.

What happened was this: Months earlier, Carrel, who'd already amazed the world by keeping a piece of chicken heart alive in a jar for twenty-three years, agreed to give a lecture on "The Mystery of Death" at the New York Academy of Medicine, an elegant building on the corner of Fifth Avenue and One Hundred Third Street, in Manhattan. His lecture, part of an ongoing series of talks at the academy given by scientists on "The Art and Romance of Medicine," was open to the public and thus was announced in the mainstream press. More than 5,000 New Yorkers tried to squeeze into the 700-seat auditorium to hear Carrel, creating a frenzy that the *New York Times* likened to "a fight crowd at Madison Square Garden." According to Sergeant Cornelius Link of the New York police, 3,000 insistent citizens jammed themselves into the hall, most of them ending up on their haunches in the aisles, while a few hundred others stood and listened to loudspeakers in the lobby outside. At least 2,000 more were sent home by policemen waving nightsticks.

Those frustrated citizens had to wait till the next morning, when the *Times*, in a story beginning on page one, published a summary of Carrel's lecture. To the surprise of many, Carrel began by praising death as the builder of civilizations.

"The weak, the diseased, and the fools were not capable of resisting its attack directly, or of learning how to protect themselves against illness," Carrel said. "Through natural selection, the strong and the intelligent persisted, and the great races and their civilizations developed." For the time being even the "strong and intelligent seem powerless before death," but "perhaps someday," Carrel said, "a pleiad of geniuses greater than Galileo, Newton, or Pasteur may rise and explore the abysses of our body and unveil the mystery of death."

An important step in that direction, Carrel said, had just been achieved by the perfusion apparatus designed by another man of great skill, Charles Lindbergh. His device gives scientists a "new method of opposing death," Carrel said, by "analyzing the conditions responsible for the aging of tissues" while studying them in whole organs, outside the body, in a "glass chamber of life" where they can "live indefinitely." Subsequent findings by Lindbergh and Carrel, the surgeon predicted, will "open new possibilities." But here Carrel issued a warning: "The fight of man against death will perhaps succeed too well," he said. "For the artificial postponement of death of a large number of individuals would be a far greater calamity than death itself."

This was the closest Carrel had yet come to revealing that his search for a biological path to immortality was not intended as a public health initiative. If Carrel ever did find the secret of eternal life, he was certain of one thing: it would remain just that—a secret—to all but a special few.

Carrel was even more energetic than usual after his lecture. He returned to the Rockefeller Institute early the next morning, after his customary walk around the Central Park reservoir, intent on moving his organ culture research to the next level. Perhaps Lindbergh could design a bigger perfusion device there, one that could perfuse large human organs. That would be a significant step forward in their shared quest, one with immediate and historic applications. They'd start quietly with organs from primates; then use fresh human cadavers; and then, if all went well, make even quieter studies with organs from live humans.

A letter in Carrel's archives suggests that he and his protégé were already laying the groundwork for such research. Sent to Carrel in 1935 by E. P. Earle of the New Jersey Board of Institutions and Agencies, the letter asked: "When are you and Col. Lindbergh going to Vineland"—a state mental institution—"with Commissioner Ellis and me to look over some of our feeble-minded 'prospects'?" As Carrel had no training in psychiatry, one must assume that those "prospects" were being "looked over" for experiments taking place in his operating rooms. This is chilling, if not conclusive, evidence that Carrel and Lindbergh were considering a program of perfusion experiments on live humans, maybe even without their informed consent.

But no matter what those "prospects" were for, there's no doubt that, from where Carrel was sitting, the future looked very exciting indeed. He was thrilled by the attention his lecture at the Academy of Medicine received. In his view the public acclaim of his genius wasn't flattery; it was evidence of a great truth confirmed. Not everyone saw it that way, however.

One who didn't had just become director of the Rockefeller Institute. This was Dr. Herbert S. Gasser, who replaced Dr. Simon Flexner in that post, after Flexner's retirement, on October 1, 1935. For Carrel this was more than a change in the institute's letterhead. Flexner had hired Carrel in 1906 and for the next three decades had been his great friend and patron, allowing Carrel to run his Experimental Surgery Division as Carrel saw fit. In fact, Flexner was one of the few people aware, right from the start, that the ultimate goal of Carrel's research was a medical path to immortality. This, too, Flexner supported by noninterference, even though he wasn't interested in reaping the rewards for himself. "I have no desire to be immortal," he wrote to Carrel. "I shall enjoy the long sleep, if enjoy it I can. But I should like to wake up once in a hundred years to come back with you to see what has happened and let you explain to me its philosophy."

Gasser didn't give a damn about Carrel's philosophy. The new director made this clear in a letter he sent to Carrel shortly after the lecture at the Academy of Medicine. "Had I seen the manuscript before it was read, I should have counseled caution with regard to certain statements." Those incautious remarks, in Gasser's view, were Carrel's assertion that the immortality of the human soul had been "demonstrated by spiritualists," and Carrel's even bolder claim that "we know positively that clairvoyants are capable of perceiving past and future events." Gasser wrote: "The experiences of spiritualistic séances do not pass the test for valid data.... The exactions which the world places on science make me hold a most conservative attitude about the making of such statements." Clearly, Gasser was urging Carrel to adjust his own attitude in that same direction—and soon.

Carrel shrugged off the criticism. But the publicity he'd generated since his return from France in the fall, coming after the wave of stories during the summer announcing the invention of Lindbergh's perfusion pump, had another impact he couldn't ignore. This is because it involved Charles Lindbergh. Those ubiquitous press accounts had the effect of robbing Carrel's press-shy protégé of his only remaining safe haven on terra firma: the Rockefeller Institute. Reporters and photographers now staked out the institute buildings on York Avenue, hoping for a sighting of Lindbergh; other journalists did the same outside Next Day Hill, where Lindbergh was living with his wife and young son, Jon. For Lindbergh there seemed no peace—and no prospect of finding any.

Lindbergh's contempt for the press hadn't abated over the years, and his animosity intensified when the appeals raised by Hauptmann's defense team got a hearing in the fall of 1935 from the governor of New Jersey, Harold S. Hoffman. (Hauptmann's execution had been set for January 17, 1936.) LINDBERGH CASE REOPENED, said the *New York Daily News*. The *New York Daily Mirror* angered Lindbergh even more by publishing installments of HAUPTMANN'S OWN STORY! That series, in turn, produced a storm of hate mail from Hauptmann's supporters, most of whom were convinced that

Lindbergh had perjured himself on the witness stand. "There are millions of people all over the world waiting for you to save this man that you condemned with your lying words," one letter said. Another anonymous writer claimed the role of Lindbergh's judge and executioner: "Well Lindy . . . here is hoping when you and your China-faced wife go up in a plane you will both come down in flames."

Threats to his wife and himself were bad enough, Lindbergh thought, but threats to his son were completely unacceptable—and these too were arriving in envelopes almost daily at Next Day Hill. And now the press was harassing the boy. Jon was a student at the Little School, a progressive nursery established in 1930 in Englewood by Anne's late sister, Elisabeth. (The school, still in existence, is now called the Elisabeth Morrow School.) One fall afternoon, as Jon's nurse was driving him home from school, a car full of strangers sped by and forced the nurse's car off the road, leaving Jon terrified and bawling in the arms of the equally frightened nurse. Then a man from the other car jumped out and ran over to the nurse's car. He thrust a large camera in the boy's face, scaring him all the more; as the photographer snapped away, each burst of his flashbulb seemed like gunfire.

The press was everywhere—at the Rockefeller Institute, outside the house at Next Day Hill, and now on the public roads. Lindbergh told his mother in Detroit that "between the politician, the tabloid press, and the criminal, a condition exists which is intolerable for us." In early December he made a decision: he told his wife to be ready to move to England for the winter (or longer), on twenty-four hours' notice. Lindbergh chose England because he believed that "the English have greater regard for law and order in their own land than the people of any other nation in the world."

Lindbergh informed his mother of his plan, then the Morrow family, and then the other most important person in his life: Alexis Carrel. Carrel wasn't happy to learn that his research partner was about to leave the institute, especially now, after they'd just achieved their first great breakthrough in their organ perfusion project. But he knew better than to ask Lindbergh to change his mind. Great men, Carrel understood, must trust their own counsel. Besides, Lindbergh planned to build a small laboratory of his own in England, once he got settled there. He'd keep Carrel abreast of his perfusion research by mail or telephone. Then, come summer, the two would re-join forces at Carrel's vacation home on Saint-Gildas, less than an hour's flight from London.

To help Lindbergh gain access to the British medical community, Carrel gave him a handwritten letter of introduction to the most prominent ex-perimental surgeon in England, Lord Moynihan. "Dear Moynihan," Carrel began, "I have the pleasure to introduce to you Colonel Charles A. Lind-bergh":

> Colonel Lindbergh is very much interested in biology. He has, as you know, invented an apparatus which allows us to keep entire glands alive in vitro, completely protected against bacterial infection. He is going to spend some time in England and would like to become fa-miliar with the work made, under your direction, by the Royal Col-lege of Surgeons. Perhaps you will be so kind as to help him to learn, in England, the subjects in which he is interested.

LINDBERGH HAD ONE MORE telephone call to make—to Lauren D. Lyman of the *New York Times*, perhaps the only reporter in the United States that Lindbergh trusted. He invited Lyman to Next Day Hill and told him of his plan and the reasons for it: threats to his family were forcing Lindbergh to leave America for Britain. He wasn't changing his citizenship, Lindbergh pointed out; he would change only his residence. Lindbergh then offered Lyman an exclusive, if he agreed to hold the story for twenty-four hours after the Lindberghs' departure. Lyman gave his word.

On Saturday, December 21, 1935, Lindbergh called Lyman to say his ship was sailing that night at midnight. At 10:30 p.m., Betty Morrow's chauffeur drove Charles, Anne, and Jon from Next Day Hill to the end of West Twentieth Street in Manhattan, where the freighter *American Im-porter* was docked. The Lindberghs would be its only passengers. When Lyman got word that the ship had departed, he went into the newsroom and wrote his story. When it was finished, the story was carried by a top editor directly to the paper's most trustworthy typesetter. While the text was being set, the *Times* banned all outgoing telephone calls. Monday's paper hit the streets with a four-column headline spread across the front page: LINDBERGH FAMILY SAILS FOR ENGLAND TO SEEK A SAFE, SE-CLUDED RESIDENCE / THREATS ON SON'S LIFE FORCE DECISION.

Ominous letters—too many to ignore, the article said—had been mailed to the Lindberghs, many targeting young Jon for murder. "So the man who eight years ago was hailed as an international hero and a good-will ambassador between the peoples of the world," Lyman wrote, "is taking his wife and son to establish, if he can, a secure haven for them in a foreign land."

—

CHAPTER 7

MEN OF GENIUS ARE NOT TALL

A S CHARLES LINDBERGH SOUGHT A REFUGE ON ANOTHER CONTI-
nent, Alexis Carrel found a home in a place he never expected to visit at all:
the top of the best-seller list. *Man, the Unknown*, the book he wrote at the
urging of his friends in the Philosophers Club, was published in the fall of
1935. It was part physiology, nearly all philosophy, and completely ferocious
in its condemnation of democracy's impact on the physical and psychic
health of the white race. In more than 300 pages, Carrel diagnosed the
disease he saw festering within western civilization, and then proposed a
cure. His analysis was a lumpy blend of science and spiritualism, self-denial
and sexism, nostalgia and authoritarianism, hubris and eugenics—frequently
seasoned with personal prejudice—that found expression in declarations
such as these:

- Men of genius are not tall.
- The feeble-minded and the man of intelligence should not be
  equal before the law.
- The Caucasian nervous system is superior to that found in any
  other race.
- Childless women are more nervous than mothers.
- The abnormal prevent the development of the normal.

- Erotic activity impedes intellectual activity.
- Certainty derived from faith is more profound than certainty derived from science.
- Clairvoyants can see the future and hidden objects at great distances.
- Democracies thwart citizens of imagination and courage.
- Natural selection no longer plays its part because the weak are saved as well as the strong.

Many of the ills facing western civilization, Carrel wrote, could be remedied by a eugenics program that encouraged the fit and eliminated those who weren't. Of all the errors propagated by democracy, the worst, he said, was its assertion of equality and liberty as the organizing principles of society. In truth, both those goals are ideological constructs that deny reality: every individual is different in intelligence, strength, and productivity—and liberty is, in fact, an illusion. An individual may think himself free, Carrel wrote, but that belief merely shows his ignorance of the true mechanism guiding his actions: heredity.

Only eugenics recognizes that mechanism. And for this reason, Carrel wrote, society must single out children with high potential, develop them as completely as possible, and then have them mate to create an aristocracy of merit. But all children, Carrel was quick to add, need the "advantages" of "privation and hardship." Those who murder or kidnap children—a clear reference to the fate of Charles Lindbergh Jr.—should be "disposed of in small euthanistic institutions supplied with the proper gases." The same treatment should be applied to the criminally insane and "those who have murdered adults, robbed while armed with automatic pistol or machine gun, despoiled the poor of their savings, or misled the public in important matters." "Why," Carrel asked, "preserve useless and harmful beings?"

The work of creating useful and beneficial beings, he said, should be directed by a "high council of experts" who would teach the rest of us how to live. Most of these experts would be physicians, professionals uniquely trained, in Carrel's view, to understand human nature. The surgeon thought himself a strong candidate for such a position, of course. In the preface to *Man, the Unknown*, Carrel described "the author" as a man who has observed "practically every form of human activity":

He is acquainted with the poor and the rich, the sound and the dis-eased, the learned and the ignorant, the weak-minded, the insane, the shrewd, and the criminal. He knows farmers, proletarians, clerks, shopkeepers, manufacturers, politicians, soldiers, professors, school-teachers, clergymen, peasants, bourgeois, and aristocrats. The cir-cumstances of his life have led him across the path of philosophers, artists, poets, and scientists. And also geniuses, heroes, and saints. . . . In addition, he has studied the most diverse subjects, from surgery to cell physiology, and metapsychics.*

The members of Carrel's high council would live together, like medieval monks, in ascetic seclusion. Such men—and in Carrel's utopia they would all be men—would function, he said, as the white race's "immortal brain." They would be the salvation of that endangered race, dedicating their lives to the study of natural and social phenomena in order to acquire the knowl-edge necessary to prevent the race from drowning in a sea of inferiors. These researchers wouldn't be polite tenured professors; they would be audacious men of unquestioned eminence, unafraid of giving offense. Through sheer intellectual force, Carrel wrote, this elite would acquire "an irresistible power" over the "dissolute majority" outside its walls.

Carrel also called for a new kind of politics in *Man, the Unknown*—not an improved democracy, but a biocracy: a system in which scientists sorted individuals according to their aptitudes and abilities, thus creating universal efficiency. Social classes, which Carrel saw as artificial, would be replaced by biological classes, which are authentic. Aristocracy would yield to meritoc-racy; the past would yield to the future. The goal of human history—the establishment of a harmonious, well-ordered world where everyone knows his place—would be achieved.

But this can occur, Carrel warned, only after "the fall of the idols [and] the collapse of the windbags." The world must "witness the disintegration of the absurd civilization constructed by the eighteenth century," a period Carrel despised as the nursery of democracy. "This catastrophe will be of the greatest magnitude," he wrote, "but it is entirely indispensable." The

* "Metapsychics" was Carrel's term for extrasensory perception and the like.

"old order" must die so that the "new order" may live. After that tumultuous upheaval, the culture-building prescriptions issued by Carrel's "high council of experts" would be applied in the only meaningful way: ruthlessly.

"Man cannot remake himself without suffering, for he is both marble and sculptor," Carrel wrote. "In order to recover his true visage, man must shatter his own substance with heavy blows of the hammer."

Carrel's cultural prescriptions were familiar to Lindbergh, who'd heard and agreed with nearly all of them—especially Carrel's critique of democracy, racial equality, and the welfare state—in countless early-morning discussions at the Rockefeller Institute. Carrel's views were new to the public, however; and the public couldn't get enough of them. Within weeks of its American publication, *Man, the Unknown* was selling roughly 1,000 copies per day, no small feat in year six of the Great Depression—in fact, no small feat even now. A year later *Publisher's Weekly* declared *Man, the Unknown* the top-selling nonfiction work of 1936—and second in total sales only to a novel called *Gone with the Wind*.

Carrel's book was translated into nearly twenty languages, and its total sales reached over 2 million copies. In France, *L'homme, cet inconnu* was sold with a special jacket band describing it as "The fascinating discovery of our true nature by the remarkable scientist who has devoted his life to the study of the mysteries of our bodies, our minds, and the universe—and who is now about to perfect a machine that will prolong human life." Carrel's German publisher asked the Frenchman to adapt his manuscript to refer to new laws in Germany, put into effect by the Nazi government, requiring the sterilization of citizens with inheritable diseases and the castration of "incurable" sexual criminals. Carrel was no Germanophile—he had spent World War I treating French soldiers blown apart by German bombs and bullets—but he agreed, writing this new paragraph for the German edition: "Germany has taken energetic measures against the propagation of retarded individuals, mental patients, and criminals. The ideal solution"—an unnerving phrase, considering the eventual "final solution"—"would be the suppression of each of these individuals as soon as he has proved dangerous. Criminality and madness can be prevented," Carrel wrote, "only by the rejection of all sentimentality."

Ironically, Carrel was rejected by the first publisher he approached—

Plon, in Paris—a fact gleaned by Madame Carrel when she found her husband's manuscript in a wastebasket in their apartment in Paris, where Carrel wrote most of the book during the winter of 1934. That Carrel was in Paris at all in the winter is testimony to the persuasive powers of his friend Frederic Coudert. According to the latter's unpublished memoir, Coudert convinced the director of the Rockefeller Institute, Simon Flexner, that Carrel should be ordered to vacate his operating rooms for three months in early 1934 so its famous walls could get a fresh coat of black paint. The painting normally would have been done during the summer recess; it was done in the winter, Coudert wrote, to trick Carrel into leaving the institute so that he could work on the book Coudert and his fellow philosophers had been urging him to write for years.

The ruse worked, in the sense that Carrel completed his manuscript. But it also failed, because the editors at Plon rejected the book. Perhaps Monsieur Carrel could take no for an answer, but not Madame. After rescuing her husband's manuscript, she telephoned Plon's chairman, Maurice Bourdel, to demand a meeting. This, according to the French engineer André Missenard, is what happened next:

> Madame Carrel asked Bourdel [at their meeting] why the manuscript had been rejected. He told her that his editorial board had judged the book would interest no one. Madame asked for a list of the members of this panel, then [upon receiving it] criticized them one by one: so-and-so was senile, another an idiot, yet another understood nothing. Moreover, Madame insisted the book would be a great success, which would certainly help Plon's finances, which everyone in Paris knew were in terrible shape. Finally, Madame threw her husband's manuscript on Bourdel's desk. "If you don't publish this book," she said, "it will simply prove that you are a stupid little shit."

There's no evidence that Cass Canfield, who acquired Carrel's book for Harper in the United States, had to endure such a meeting. Like most Americans, Canfield knew Dr. Carrel through his work as a Nobel-winning surgeon and a pioneer in tissue culture. Canfield published his book because of that renown, hoping that Carrel's nonmedical ideas might find a respectable audience of say, 10,000 readers. An excerpt published in *Reader's Digest* surely helped *Man, the Unknown* find a larger readership. So did a positive review in the *New York Times*. "For probably the first time in history," wrote Raymond Pearl, a professor of biology at Johns Hopkins, "the Soul has taken a duly appointed place in a first-rate professional treatise on medical science." Carrel, Pearl wrote, "goes all the way: Telepathy, clairvoyance, miraculous healing and other unusual things are not only subscribed to, but discussed seriously as integral and important parts of human biology."

Other scientists thought "all the way" was way too far. Reviewing Carrel in the *British Medical Journal*, the physiologist Sir Arthur Keith warned that "those of us who believe that the art of healing can be advanced only by careful observation, clear-cut experiment, and sound reasoning will have Dr. Carrel cast in our teeth by charlatans who think there is a shorter path." But Keith's ultimate criticism of Carrel's book was less medical than political. That Carrel called for society to be remade along eugenical lines by "a junta of superphysicians," segregated in a "superinstitute," filled Keith with horror. *Man, the Unknown*, Keith concluded, "reveals far more concerning Dr. Alexis Carrel than it does about Man."

But nonscientists around the world were thrilled by those revelations. Despite his own elitism, Carrel wrote a book that appealed to a wide range of groups, some intellectual, most not, who found confirmation of their own ideas in his. *Scouting*, the journal of the Boy Scouts of America, praised Carrel's advocacy of hiking in *Man, the Unknown*. *Psychic News*, a British magazine for séance leaders, was similarly impressed, if for different reasons. The left praised Carrel's critique of the alienation caused by industrialism; the right praised his contempt for parliamentary democracy. Radicals celebrated his call to remake society from the ground up; conservatives celebrated his acknowledgment of elites. Hundreds of stay-at-home mothers on both sides of the Atlantic wrote to Carrel to thank him for praising their contributions to society.

Clerics were among Carrel's loudest champions. The Reverend George A. Buttrick of the Madison Avenue Presbyterian Church in New York delivered a sermon on *Man, the Unknown* on March 1, 1936, a few weeks after another New York cleric, Rabbi Stephen S. Wise of the Free Synagogue, called Carrel's book an "oasis of beauty in a world without beauty or spaciousness." "Like all true things," Wise said from his pulpit, grasping a truth about Carrel more accurately than perhaps he knew, "this book, and the man who wrote it, are testing for immortality."

Actually, *Man, the Unknown* had already achieved a kind of immortality for its author. On September 16, 1935, Carrel was accorded what was in that decade, and for many decades to come, the highest recognition one could hope for in western popular culture: his round and somewhat fierce visage, emerging from a flowing black robe and topped by a little white cap, was put on the cover of *Time* magazine. Carrel was delighted to be in the same company as Benito Mussolini, General Douglas MacArthur, and J. Edgar Hoover, men he admired greatly. What gave the first-time author even more pleasure was knowing that he was one of only two writers so honored by *Time* that year.*

CARREL WAS NO LONGER the only Nobel laureate at the Rockefeller Institute, but he was the only member to have written a best seller, to have his face on the cover of *Time*, and to be known as the scientific mentor to the world's most famous man—feats several of Carrel's colleagues deemed vulgar, which he noticed: "For me, personally, [*Man, the Unknown*] has been something of a calamity," Carrel wrote to a friend after his appearance in *Time*. "[My] colleagues are offended at its great success . . . and accuse me of writing it out of a love for publicity!" That Carrel's new acclaim arose not from serious laboratory research, the clearly stated mission of the Rockefeller Institute, but from a rumination on the failings of western civilization that proposed a kind of biological fascism as its cure, was especially troubling to many of Carrel's peers. Several thought he'd crossed an important line. Was Carrel still a scientist? Or was he a fame-seeking fascist mystic?

There had always been resentment of Carrel's special status at the institute—the huge budget for his mousery, the black uniforms he demanded that the institute provide for his staff, and so forth—and now this resentment began to turn malignant. It didn't help that Carrel often held himself aloof from his colleagues. He typically ate alone in the lunchroom,

* The other was the novelist Kathleen Norris, by coincidence a member of the team sent by the *New York Times* to report on the trial of Bruno Richard Hauptmann.

or only with members of his staff; and he discouraged other members of the institute from observing his work on organ cultures, which continued even though Lindbergh was no longer there to provide on-site technological guidance.

Carrel's door was still open to celebrities, however. The English literary star Aldous Huxley, author of *Brave New World* and *Point Counter Point*, visited Carrel's laboratory not long after the publication of *Man, the Unknown*. Huxley was so taken with Carrel that he modeled a character after him—the mysterious Dr. Obispo—in his next novel, *After Many a Summer Dies the Swan*. Published in 1939, this book told the story of an American multimillionaire so obsessed with the idea of living forever that he hired a brilliant physician, Dr. Obispo, and set him up in a private laboratory to find a medical path to immortality. Obispo accepted that seemingly impossible challenge as a liberation from the ordeal of caring for normal patients. Besides, the doctor believed that if he succeeded he would have done a great service to humanity—or, more precisely, to the very select segment of humanity worthy of eternal life.

The multimillionaire who really signed Carrel's paychecks, John D. Rockefeller Jr., kept his opinions of Carrel's extracurricular career to himself. Not so the man he entrusted to run the Rockefeller Institute, however. Herbert Gasser was appalled by the role Carrel claimed for himself as a cultural critic. Nor was Gasser impressed that the public seemed all too willing to embrace the surgeon as such. That some of Carrel's readers showed up now and then at the institute demanding to meet Carrel was a distraction from the real work of the institute which Gasser refused to countenance.

The director's patience almost snapped when one of those frustrated visitors—obviously deranged—killed himself in the institute's lobby. "His name was Yulian Syzygy," Carrel wrote, shortly after the incident. "He walked into the main office and, when asked whom he wished to see, gave my name. He was told to wait in the Reception Room." When Carrel's office sent word that the scientist was too busy to come down, "the man swallowed cyanide and died twelve minutes later." Carrel was shocked and horrified. Gasser was nearly apoplectic.

Gasser had already warned Carrel, after the "Mystery of Death" speech at the New York Academy of Medicine, that the institute was not well served by having one of its most prominent members speak approvingly of clairvoyance and the quest for immortality. Gasser was furious that Carrel had forced him to restate this position. The proper work of an institute member, Gasser reminded Carrel, was done quietly and professionally in the research laboratory—and only there.

The two men had never been close, and now they were farther apart than ever. It was not in Carrel's nature to take direction easily, especially from someone he didn't consider his intellectual equal. Gasser would later win a Nobel Prize for his own work on nerve cells, but he wasn't, in Carrel's view, a first-rate thinker, at least not on the big subjects close to Carrel's heart. Gasser was, however, a skilled political player. Whether Alexis Carrel realized it or not, he'd made a powerful enemy. It wasn't long before he'd know it all too well.

CHAPTER 8

A TINY PUFF OF SMOKE

LINDBERGH WAS THRILLED BY CARREL'S PUBLISHING BREAK-through. The "demand for your book . . . seems to increase with every month," he wrote to Carrel from Long Barn, the house Lindbergh had moved into with Anne and their son, Jon, in the Weald of Kent, some thirty miles south of London. "Your book's success," he added, "is very, very significant."

This was not as easy for Lindbergh to say as you might think: the book Carrel had knocked off the top of the best-seller list in the United States was *North to the Orient*, Anne's account of the Lindberghs' 1931 aerial expedition from New York to China and back, a trip that included stops at Inuit villages above the Arctic Circle where Anne was the first Caucasian woman the locals had ever seen. ("I wouldn't take *my* wife into that territory," a Canadian official had told Lindbergh. "You forget," he replied, "she's crew"—a moment Anne later described as "the proudest moment of my life.") The literary glory achieved by Lindbergh's blunt-speaking mentor and the pilot's shy but courageous wife wasn't just gratifying to Lindbergh, it was confirming: their successes intensified his feelings of separateness and superiority. The three of them, individually and together, had reached the pinnacle of the most exclusive elite of all: the elite of genuine achievement.

And no one—not the press, not a small-minded scientific bureaucrat, not even a semiliterate kidnapper—could take that away.

Not that the British press didn't try, at least at first. Lindbergh brought his family to England to find peace, only to be met by chaos. The dock in Liverpool where *American Importer* moored on the final day of 1935 was "besieged by the largest international gathering of newspapers and photographers ever assembled" in that city, according to the *Liverpool Daily Post*. After policemen wielding nightsticks cleared a path down the gangplank, Charles emerged to a meteor storm of flashbulbs, carrying his frightened three-and-a-half-year-old son, who buried his face in his father's shoulder. A waiting limousine took the Lindbergh family to a nearby hotel, where they barricaded themselves in a suite guarded by a private detective.

Lindbergh was happy to see that the British soon lost interest. The press didn't camp out in front of the eccentric home he'd rented from the British writers Harold Nicolson and Vita Sackville-West. In truth, Long Barn was less a house than three cottages and a barn pressed together: the oldest parts, built of oak beams salvaged from sailing ships, dated from 1380. The roof sagged, the floors sloped, and the walls slanted—and both Charles and Anne found all this charming. Lindbergh was also pleased that none of the locals bothered him as he walked around Sevenoaks, the nearest town with a railroad station and post office, to do errands and mail his letters.

But Lindbergh hadn't come to England for a holiday. Though he was thousands of miles away from Carrel, he had no intention of abandoning their scientific research. Lindbergh claimed a wing of Long Barn as his laboratory and library, where he quickly set up his pumps, gas canisters, microscopes, and so forth, along with his physiology and biology textbooks. Carrel gave Lindbergh a big job to do in England: he was to design and build a mechanical kidney to work with his perfusion pump.

Lindbergh's current device used sand and screens to filter out impurities in the perfusion medium and waste products emanating from the perfused organ. But those simple screening tools were not sufficient to clean the perfusion fluid over long periods of time; as a result the fluid had to be replaced at regular intervals. If Lindbergh could make a mechanical kidney to filter and drain the wastes sent into the medium by the organ inside his pump, the need for replacement fluid (and the labor and observation that it entailed) would be greatly diminished. This would also move Lindbergh and Carrel a significant step closer to their next goal: experimentation on larger organs—including those from humans.

Designing a more proficient filtering system didn't strike Lindbergh as an insolvable problem. He was far more worried about preventing contaminants from entering his device as the mechanical kidney was draining wastes out. Lindbergh started his research in his library, reading everything he could about the form and function of the human kidney. Then he began studying even more detailed physiology texts and journals at the Royal College of Physicians library in London. Before long, however, this research was momentarily halted by a distraction from an all too familiar source.

Shortly after 8:40 p.m., on April 3, 1936, Bruno Richard Hauptmann, his head shaved and his feet encased in bedroom slippers, shuffled into the execution chamber at the New Jersey State Prison in Trenton, where he was strapped into the electric chair. As fifty-five witnesses (including Colonel H. Norman Schwarzkopf of the New Jersey state police and thirty newsmen) looked on, the executioner's assistant attached the first electrode to Hauptmann's right calf, just beneath the split in his trousers that had been put there for this purpose. Then the executioner took a rubber skullcap that had been dipped in a nearby pail of salt water and secured it to Hauptmann's head by tightening a chin strap. Inside the cap was the second electrode. The reporters held pencils in their hands, in case Hauptmann had something to say.

He didn't. At 8:44, Warden Mark O. Kimberling signaled the executioner to proceed. Standing behind the electric chair, his back to the prisoner, the executioner turned the wheel on his generator's control panel. The generator made a whining noise, after which witnesses heard a sudden creaking of the leather straps restraining Hauptmann. Seconds later a tiny puff of smoke materialized above his head.

The United Press had cabled Lindbergh earlier that day: "WHILE APOLOGIZING FOR THIS INTRUDING YOUR PRIVACY WOULD LIKE TO OFFER FULL FACILITIES OF UNITED PRESS OF AMERICA IF YOU SHOULD DESIRE TO MAKE ANY STATEMENT WHATSOEVER IN CONNECTION HAUPTMANNS EXECUTION." He made no reply. One can only wonder if the date of the execution rang a bell: it was precisely one year after Lindbergh's perfusion pump was used in the first successful organ culture experiment in Carrel's laboratory—the first successful long-term organ culture experiment in history.

If Lindbergh did notice the coincidence, he never said so. Instead, he wrote to friends indicating his surprise that Hauptmann's execution led so many American editorial writers to assume he would now be bringing his family back home. "There has never been any question about returning now," Lindbergh wrote to Dr. Abraham Flexner, Simon's brother. "The crime situation [in America] is a little better, but the newspapers have improved little, if at all. . . . [Anne and I] are very happy in England."

Lindbergh was happiest in his laboratory, making design sketches for his mechanical kidney, then checking those plans against notes he'd taken at the Royal College of Physicians. Would a glass chamber be sturdy enough to hold the larger organs he hoped to perfuse? Would a metal chamber be better? What sterility issues would that change raise? Was his oil flask mechanism powerful enough to send sufficient oxygen and nutrient medium into those larger organs? Would he have to design a new pumping system? Was there a physiological issue he'd overlooked?

Questions like these swirled through Lindbergh's mind. He shared most of them with Carrel. The two men corresponded two, three, or sometimes even five times a week. Some of these letters detailed the problems—many of them unexpected—Lindbergh encountered in his new home. Designing an artificial kidney was hard enough; getting it made in England, Lindbergh learned, was even harder. Now that he was on the other side of the Atlantic, Lindbergh no longer had easy access to the glassblowing wizard Otto Hopf or the other technicians he'd collaborated with in New York. He tried to partner with British manufacturers but was consistently frustrated, he told Carrel, by their "different conception of time." When a gasket in one of his early prototypes needed fixing—a job that would have taken one day in America—Lindbergh's British manufacturer asked for two weeks.

Though Lindbergh was still angry at America, he felt only admiration for the Americans he'd worked with there. "I have missed working in [Carrel's] department more than anything else since I sailed for Europe," he wrote to Dr. Irene McFaul, one of Carrel's top aides. "You have an exceptionally fine organization there . . . [and] I hope very much that I will be able to work [there] again in the future. . . . You must have been very busy since I left, as I understand the number of completed [organ] cultivation experiments is now over 200." Most of those experiments tested different recipes for the perfusion media; Carrel wanted to find the safest and most efficient substance before he began attempting to treat diseased organs perfused inside Lindbergh's machine.

In England, Lindbergh hadn't been able to complete even one artificial kidney experiment successfully. This meant that his chance of achieving the goal Carrel had given him was next to nil unless he were willing to do the one thing he wasn't yet willing to do: return to America. In the end, Lindbergh dealt with his frustration by making a decision that was one part expedience, one part stubbornness, and all self-confidence—a typically Lindberghian formulation. He filed away his design sketches for the mechanical kidney and moved on, working on the assumption that he *had* built a working prototype. His new goal, Lindbergh decided, was to prepare for the next round of studies, which would take place in a perfusion pump equipped with that as yet unbuilt and untested artificial kidney, even if those studies wouldn't occur until he returned to New York.

To advance his immortality research now, without going back to America, Lindbergh knew he'd have to learn all he could about organs more complex than the thyroids and ovaries from cats, dogs, and chickens that Carrel and he had already worked with. These new organs would be extracted from beasts as close to *Homo sapiens* as possible. With that huge step in mind, the man who'd already transformed himself from a failed farmer into the world's most admired aviator, and then from a college flunkout into the scientific partner of a Nobel Prize–winner, reinvented himself again: Lindbergh the perfusionist became Lindbergh the primatologist.

He began this transformation by initiating a correspondence with Dr. Hayden B. Harris, one of America's leading experts on apes and monkeys. Before long the two men developed a plan titled "Research Laboratory for Using Subhuman Primates." Several sites were discussed as a potential home for this research center, including eastern Maryland, Cuba, and Jamaica. (The advantage of the Caribbean locales was that there the animals could be raised in different environments: at sea level and in the mountains.) Different primate species, Lindbergh learned, presented different research opportunities. Baboons thrived in temperate climates and were even-tempered—at least, as even-tempered as a baboon in semi-captivity could be. Cebus monkeys had high intelligence, as well as a record of living ten years under close human supervision. Spider monkeys had the most pliant dispositions. Squirrel monkeys were the smallest. Chimpanzees were the easiest to procure in large numbers.

Lindbergh kept Carrel abreast of this new research. "I have been thinking more about Anthropoid Apes," he wrote. "It would be desirable [for our experiments] to have animals with a minimum of hereditary difference. It takes too long to inbreed (years to a generation), but it has occurred to me that we might be able to find already inbred gibbons on some of the small islands off the coast of Indo China or the Malay Peninsula. If they . . . have no access to the mainland, they must be fairly well inbred."

When Lindbergh heard, in the spring of 1936, that several gibbons were kept on a tiny man-made island surrounded by a moat, at a château near the French city of Rouen, he flew there with Anne to see how the apes were cared for, to learn what vegetation and other ecosystems had been provided for them, and to assess their health in captivity. At Carrel's suggestion, Lindbergh also went to Paris on this trip, where he met with Carrel's former assistant at the Rockefeller Institute, the biophysicist Pierre Lecomte du Noüy, now a member of the Pasteur Institute in Paris, which maintained a chimpanzee colony for medical research purposes in Africa.

While visiting with the du Noüys, Charles and Anne met Madame Carrel for the first time. "Mme. Carrel is my idea of a very wonderful woman," Anne wrote in her diary. "She has a woman's emotion . . . and yet a man's breadth of mind, breadth of view, clarity of vision, and impersonality of attitude." Anne called it the "scientific attitude," perhaps with some envy, knowing that her husband wished there were more of that attitude in her.

Charles, too, was smitten with Madame Carrel, especially when he learned that her "clarity of vision" extended to things that couldn't be seen. While walking in the du Noüys' grounds after lunch, Lindbergh encountered Madame Carrel after she'd inadvertently dropped her wedding ring. He watched her draw a map of the grounds; then, as his fascination increased, she took out a small pendulum, which she held over the map. The weight remained still until it hovered over a particular spot, when it started to swing. She and Lindbergh then walked to that section of the grounds. "Mme. Carrel took the lead," holding the pendulum in front of her, Lindbergh later wrote. "About 200 yards from the house, and in the area which she had located [on her map], the pendulum began to swing in circles." Moments later, as they kept walking, the pendulum started to "swing violently." The two of them looked down: there was the ring.

AFTER RETURNING TO ENGLAND, Lindbergh decided to merge his new interest in paranormal phenomena and the occult with his preexisting interest in immortality. At the library of the National Institute for Medical Research in London he read detailed reports of holy men in India who'd extended their lives by lowering their pulse rate and body temperature. Lindbergh was enthralled. "Had Yogic masters found ways of sharpening perception until it could pass through walls enclosing the normal mind and senses?" he wondered in *Autobiography of Values*. "Was it true that mystics sometimes sat in mountain snows for hours without any sense of being cold?" "Did their body temperature decrease?" "What life-extending benefits could be obtained by such procedures?"

Lindbergh asked these questions of Sir Francis Younghusband (a Briton born in India), at Younghusband's country home, not far from Long Barn. The living embodiment of a Kipling character, Younghusband had led a life of extraordinary adventure in the farthest reaches of the British Empire. It was said that he'd single-handedly invaded Tibet, discovered the source of the Indus River, made one of the earliest attempts by a westerner to climb Mount Everest, and had been the first European since Marco Polo to travel from Peking (now Beijing) to central Asia, only to abandon his life of physical heroism to become a mystic. Lindbergh was mesmerized by his host— "beneath his white lashes, deep-set eyes held you"—but was disappointed to learn that Younghusband didn't know if anyone had ever inserted a thermometer into the body of a meditating yogi or performed similar biological tests. Younghusband would be returning to India the following spring, however, and offered to introduce Lindbergh to yogis there, so that he might perform such tests himself. Lindbergh immediately accepted the invitation.

He was distracted, however, by two letters that arrived weeks later at Long Barn. One was from Carrel. A former tissue culture assistant of his—Dr. Albert Fischer, now chief of the Carlsbad Biological Institute in Copenhagen—was organizing the International Congress of Experimental

Cytology (the branch of biology dealing with the structure, function, and pathology of cells) in that city in August. Fischer hoped Lindbergh would demonstrate at that Congress the perfusion pump he'd invented. Carrel urged Lindbergh to do it. Lindbergh agreed, telling his mentor he'd get to Denmark several days before the event, to ensure that the device, once it arrived from New York, was properly set up.

But Lindbergh told Carrel he'd be stopping someplace else before he got to Copenhagen. He'd received another invitation about the same time Carrel's letter arrived. This second letter came from an American military attaché stationed in a European country quite close to Denmark. The attaché was asking if Lindbergh would consider traveling to that country to inspect that nation's rapidly growing aviation industry. The host country's government, the attaché wrote, was eager for someone of Lindbergh's stature to see the progress they'd made. Lindbergh said he would be happy to; he'd fly there with his wife. All he asked was that they be treated like any other private citizens once they arrived—with no parades or special honors.

The reason he agreed to the attaché's request, Charles said, was that it meant going to a place he'd always wanted to see, especially now, when that place was experiencing a political and cultural rebirth unrivaled in Europe, led by a man who had defeated unemployment and crushed communism— a man who seemed to be the most virile and dynamic leader on the planet. That fortunate and special place, Lindbergh told Carrel, was Adolf Hitler's Germany.

CHAPTER 9

THE MOST INTERESTING PLACE
IN THE WORLD TODAY

T HE INVITATION WAS SENT BY U.S. ARMY MAJOR TRUMAN SMITH, whose job in Berlin was to monitor the German military, then assembling a huge well-equipped air force, the Luftwaffe. Smith was certainly aware of that, but he was an infantry specialist; he needed an aviation expert to be his eyes and ears, and then explain to him what the details meant. Was there an American anywhere more qualified for that mission than Charles Lindbergh?

When Smith sounded out the Germans about having the world's most famous pilot tour their aviation sites, approval came down almost immediately from Hitler's air minister and chief confidant, General Hermann Göring. Before moving ahead, though, Smith asked for a list of the factories and bases that would be made available to Lindbergh for inspection. When Smith saw that this list included places never before seen by an American, he wrote to Lindbergh, on May 25, 1936:

Although I have not had the pleasure of your personal acquaintance, I feel free on account of my position in corresponding with you with respect to a possible desire on your part to visit Germany. . . . In a recent discussion with high officials of the German Air Ministry, I was requested to extend to you in the name of General Göring an invitation to visit Germany and inspect the new German civil and military establishments. . . . I need hardly tell you that the present German air development is very imposing and on a scale which I believe is unmatched in the world. Up until very recently this development was highly secretive, but in recent months the [Germans] have become extraordinarily friendly to American representatives. . . . From a purely American point of view I consider that your visit here would be of high patriotic benefit. I am certain that they will go out of their way to show you even more than they show us.

Lindbergh accepted Smith's invitation after the dates of the trip—the last week of July—were rearranged to accommodate two previous commitments: the perfusion demonstration with Carrel at the Cytology Congress in Denmark; and a meeting in France with a man said to have amazing abilities with a pendulum—a friend, no doubt, of Madame Carrel's. A letter Lindbergh sent to his mother in Detroit, telling her of his decision to go to Germany, is revealing, not so much for what it said as for what it didn't say: "Comparatively little is known about the present status of Aviation in Germany, so I am looking forward, with great interest, to going there. . . . Germany has taken a leading part in a number of aviation developments, including metal construction, low-wing designs, dirigibles, and Diesel designs. If it had not been for the war she probably would have produced a great deal more."

What Germany also had produced after World War I, and before Lindbergh's visit, was the Nuremberg code of "racial purity" laws passed in September 1935, which stripped German Jews of their citizenship rights, another civic insult in a two-year process that already barred them from public office, the civil service, journalism, farming, and professorships in German universities. These laws were neither passed nor implemented in secret; they were written about in newspapers around the globe and prompted several large protest demonstrations, including one attended by 20,000 Americans at Madison Square Garden, when Lindbergh was working less than three miles away at the Rockefeller Institute. It's simply not possible that Lindbergh was unaware of the worsening human-rights situation in

Germany before he went there. Once again, however, Charles Lindbergh was less interested in people than in machines.

On July 22, 1936, the Lindberghs landed their plane at a military airport outside Berlin. The first words they heard in Germany were "*Heil Hitler!*" barked by heel-clicking soldiers standing beside fifteen Luftwaffe bombers. The Lindberghs were driven into Berlin in two cars—Charles with Major Smith, and Anne with Smith's wife, Kay, and the Smiths' daughter, Katchen. Anne recorded her impressions of that drive in her diary that night: "The neatness, order, trimness, and cleanliness [of Berlin]. . . . The sense of festivity, the Nazi flag, red with a swastika on it, *everywhere.*"

The Lindberghs would be separated for much of their stay. Charles's days typically began with the arrival of Major Smith and his aide, Captain Theodore Koenig, who drove Lindbergh to the sites on that day's schedule. One morning Lindbergh flew two German planes: the Junkers JU 52, the mainstay of the Luftwaffe's new bomber force; and the Hindenburg, an experimental plane that had four of the biggest engines Lindbergh had ever seen. At a Junkers factory he saw the new liquid-cooled JU 210 engine, a machine far more sophisticated than anything he had anticipated. An equally eye-opening experience occurred when he was shown the dive-bomber and fighter plane made for the Luftwaffe by the Heinkel Corporation—both, Lindbergh concluded, superior to anything made in the United States. The next day he met pilots from the Richthofen Geschwader, the Luftwaffe's elite fighter squadron. Lindbergh's respect for German aircraft was now matched by his awe at the national pride and disciplined focus of the men who designed and flew those planes.

Anne, meanwhile, was busy experiencing Berlin, a city marching to a military beat in peacetime. "The sound of drums, music. What is it, a parade?" she wrote in her diary on July 23. "Oh, no, [just] the changing of the Air Ministry Guard. Is it something special today? Oh, no, it happens every day—the clomp clomp of the feet, the goose step." Then, a few days later: "With the Smiths in the car. . . . All kinds of uniforms in the streets. S.A. in brown, S.S. in black—Hitler's special shock troops (one is always guarding his entrance). Also Hitler Jugend. . . . The boys wear black corduroy shorts and brown shirts. The girls wear long black skirts. (Hitler likes modesty in women!) The crowds are awe-struck. '*Wunderbar!*' It is a great show."

An even greater show took place on July 28, at Göring's palatial home on Wilhelmstrasse, where the air minister, wearing a crisply pressed uniform covered with medals, hosted a lunch in Charles's honor. Though Anne had grown up in a large house and lived for a while in the American ambassador's residence in Mexico City, she found herself, to her surprise, feeling "small and out of the picture" in Göring's home.

Not her husband: Göring's focus was directly on him. The air minister, who'd recorded twenty-two "kills" as a fighter pilot in World War I, asked Lindbergh about his celebrated flight to Paris, then about what he was doing in England, then specifically about the scientific work he'd started in New York with Carrel. Was Lindbergh working on a new perfusion pump? "Why don't you work here, in Germany?" Göring asked. "There are better scientists here." Once the remarks were translated, Lindbergh just smiled.

After the meal an aide brought in Göring's pet lion, who sprawled on his master's lap. Then there was an awkward development: the lion urinated on Göring's leg, but the air minister was the last to notice. "Just like a child," Göring said, when he finally did. The aide took the animal away; Göring left as well. When he returned he was wearing civvies, rouge, and the sweet stink of cologne. Now the Lindberghs and the Görings moved to the general's office, a room lined with books, paintings (including one by Hitler), and statues, where Göring continued to expound on Germany's scientific superiority to Britain, France, and the United States. A photographer was called in to record this moment—not just any photographer, but Hitler's personal photographer, Heinrich Hoffman. This photo would be seen around the world: General and Mrs. Göring, Colonel and Mrs. Lindbergh, and the interpreter, standing beside Göring's massive desk, on which sat another famous photo by Hoffman—a framed portrait of *der Führer*.

Lindbergh would never meet Hitler, though the two sat not far from each other at the Olympic Games of 1936, in Berlin. Even so, Lindbergh was convinced he knew what he needed to know about the man and the country Hitler had transformed. For Lindbergh, Germany seemed everything that America was not and probably never could be: a country composed of one virile, morally and ethnically pure race committed to science, and united in a vision of national greatness. That such unity came at the cost of democratic institutions, individual rights, and a free press didn't alienate him. Democracy was a noble ideal, Lindbergh believed, but the reality was quite different. This was especially true, his own history taught him, in the United States, where social and political equality, together with a free (and, to Lindbergh's mind, irresponsible) press, produced a climate of degeneracy that led to the murder of his firstborn son. Only a strong, visionary, and, yes, even fascist leader was best equipped to restore moral order to western civilization. (Hadn't Carrel been telling him that for years?) Whatever price was required for that transformation, Lindbergh believed, would be worth it.

Lindbergh committed these views to paper in a letter-writing spree. One of the first letters was a handwritten note to Carrel, who was then with his wife in Paris. "I had intended to write you before this," Lindbergh began, "but it is difficult to take time to write when there are as many interesting things to see and think about as we find in Germany":

> You must spend a few days in this country before going back to America. I can promise you that you will find it worthwhile. I believe that Germany is in many ways the most interesting place in the world today, and some of the things I see here encourage me greatly.

A few days later Lindbergh wrote to Harry Davison, his financial adviser in America:

> [Hitler] is undoubtedly a great man, and I believe has done much for the German people. He is a fanatic in many ways, and anyone can see that there is a certain amount of fanaticism in Germany today.... On the other hand, Hitler has accomplished results ... which could hardly have been accomplished without some fanaticism.

Anne, too, was swept away by her ten "thrilling" days in Berlin, urging her mother to reject the "puritanical view" that "dictatorships are of necessity wrong, evil, unstable, and no good can come of them," a falsehood spread, Anne wrote, by "Jewish propaganda in the Jewish owned papers." "There is no question," she added, "of the power, unity, and purposefulness of Germany":

It is terrific. I have never in my life been so conscious of such a directed force. . . . Hitler, I am beginning to feel, is like an inspired religious leader, . . . a visionary who really wants the best for his country.

THE LETTER-WRITING CAMPAIGN INTENSIFIED when the Lindberghs got to Copenhagen on August 2. After installing his perfusion pump, and while waiting for the Cytology Congress to convene, Charles mailed twenty-two thank-you notes to Nazi officials, including one to Göring. If Lindbergh expected Carrel to embrace his growing Naziphilia, however, he was disappointed. Carrel was delighted to see his protégé again, and showed his pleasure with kisses on each side of Lindbergh's face, when they met at their hotel. (Carrel usually avoided such French mannerisms when he was in the United States.) Though Carrel supported iron-willed dictatorships, especially those that promoted eugenics, he dismissed Lindbergh's enthusiasm for recent political events in Germany, telling Lindbergh in no uncertain terms that Hitlerism was a "pagan cult" threatening the very survival of Christianity. In reality, the Germans, no matter who led them, would always be *les boches* to Carrel, the gruesomely methodical soldiers who killed so many Frenchmen in World War I. Lindbergh knew better than to argue.

Besides, Carrel hadn't come to Denmark to talk about German politics. He'd come to talk about organ perfusion. Even Anne, who knew next to nothing about cell biology, marveled in her diary at how Carrel's enthusiasm transformed the "musty, beardy atmosphere" of the Cytology Congress into something lively and dramatic. Carrel had Lindbergh move the pump to a small room in the exposition hall, then decreed that only ten scientists at a time could see the device in action, as it clucked and hissed, sending a watery red nutrient into a cat thyroid that now lived and functioned, outside the body that created it, in a transparent glass perfusion pump. Other scientists had to wait their turn in line outside. The details of Lindbergh's death-defeating machine were explained to each batch of ten—by Carrel, in French; and Lindbergh, in English—as they stood behind it: the two high priests of perfusion in identical flowing black robes.

Actually, Carrel made sure the drama in Denmark began even before those demonstrations. Several biologists from the congress were invited to watch him as he removed the thyroid to be cultured in Lindbergh's pump from the donor animal. "Carrel and an assistant were in black blouses and wore black rubber gloves," a witness told the Associated Press. "It was a thrilling moment when those two black-clad scientists sat down at the black table to perform the operation. [It] reminded me of Rembrandt's anatomical paintings."

There is another account of Lindbergh's and Carrel's activities in Denmark, the only one available from a living eyewitness—indeed, from the only man still alive who worked with both Lindbergh and Carrel. This man is the research cardiologist Richard Bing, now ninety-five and living (as this is written) in retirement in California. In 1936 Bing was a twenty-six-year-old German physician, recently graduated from the University of Munich, working as a research fellow at the Carlsberg Biological Institute in Copenhagen. The chief of that institute, Albert Fischer, had organized the Cytology Congress then under way in Copenhagen. Just before the congress began Fischer asked Bing—who spoke Danish, English, French, and German—to work as a translator and technical assistant to the two most famous exhibitors.

"I immediately agreed," Dr. Bing said in a telephone interview. "After all, here was a chance to work with one of my heroes." The hero Bing was eager to work with wasn't Lindbergh, but Carrel. "I knew who Lindbergh was, naturally," Bing said, "but I was only a teenager in Germany when he made his flight. I knew he was famous, but it wasn't something I thought about much. Carrel, on the other hand, was the world's leader in the field I was working in." (That field was tissue culture.) "To me, he was very famous and probably a genius. I was thrilled by the prospect of learning from him. You have to remember, I was educated in the European system, where men who made important scientific breakthroughs were thought of as gods."

Bing would find that belief put to an early test. Carrel, he learned, "was a man of contradictions, a scientist who believed in nonscientific things." One night, Bing said, "we were having dinner at Dr. Fischer's house"—Dr. and Mrs. Fischer, Dr. and Madame Carrel, Colonel and Mrs. Lindbergh, and Bing—"when Carrel began talking about the power of the human mind, and how science knows so little about it. He believed in ESP and mind-reading. I was surprised to hear this."

Suddenly Carrel insisted that everyone at the dinner table join him in a demonstration of one of those little-understood powers. "He wanted us to each put one hand under the table, which was extremely heavy, and, on the count of three, while we were still sitting there, lift the table off the floor, using our powers of concentration." Everyone agreed to try, though most were skeptical. Carrel had been speaking in English, but Bing remembers him counting in French. " '*Un, deux, trois,*' and then we all pushed our palms upward." The heavy table, much to Bing's surprise, moved six inches into the air. Everyone laughed but Carrel, who exulted in his proof. "I suppose this proved something," Bing said. "What, I don't know."

What was perfectly clear to Bing was the awe in which Lindbergh held his mentor. "Carrel was more than a hero to Lindbergh. He was a father figure. I'm no psychoanalyst, but I think Lindbergh's relationship with his own father must have been not quite satisfactory." The public didn't realize something important about Lindbergh, Bing said: "He was a romantic. People thought of him as a pilot and engineer, a brave man good with machines. He was those things, but he was also very much interested in metaphysics. Carrel would talk for hours about 'the meaning of life,' which I, as a laboratory scientist, felt inadequate to discuss. Lindbergh loved to talk about such things with Carrel. One reason, I think, is that Lindbergh was not a sophisticated person, so he looked up to Carrel as the epitome of sophistication."

But "a truly sophisticated person," Bing said, "wouldn't need to wear black robes when demonstrating his work. He would put the focus on the work. But Carrel was a showman." There were many scientists at the congress who didn't like Carrel—"one could see that," Bing said. "Most of them were angry about the political things he wrote in his book" (*Man, the Unknown*). But, as Bing also saw, Carrel's charisma and the results he got with Lindbergh's machine made most of those critics at the congress back down.

Lindbergh's lack of showmanship made a vivid impression on Bing. "He could be aloof, especially with newspapermen, but he had every right to be. If you were a scientist or a technician, he was very friendly and respectful. There was not a bit of conceit in him." That modesty, Bing said, was all the more remarkable because the machine Lindbergh invented really was significant. "Lindbergh would be out of his depth now, of course. His approach was mechanical; now everything is molecular and genetic—about DNA and such." But seventy years ago, Bing said, Lindbergh's machine was a genuine breakthrough. "It maintained isolated organs in a sterile environment with pulsatile perfusion. No one had ever done that. I [later] used his machine to study the metabolism of cholesterol in perfused arteries. It was extremely helpful."

Apparently, Carrel and Lindbergh thought Bing was so helpful to them in Denmark that they decided to help *him*. Bing, you see, was half-Jewish, and in 1936 he found his scientific career in Germany—along with most of his basic rights—blocked by the Nazis' "racial purity" laws. With Lindbergh's and Carrel's assistance, Bing got a visa to the United States, where, in November 1936, he joined Carrel's perfusion team at the Rockefeller Institute.

Bing would later join the cardiology staff at Columbia-Presbyterian Hospital in New York; direct the cardiac catheterization laboratory at Johns Hopkins Hospital in Baltimore; become chief of experimental cardiology at the Huntington Medical Research Institutes in Pasadena, California; and publish more than 400 articles in major medical journals. In 2001, sixty years after he became an American citizen, Bing was awarded the President's Citation from the American College of Cardiology, and the Medal of Merit from the International Academy of Cardiovascular Sciences, each for his lifetime of service. The award given annually to the Best Young Researcher by the International Society for Heart Research is now called the Richard Bing Award.

It's possible Bing would have achieved all that without Carrel and Lindbergh's help, but we'll never know. Does the timely aid given Bing in 1936 by Carrel (who once praised the Garden City Hotel on Long Island, where he briefly lived, for banning Jews) and Lindbergh (who praised Hitler's leadership) prove that those two men weren't Nazis committed to the extermination of the Jewish "race"? It probably does. Does it prove that they were incapable of anti-Semitic beliefs? It does not.

AFTER LEAVING THE CYTOLOGY Congress in triumph, Carrel ignored Lindbergh's suggestion that he see with his own eyes the "thrilling" revival then under way in Nazi Germany. He also declined a written solicitation from the Allied News Agency in London to write "six articles in a simple yet arresting style" on the "secrets of everlasting life." Instead, Carrel and his wife returned to their apartment on rue Georges-Delavanne in Paris, where they unpacked, rested, and then repacked a few days later before traveling to their favorite destination on earth: the rustic summer house they owned on Saint-Gildas.

The Carrels owned more than the house; they owned the whole island. "My Nobel Prize was used to buy beautiful little Saint-Gildas, in the English Channel, where Gildas the Wise landed when he went from Ireland to preach the Gospel to the inhabitants of Brittany," Carrel once told a Swedish journalist who had asked him how the prize money changed his life. "It is of immense advantage for a man of science to spend some time every year in solitude, and to meditate in complete peace," Carrel said. "There is no doubt that the Nobel Prize," by making that possible, "has facilitated my work and enlarged its scope in a very marked way."

Saint-Gildas was a spectacular place to do that meditating. Roaring waves broke on the shore at high tide, smashing into huge boulders, coating them with white foam before retreating with a current so fierce that loose stones the size of automobile tires rolled on the beach like marbles. Low tide brought its own sense of wonder: you could leave the island and walk on the sea bottom, around reefs choked with seaweed, by tiny pools that trapped squid and other fish, and by small boats that hours earlier were floating but now lay keeled over on the damp sea floor. Brittany was just 1,000 yards away; other Channel islands were equally accessible by foot. But you had to be careful. A mailman once lost his way in a fog after leaving Saint-Gildas, walked in the wrong direction, and soon drowned in the fast-rising surf.

The Carrels' house had been built in Napoleonic times, in the "old Breton" style. There was a large living room with ochre-colored plaster walls, a beamed ceiling, a somewhat bumpy floor made of unjoined flagstones, a wide fireplace, and simple wood furniture. There was no electricity or indoor plumbing. A gardener's storehouse had been annexed to the house, serving as a library and a suite of bedrooms.

Normally, the Carrels spent all of July and August on their 100-acre island, joined only by a small staff of domestic help, animal wranglers (cattle, goats, and chickens were raised there) and groundskeepers. The Cytology Congress had already altered that routine. Now another interruption was on the way: the Lindberghs, after making a quick return to England, would be visiting Saint-Gildas in the last week of August.

Charles was eager to get there. He was gratified by the reception given to his pump by the scientists in Denmark, but the nearly constant presence of newspapermen there infuriated him. "It is not pleasant to be surrounded by a group of human monkeys with flashbulbs and motion picture cameras, attempting to record every movement you make, and hoping that by some favor of a photographer's god, you will fall and break your neck while you are having your picture taken," he wrote to Dr. Raymond Parker, one of Carrel's chief tissue culture aides, after the Cytology Congress. The idea of meeting his mentor on a private island, without any photographers, filled Lindbergh with a sense of possibility he hadn't felt since he left the Rockefeller Institute.

He was even more inspired once he arrived on Saint-Gildas. The fierce grandeur of the place took his breath away. "You felt the forces of earth and cosmos as though God himself exposed his hand," Lindbergh wrote. "You sensed the beauty and danger of existence." The possibility of extending human existence, maybe forever, was the subject Lindbergh came to discuss with Carrel. Lindbergh was upset by his inability to advance that secret project in a tangible way on his own in England, and especially by his failure to build a working prototype of a mechanical kidney. Still, he was sure Carrel would teach him something on Saint-Gildas that would end his frustration.

But Carrel preferred to talk about the precarious state of western civilization. Many of these discussions—on the need for the intelligent few to rule the ignorant many, the promise of eugenics, the vulgarity of proletarian movements, and so on—took place at breakfast, over bowls of steaming coffee and thick slices of black bread. Others occurred during hikes along the island's rocky, windblown shore, or during treks through its hilly, gorse-covered interior, Carrel in a white windbreaker and a black beret, peering through his pince-nez, Lindbergh in chino pants and a flight jacket, happy not to need a disguise, or even to worry about being recognized by anyone, save one of Carrel's friendly Breton field hands.

And so they'd walk and talk, these two men, on this tiny island that seemed so far from the chaos and decadence of modern civilization. Though Carrel had little interest in the ongoing Nazi experiment in racial superiority, he still believed in biological elites. He'd tried to create one on Saint-Gildas: German shepherds, specially mated, specially fed, and even given special hypnotic suggestions by Madame Carrel, to exhibit clear signs of superiority.

The results so far, Carrel admitted to Lindbergh, were terrible: whatever his animals gained by his efforts in form, they'd lost in function. He'd created "beasts that are superb yet stupid," he said. What went wrong with his dogs, Carrel added, had already gone wrong with humans in America—without any planning or direction. They've produced a race of giants there, the tiny Frenchman said, but lost nearly all intelligence and sensitivity in the process. The towering Lindbergh nodded in agreement. Men of genius are not tall.

Anne Lindbergh wasn't finding her visit nearly as enthralling. Madame Carrel, who'd made such a positive impression on her when they met in Paris, now seemed a different person: much more judgmental, for starters. Dr. Carrel, who took pride in his wife's ability to see auras emanating from people's bodies, insisted that she make psychic examinations of the Lindberghs, all of which found Anne deficient. Charles's aura was a strong violet, Madame Carrel announced, while Anne's was a weak blue. Charles could see infrared rays, Madame Carrel said; Anne couldn't. Charles was especially sensitive to touch; Anne was virtually numb. On and on it went. These pronouncements made Anne feel even more inferior to her husband than she already did. It's possible that Anne would have felt this way about any man she married. That she married the world's most famous man only exacerbated her self-doubt, which continued, undiminished, even after her success as a writer.

Anne's diaries suggest she was intimidated by the Carrels' ready categorization of human types. Not that Charles noticed. He could barely separate himself for a moment from the Carrels, least of all from Dr. Carrel. The success of *Man, the Unknown*, the surgeon-turned-philosopher told Lindbergh, didn't just prove there was an audience for his ideas; it gave him a mandate to transform those ideas into reality. The first step, Carrel said, was to create the organization he called for in his book to save western civilization—he'd decided to call it the "Institute of Man"—along with a "high council of experts" to staff it. Carrel received dozens of invitations each year to speak at medical school commencements and the like. Typically, he declined. But he told Lindbergh he would accept some of these invitations in the coming year and use those opportunities to argue for the implementation of his ideas. Lindbergh pledged his support.

What Carrel didn't tell his protégé was that his zeal to create the "Institute of Man" was partially fueled by growing doubts about the ultimate success of their immortality project. The origin of those doubts was this: After Lindbergh left America, Carrel did numerous experiments on different types of organs at the Rockefeller Institute with Lindbergh's perfusion device. Nearly all were successful, but one type of tissue resisted efforts to be cultured, and almost always died. This tissue was nerve tissue.

That was significant because in Carrel's original theory of immortality by means of organ perfusion, the idea was to turn the body into a machine with constantly reparable or replaceable parts. The only parts Carrel focused on, however, were those he called, with poetic license, the body's "moving parts"—internal organs such as the heart, liver, and kidney that wear down over time. Carrel wasn't thinking about the brain, because he didn't think it encountered as much physical wear. In Carrel's initial conception, immortality could be achieved in man by replacing his internal organs, but retaining the original brain and nervous system.

This idea seems absurd now, but a century ago it didn't. Median life expectancy in 1900 was forty-nine years; most people didn't live long enough to show clear signs of mental deterioration. And those who did, Carrel believed, shouldn't live longer. But the medical advances achieved in the early twentieth century extended life expectancy dramatically. This longer life span then created large populations with symptoms of memory loss, senility, and other age-related mental deficits—even in people of undeniable intelligence. Apparently, the human brain did wear down. Hence Carrel's frustration at his inability to culture nerve tissue. A vexing conclusion took hold in his mind: immortality was still possible—but only if humans didn't have a brain.

This is why Carrel was so intent on transforming his "high council of experts"—the "immortal brain" of the white race, as he called it in *Man, the Unknown*—from a figure of speech into a reality. The world needed an "Institute of Man" where the study of man could go on, uninterrupted, for centuries, an institute whose experts, on reaching their conclusions after a long, peaceful period of study, would issue edicts telling the rest of us how to think and live. Carrel and Lindbergh would contribute to that civilization-saving process not only as members of the "high council," but also with their organ perfusion experiments. If successful on humans, those experiments might enable members of the "high council" to live, if not forever, for a span of years, maybe even centuries, never before possible.

AS AUTUMN BEGAN IN 1936, the Lindberghs were back at Long Barn, Madame Carrel was in Paris, and Dr. Carrel was in New York, where his mood brightened. "An improvement in the way of filtering the fluid [in your machine] has suppressed the areas of necrosis which we observe from time to time," he wrote to Lindbergh in November. Richard Bing arrived at the Rockefeller Institute that same month and acquired a quick mastery of Lindbergh's pump. "Bing has learned the technique quite easily," Carrel wrote. "He can cultivate without accident a thyroid, a suprarenal, an ovary, etc."

Lindbergh was delighted to hear this. He started sketching plans for a pressure tank to enable his pump to achieve the higher perfusion pressures required by larger organs. Lindbergh asked for advice on this project from the one person who, after Carrel, was his favorite correspondent now that he lived in England. This was Dr. Robert H. Goddard, the brilliant American physicist who designed, built, and fired the world's first liquid fueled rocket, in 1926, in Auburn, Massachusetts. Lindbergh was nervous about one possibility in particular: "I am designing a tank which will contain oxygen under 10 atmospheres pressure," he wrote to Goddard, who was then continuing his research on rocketry in Roswell, New Mexico. "I would like to use mineral oil in some of the instruments the tank will contain. Do you believe . . . an explosion would take place if a spark should come in contact with the surface of oil under this pressure?"

Apparently, there were no explosions. (Goddard advised against the use of mineral oil.) The finished tank, made by an English firm that specialized in diving helmets, was "built with the substantiality and beauty of an English beef roast," Lindbergh wrote to Carrel. "It will serve our purpose well and it should last forever." This was news Carrel was happy to hear. "It is quite gratifying that you have succeeded in building the pressure tank," he wrote back. "As you know, kidneys and testicles degenerate rapidly when cultivated in serum. . . . It is possible this is due to lack of oxygen. I hope that serum under pressure from your new tank will be able to maintain kidney, testicle and possibly even nerve cells in normal condition. I am very excited by the possibilities of this new apparatus."

That Lindbergh's new apparatus might be usable on apes made Carrel even more excited. "[Robert] Yerkes," the founder of the Yale Laboratories of Primate Biology, "came to see me," he wrote a few weeks later. "He has about 40 chimpanzees. . . . You are doing exactly what should be done," Carrel told Lindbergh. "If you go to India, you will likely get more information."

—

CHAPTER 10

THE EXPLORATION OF THIS REALM IS A GREAT NEW ADVENTURE

LINDBERGH TOOK OFF FOR INDIA IN FEBRUARY 1937, ACCOM-
panied by Anne, then more than six months' pregnant with her third child
(counting Charles Jr.). Stopping in Rome, once capital of the world's greatest empire, Lindbergh saw the ruins of the imperial forum; in Athens, he
walked through the chipped, broken columns of the Acropolis; in Tunisia,
he saw the dust of once mighty Carthage. In these beautiful remnants, he
decided, lay a terrible warning. "One realized how easily strength was per-
verted into decay," Lindbergh wrote. "Western Civilization—how I had
taken it for granted! It had seemed immortal."

Lindbergh's stay in India confirmed the "truth" of one of Carrel's core
beliefs: Civilization means *white* civilization. The dark, servile Indians
Lindbergh encountered in the narrow streets of Calcutta reminded him of
"farm animals." Human life there, he wrote in *Autobiography of Values*, "had
sunk to levels [I] had never seen":

> Ragged, hungry people milled back and forth on filthy streets.
> Cripples sat on curbstones. Scabby, thin-legged children followed
> us with outstretched hands. At night, we stepped around sleeping
> bodies on the sidewalks close to our luxurious European-style hotel.

"You never know the difference," we were told, "but sometimes one of them is dead."

Lindbergh saw what he called the "barrier of race" with new clarity. The ongoing supremacy of the white race—something "which previously seemed secure beyond the need of questioning," he wrote—now seemed a very real question. Suppose, he asked himself, modern western civilization "destroyed [itself] with politics and wars, as ancient Rome had done? . . . Would other races do to us as we had done to them?" That frightening thought led Lindbergh to think of moving his family yet again, this time to the "distant white men's frontier" of Australia. Maybe others thought racial integration was a noble goal; for Lindbergh it defied reality. As a child hunting with his father in Minnesota, he saw that "wild birds never mixed. Partridge flew with partridge and mallard never crossed with teal." That humans were less particular in their mating patterns was obvious, but the results, particularly in the United States, created what Lindbergh saw as a kind of schizophrenia: "The American takes pride in his genetic internationality" and "boasts of his diversity," he wrote. But the American is also "alarmed by the rising Negro population throughout his United States."

The goal of his trip to India, Lindbergh reminded Carrel in a letter from Calcutta, was to meet Sir Francis Younghusband at the Parliament of Religions convened there to mark the centennial of the birth of Sri Ramakrishna, the great Indian holy man born in that city, then to use Younghusband's connections to meet Himalayan yogis who'd extended their lives, sometimes several decades past 100 years, through body-temperature control. Lindbergh's other areas of interest, he told Carrel, were "Hypnosis, Levitation, Fire Walking, Control of Bleeding, Immunity to Poisons and Diseases, and Control of Various Portions of the Body." The Lindberghs attended the religious conference, where, to Charles's great shock, he heard an Indian poet, Sarojini Naidu, compare him from the podium to "Buddha, Galileo, and other leading spiritual figures of the world." This praise was doubly bizarre: Lindbergh was an agnostic who belonged to no church; and Galileo was tried for heresy before the Inquisition. Lindbergh's embarrassment was so visible it was noticed by the press. "What other man on earth today could have his blush reported on five continents?" the *New York Times* asked.

Lindbergh's distress was assuaged when Younghusband introduced him to an Indian yogi who agreed to speak with Charles and Anne in their suite at the Great Eastern Hotel. "He would not sit on a chair," Lindbergh wrote to Carrel, "but sat on the floor, with his legs crossed in yoga posture. . . . He talked of weeks spent without taking a breath, and of oranges being changed to stone by another, even greater 'Holy Man.' We thought he was about to go into a trance while we were talking to him."

A few days later, Charles met a professor at the University of Calcutta, Dr. H. C. Mookerjee, who was one of the city's most experienced and articulate yoga adepts. "He classed his statements under three headings," Lindbergh told Carrel: "What he knew from personal experience, what he believed to be true, and what he heard only as claims and stories":

Dr. Mookerjee said that he had seen a yogi buried underground for four hours, and believed that authentic records existed of yogis being buried for as much as forty days. He had seen a yogi sit in a bowl of goat's milk, draw the milk up through his intestines, and spit it out of his mouth. He had seen a man drink H2SO4 (sulfuric acid). He had seen another yogi take his intestines out through his anus, wash them, and replace them again.

As Lindbergh's acceptance of these stories attests, being world-famous did not make him very worldly. The goat's milk trick is surely one that even an amateur magician could duplicate. The claim that someone imbibed a large quantity of sulfuric acid—one of the most toxic and corrosive substances known to man—and lived is preposterous, as is the story about removing and washing intestines.

Carrel, a trained scientist, was more skeptical. "There is probably some relation between the condition of latent life observed in India and certain states of hysteria," he wrote back. But Lindbergh swallowed these tales whole. That is why he was all the more disappointed when engine problems in his plane forced him to cancel his flight to meet the life-extending yogis of the Himalayas.

Even so, Charles came back from India with a sharpened focus. "We whites [were] so accustomed to dominating that it was difficult to realize that we were a minority in a world of yellow, brown, and black," he wrote in

Autobiography of Values. Suddenly the dire future facing white western civilization seemed clearer than ever to him. He returned to his laboratory at Long Barn, working to advance the project for saving white civilization that Carrel and he started in secret at the Rockefeller Institute five years earlier. The stakes, Lindbergh understood, were even higher than he previously thought. He took some comfort, however, in knowing that there was one western nation aware of the crisis—and taking vigorous steps to confront it. Lindbergh had a feeling he'd be going there again.

UNTIL THEN THERE WERE experiments to do. Lindbergh hadn't met any temperature-lowering yogis in the Himalayas, so he decided to try to recreate their body-altering experiences at Long Barn, using animals. The key to the yogis' experiences, Lindbergh believed, was their slower breathing, which altered the composition of the gases in their lungs—elevating the carbon dioxide and lowering the oxygen. Years earlier in New York, when redesigning tissue flasks for Carrel, Lindbergh saw the dramatic effect even a small change in oxygen supply could have on living cells. Suppose, he wondered, he took some air, lowered its oxygen pressure and increased its carbon dioxide pressure, designed a simple closed system for this air to circulate in, and then placed a rodent in that environment. Would its respiration rate slow down? Would its body temperature fall? Would those changes, working in tandem, create a death-delaying state of "suspended animation"?

Lindbergh drove to London, a city where, much to his pleasure, he could walk around without a disguise and not be harassed by passersby, even when they recognized him. There he purchased a vacuum pump, a glass bell jar, and a few dozen guinea pigs and white mice. These rodents were presented to his son Jon as pets, and housed in cages Charles built in Long Barn's backyard. The reason Lindbergh bought so many animals was that he didn't want Jon to notice the absence of several of them once he started his experiments. Death, after all, was a very possible outcome.

A few days later Lindbergh placed a healthy young rodent inside the bell jar, used the vacuum pump to alter the pressure inside, and removed the animal after different lengths of time. Then he repeated the tests on other animals. A number of these test subjects—those exposed to high atmospheric pressure for several days—did die. "I wondered whether [this] was due to the phenomena connected with the transmission of gas through the membranes of the lungs to the blood, or to some effect on the blood corpuscles, or to an effect on the tissues, or to a combination of these factors," Lindbergh mused in a letter to Carrel.

But rodents exposed to lower pressure survived unharmed. "I found that rectal temperature [in the rodent] would drop from 37 degrees centigrade to 23 degrees within several hours," wrote Lindbergh, who took those rectal readings himself. Once returned to the open air, the rodent slowly left the trance state it had entered inside the bell jar and eventually returned to normal activity—"showing no effect," Lindbergh noted, "of the strange experience it had just undergone."

Questions filled Lindbergh's mind: How long could a human survive with a lowered respiration rate? What changes would occur inside the brain as the body's temperature went down? Would such an experiment create a viable state of suspended animation, or merely induce a type of temporary coma? And most important of all: what life-extending possibilities would be raised by such thermal and respiratory changes? To find out, Lindbergh decided to do the experiment on himself. This would involve not only replacing the small glass jar with a much larger chamber, but placing the experimental subject—Lindbergh—in real physical jeopardy. When Lindbergh told Carrel of his plan, the surgeon was horrified. Such an experiment, he warned, could lead to Lindbergh's next appearance on the front page in newspapers all over the globe—in an obituary. He strongly advised Lindbergh not to do it. Lindbergh backed down.

But Carrel hadn't lost interest in all hypothermia experiments. He asked Lindbergh to build a chamber that could control temperature and oxygen levels in Carrel's operating rooms. The potential benefits of that device, Carrel wrote, were worth pursuing: "If it is possible to reduce greatly the temperature of a homeothermic (warm-blooded) animal, a new field is open to surgery. The lowering of temperature will decrease pain. Operations may be performed without anesthetics." (And, indeed, today some are, in a method called cryosurgery.) Carrel also thought such a chamber would be useful for new life-extending experiments on primates, studies that, if successful, could be tried on humans. But, he reminded Lindbergh, only *after* they'd been tried on primates first.

Carrel was already doing new experiments in America—not biological but oratorical. In February 1937 he gave the first of the talks he had told Lindbergh about the previous summer. Carrel gave this talk at the University of Illinois, where he accepted the Cardinal Newman Award, presented annually to the individual "who has made an outstanding contribution to the enrichment of human life." If the students, faculty, and Catholic clergy in the audience expected to hear uplifting platitudes from their speaker, they were in for a shock. Talking rapidly in his French-inflected English and holding the lectern so tightly that it sometimes shook, Carrel warned them that western civilization was "crumbling," and that the remedies put forth by "economists, psychologists, and politicians" were mere "abstractions" that were doomed to fail.

This was because the persons offering those remedies understood only a fragment of man. What was needed to prevent western civilization from ending, Carrel said, was "an Institute of Man," where man could be studied "in his totality." And not just studied, but shaped. "Man," the surgeon said, "must be molded into a definite organic and spiritual form by scientific, intellectual, religious, and aesthetic experts." For that effort to succeed, another step had to precede it: a national eugenics survey that would assess the intellectual and physical potential of every American citizen under thirty to determine who among them would not be allowed to propagate. The propagation of "the best strains" in America had "decreased in an alarming manner" compared with "lesser strains," Carrel told his audience. A program to reverse that dangerous demographic slide—one that would, through scientific mating principles, create a natural aristocracy of merit—"must be undertaken immediately."

This speech, met first by stunned silence, then thunderous applause, received extensive and almost universally positive coverage in American newspapers as far away as San Antonio, Texas, as did several other talks Carrel gave on the same subject in the months to come, until Carrel left the United States in late May for France. Carrel was thrilled that his warnings about the crisis facing western civilization were being taken seriously in America; he summoned the Lindberghs to Saint-Gildas so that they could hear about it directly from him.

This time, however, Charles came alone. Anne had given birth to another son, Land (named after Charles's dentist grandfather), in London in

May; she said she needed time to nurse the infant properly. In reality, Anne was reluctant to go back to Saint-Gildas because she felt the Carrels had condescended to her the first time she went there.

Flying alone from London to Saint-Gildas was simple enough for Charles; setting foot on it, he learned, was more complicated. (On their first visit, the Lindberghs had traveled by ferry and train.) Lindbergh was cruising in his black-and-orange monoplane at an elevation of 2,000 feet when he sighted the tiny island. While circling above, he looked down from his cockpit and saw several chickens scatter in fear; then he saw Dr. and Madame Carrel, who stepped out of their house to wave. Lindbergh wrote a note on a piece of paper saying he'd touch down on the mainland, then return. He wrapped the message in a long strip of cloth and threw it overboard; the cloth was weighted down with a rock Lindbergh had scooped out of his garden, back in England. The message landed not far from the Carrels' house. Dr. Carrel picked it up, read it, and waved again.

But Lindbergh couldn't find a landing strip nearby. After searching for more than an hour, he landed at an airdrome about forty-four miles away. He couldn't call the Carrels—they had no phone—and he suspected that it would be nearly midnight, and high tide, by the time the taxi he hired got him to the village of Port-Blanc on the Brittany coast, just across from Saint-Gildas. He was right, so, after reaching the shore, Lindbergh removed the plane's emergency raft that he'd stowed in the taxi's trunk, inflated it, threw on his personal bag, unfolded the collapsible oars, pushed the raft into the black water, jumped in, and started rowing. Minutes later Lindbergh reached his destination. He hoisted his raft onto his shoulder, picked up his bag, and began feeling his way, ever so slowly, in the darkness, toward the Carrels' house. He was completely alone and exultant.

In the days that followed Lindbergh exulted in the company of his mentor. It was just weeks since Lindbergh had completed his suspended-animation studies on rodents at Long Barn; he gave Carrel a full report. The time had come, the surgeon said after listening, for a round of experiments performed on primates in a larger chamber. Carrel insisted, however, that these studies be done at the Rockefeller Institute, where they could be properly monitored. Carrel used all his powers to convince Lindbergh to agree to this, but his protégé refused. He simply wasn't ready to return to America full-time. Carrel then urged Lindbergh to consider returning briefly—for the Christmas holiday, perhaps. Lindbergh said he'd think about it, but made no promises.

Carrel also asked Lindbergh to join him on another project: a book detailing what they'd already accomplished in organ perfusion. Lindbergh would describe the construction, use, and maintenance of his perfusion pump; Carrel would describe the biological results obtained. They'd made a historic leap in understanding the secret workings of the body with their research, Carrel told Lindbergh. The world required a written record of that work, Carrel said. Lindbergh agreed to do his part.

After flying back to England a few days later, Lindbergh was shocked to read an account of his visit to Saint-Gildas in London's *Sunday Express*. Apparently, the taxi driver who drove Lindbergh from the airstrip to the Brittany coast had recognized him. A hack for the *Sunday Express* named Victor Burnett took it from there. Burnett's first-person piece was headlined LINDBERGH SEEKS THE SECRET OF LIFE / LONELY ISLAND EXPERIMENTS WITH MACHINE THAT KEEPS A BRAIN ALIVE / SOLVING MAN'S GREATEST PROBLEMS / WHAT IS DEATH? / CAN IT EVER BE CONQUERED? It began this way:

I have just learned the secret behind Colonel Lindbergh's dramatic visit to the lonely island of St. Gildas, on the Brittany coast, last week. The famous airman's orange and black monoplane swooped over the isle on Sunday night and landed at a near-by airfield. Later he paddled to the island in a rubber boat.

He had arrived to help a friend, Dr. Alexis Carrel, a scientist and Nobel Prize winner, in a vital part of an experiment that may reveal the secret of life itself.

In a guarded, walled laboratory on St. Gildas, a machine is slowly being assembled. Only two men have ever seen it or know exactly what its ultimate function will be. They are Dr. Carrel and Colonel Lindbergh.

Lindbergh has helped Carrel to build the machine. It was these two men who startled the scientific world recently by perfecting an artificial heart to keep organs in the body alive almost indefinitely.

Lindbergh's visit to St. Gildas was to fit one of the most important parts of a new machine. . . . [This machine] takes living organs, kills them by drying them—and then attempts to bring them back to life.

If successful, the experiment means that an animal that has apparently been dead for years can be brought back to life in exactly the condition it was in when life was suspended—no older, in perfect health. . . . On the Carrel island there is now an improved version of this machine that can even keep a brain alive after the animal it was taken from is dead. This new apparatus, I understand, is part of a secret life machine.

Though nearly every ghoulish detail was false, the article in the *Express* did convey an important truth: it was once thought Alexis Carrel alone was privy to the secrets of eternal life; now it was believed Charles Lindbergh knew them as well.

LINDBERGH MADE SEVERAL MORE flights to Saint-Gildas that summer, each devoted to the book he was writing with Carrel. The title, the two men agreed, would be *The Culture of Organs*. Carrel suggested that the book be dedicated to his scientific hero, Claude Bernard; it was. A reporter for *Paris Soir* interviewed Carrel about this project in July and found the surgeon in a surly mood. "I am here to write a book about the work I have done with Colonel Lindbergh," Carrel told him. "If I try to keep people away from my island, as is my right, it is to be able to work quietly with my friend and collaborator. . . . I pray you, do not try to see him. He has suffered enough."

Two of Lindbergh's subsequent visits to Saint-Gildas that summer were noted by the Associated Press. Neither article, Carrel and Lindbergh were relieved to see, talked about culturing brains or restoring life to the dried out or the dead. Even so, Carrel admonished the press about intruding on his privacy when he returned to New York. CARREL, BACK, SAYS WE TALK TOO MUCH / SCIENTIST CANNOT UNDERSTAND WHY LINDBERGH'S VISITS TO HIM SHOULD CAUSE A STIR was the headline in the *Times*. Carrel wasn't about to follow his own advice, however. Ten days later he gave a provocative speech at Dartmouth College. "Those who have given their lives to the search for the prevention and cure of disease are keenly disappointed to see their efforts have resulted in a large number of healthy defectives, healthy lunatics, and healthy criminals," Carrel told members of the school's Phi Beta Kappa chapter.

Two months later Carrel spoke again, lecturing on "The Prolongation of Life" at a convention of life insurance executives. The businessmen had invited Carrel to speak to them because they were worried that their industry would die if people could live forever—a prospect Carrel's research in organ perfusion seemed to be making possible. At first, Carrel gave the executives more reason to worry: "The quality of tissues and of the regulating mechanisms" within the body "are responsible for longevity," Carrel told them. "When waste products are not allowed to accumulate, senescence and death are indefinitely postponed."

But Carrel was less concerned with negative changes in the executives' balance sheets than he was that the wrong people would live longer. "Prolonging the life of a great many people would profit neither themselves nor society," he said. Civilization is "already encumbered with those who should be dead: the weak, the diseased, and the fools." The question of immortality, Carrel said, has moved "beyond the frontiers of hygiene and medicine into uncharted territory. The exploration of this realm is a great new adventure, and requires the help of physiology, chemistry, and physics in their most organized and intelligent form."

The specific outcome of the quest is not predictable, Carrel conceded. "But we must remember that there is no example of a scientific search for truth which has not been rewarded."

—

CHAPTER 11

THE REACTION OF A MAN WOULD PROBABLY BE SIMILAR

ACROSS THE ATLANTIC, LINDBERGH WAS CONTINUING HIS own search for truth—in Nazi Germany. On October 11, 1937, Charles and Anne flew to Munich, where Charles had been invited to attend the annual Lilienthal Aeronautical Society Congress, named for Otto Lilienthal, a nineteenth-century German engineer who did some of the earliest aerodynamic experiments with gliders. It's interesting to note that five months earlier Lindbergh had refused to return to the United States to participate in ceremonies marking the tenth anniversary of his own flight from New York to Paris. Those events included a banquet in his honor at the Waldorf-Astoria organized by Governor Herbert Lehman of New York, Mayor Fiorello La Guardia of New York City, and Orville Wright, as well as a three-day air meet in St. Louis attended by nearly 200,000 aviation enthusiasts. Instead, Lindbergh spent May 20 and 21, 1937, in seclusion at Long Barn, refusing all interviews.

He was far more sociable in Munich. Lindbergh not only mingled with the Nazi aviation elite at the Lilienthal Congress, but eagerly accepted their invitations to visit factories he'd missed on his first visit to Germany a year earlier. Colonel Ernst Udet of the Luftwaffe, who had been a flying ace in World War I, took Lindbergh to the Rechlin air testing station in Pomerania. There he became the first American to see the Luftwaffe's most advanced single-engine fighter, the Messerschmitt ME 109, along with the Germans' new bomber-reconnaissance plane, the Dornier DO 17. Lindbergh also noticed that the Nazis were developing a twin-engine Messerschmitt, with 1,200-horsepower Daimler-Benz motors. At the Focke-Wulf plant in Bremen, Lindbergh saw a fantastic new aircraft that could ascend vertically, hover without significant movement, move forward or backward, and turn with unprecedented maneuverability. "I have never seen a more impressive demonstration of an experimental machine," Lindbergh wrote of this helicopter to an American friend.

Lindbergh doubted the British or French could match Germany's air strength; he was equally certain Germany would soon surpass the United States. To make sure the American government was aware of his assessment, Lindbergh collaborated on a report with Major Truman Smith. According to Smith, the "General Estimate (of Germany's Air Power) of Nov. 1, 1937"—sent to Washington under Smith's name—was written almost entirely "in Lindbergh's exact words." Lindbergh is not here to confirm that, but the tone of the report does suggest that it was written less by a career military officer than by an awestruck tourist. The size of Germany's aviation program is described as "literally amazing"; the efficiency of German manufacturing is "formidable" and "astounding"; and Germany's rebirth, coming after both the Great Depression and the country's defeat in World War I, is judged to be "one of the important world events of our time."

This last comment is significant. Years later, after Germany was defeated in World War II, Lindbergh's supporters characterized the "General Estimate of Nov. 1, 1937" as an American patriot's attempt to sound an alarm about Germany's growing air power—and, in particular, the threat this power posed to American security. It's far more likely, however, that Lindbergh's true motive was something else. Rather than presenting German strength as a danger, Lindbergh, as he made clear in *Autobiography of Values,* saw it as something positive: a bulwark against communism and other insidious forces, external and internal, threatening western civilization. Instead of urging the United States and its allies to confront German power, Lindbergh was urging them to fall in step behind Germany in a struggle to save their common cultural heritage.

Especially revealing in this context is the quiet racial analysis in the

"General Estimate." The rapid creation of Germany's powerful air force is credited to the innate courage and scientific skill of the manly German race. The link between race and courage, Anne noted in her diary, was the subject of a long conversation one night on this trip to Germany between her husband and Kay Smith, Major Smith's wife. Courage might be reinforced by military training, Charles and Kay said in their discussion, but the clear lesson of history, they agreed, was that courage is chiefly the product of a race's genetic heritage.

Kay Smith's ethnohistorical views were certainly shared by Major Smith, whom Lindbergh greatly respected. But there was an even closer source for Lindbergh's racialist views: Alexis Carrel. As a former French army surgeon, Carrel was no friend of the German race, but he was a friend of racial analyses. It's easy to believe that the report's racial slant came from Lindbergh, who'd spent countless hours listening to Carrel speak about such matters, and especially about his conviction that the white race was in jeopardy of drowning in a sea of inferiors. What we know for sure is that, days after working with Major Smith on the "General Estimate," Lindbergh wrote a letter to his friend Harry Davison in America echoing Carrel's worldview which equated certain "races" with masculinity, discipline, and strength, and others with weakness and vice. The Germans and Italians, partners since 1936 in the Rome-Berlin Axis, are without question, Lindbergh wrote in this letter, the "most virile nations in Europe today."

With questions in his mind about America's virility, Lindbergh found himself weighing Carrel's suggestion that he return to the United States for the Christmas holiday. Perhaps he could enter the country without fanfare (or at least try to), then make a tour of American aviation facilities, where he'd quietly consult with high-ranking military officials. Equally important—and even more quietly—he'd return to the Rockefeller Institute, where he'd perform important perfusion experiments on primates, studies that would advance his secret research with Carrel.

"Fanaticism and crime are probably as rife . . . and the press as irresponsible and lawless as when we left," Lindbergh said to Carrel in a twenty-six-page handwritten letter announcing his intention to return, temporarily, to America. "[But] my greatest concern is . . . to help [you] in any experiments with the pressure tank and 'artificial hibernation.' . . . I would be glad to make the trip home for that purpose alone."

—

EVEN BEFORE *PRESIDENT HARDING*, the ship on which the Lindberghs sailed from England, arrived in New York on December 5, 1937, Charles knew that his worst anxieties about the press were about to come true. A steward informed him that nearly 100 photographers were jostling for room at the foot of the first-class gangplank. Apparently, word of the Lindberghs' presence on board had leaked. So Charles grabbed Anne, ran with her to the third-class section, then walked down a gangplank and jumped into a freight elevator, which deposited them at ground level. This evasive maneuver worked, but only for a moment: when the Lindberghs were spotted moving toward the limousine sent for them by Anne's mother, reporters and photographers chased after them in a loud, unruly horde.

LINDBERGHS ARRIVE HOME ON SURPRISE HOLIDAY VISIT / TRY TO SLIP IN AS SECRETLY AS THEY LEFT U.S. 2 YEARS AGO, BUT ARE RECOGNIZED LEAVING SHIP, read the *New York Times* headline on December 6. The frenzy at the dock was nearly matched at Next Day Hill, where the Lindberghs were staying. (Charles and Anne had left their sons, Jon and Land, with servants in England.) Within minutes of their arrival at the Morrow estate, the *Times* reported, "crowds gathered at the gate and automobiles moved slowly by, their occupants craning their necks at the house half hidden behind trees." The press, just as it had years earlier, blamed the Lindberghs for the commotion: "In their anxiety to avoid publicity," the *Times* declared, "the famous couple adopt tactics that by their very mystery arouse public interest." Lindbergh vented his anger in his journal: "Rumors, lies & all the sensation of American journalism at its worst," he wrote.

One press mogul belatedly urged restraint: "Let's leave Colonel Lindbergh alone," Frank E. Gannett, owner of what was then the country's sixth-largest chain, wrote in the *Rochester* (New York) *Times*. Other news organizations declared that they, too, had learned the lessons taught by the hero's hasty departure from America two years earlier. Almost none honored their promises.

Even so, the Lindberghs' visit, originally conceived as lasting a few weeks, stretched to three months. One reason was that Charles decided that his wife and he would have their portraits painted in New York by the artist Robert Brackman, a well-known portraitist and teacher at the National

Academy of Design and the Art Students League, both in Manhattan. The time required to sit for those paintings, Charles learned, created scheduling problems that prevented his completing the extensive tour of American aviation facilities he'd hoped to make when he left England. He did get to see a few bases and factories, however, and what he saw confirmed his pre-existing view. "Germany," he wrote U.S. Army Major General Frank Ross McCoy, on January 31, 1938, "is rapidly surpassing us in air strength."

After donning lensless eyeglasses and a floppy fedora, Lindbergh was able to achieve his other goal: driving into Manhattan from Next Day Hill to enter the Rockefeller Institute, unrecognized by the press or anyone else. Once inside, he happily renewed old friendships with Carrel's support staff, some of whom were still tending daily to Carrel's chick heart, pulsing as regularly as ever a full twenty-five years after its birth in Carrel's laboratory. Lindbergh also saw ovaries, thyroids, kidneys, and other organs—surgically removed from the animals in which they'd been created—living on in his perfusion pumps, which hissed and snored like a cave full of hibernating bears.

It's unlikely Lindbergh needed to be reminded of the ultimate goal of his scientific quest, or of Carrel's special genius. If he did, Carrel's famous chick-heart cells, which Lindbergh examined shortly after his arrival, served both purposes. The ongoing existence of the pulsing tissue—an existence which was already longer than that of a real chicken—was "proof" that Carrel had challenged death, and won. Working together, Lindbergh and Carrel had achieved the next step in their quest: perfusing whole organs. The task still ahead of them, however, was formidable: taking what they'd learned and applying it to higher animals—first apes, then humans. If they succeeded, the immortality Carrel had seemingly demonstrated on the cellular level might be achieved in *Homo sapiens*.

Carrel was already testing an important step in that process: taking thyroid glands he'd surgically removed from their hosts (typically dogs or cats), perfusing them in Lindbergh's pump for several days, and then reimplanting those organs back into the host animal. The results of these operations, some witnessed on this visit by Lindbergh, were inconsistent, yet promising: infection, even death, resulted in a few instances; but other retransplantations "took," and the animal appeared to be in perfect health. Carrel and Lindbergh's basic hypothesis—that the body is a living machine with constantly reparable and replaceable parts—was approaching empirical reality. Building a larger perfusion device for similar experiments on primates, and then humans, Lindbergh was certain, would create no perceptible engineering problems. In fact, he was sure that the glassblowing expert Otto Hopf could manufacture them by himself from drawings Lindbergh would send over once he returned to England.

While in New York, Lindbergh was very eager to advance his hypothermia and pressure-lowering experiments—"artificial hibernation" was the term Carrel typically used—on primates. Before he'd left England Lindbergh had his completed pressure tank shipped to Carrel's laboratory; it had cleared customs and was now at the institute. After a young monkey was sedated by one of Carrel's technicians, the animal was placed inside this multi-riveted metal chamber—a device resembling a small vertical submarine—which had been constructed with a glass window to facilitate observation. After the monkey was exposed to lowered temperatures for several hours, two technicians removed the dazed animal, checking its vital signs and taking internal readings. Just as the temperature inside the chamber had gone down, so had the monkey's, by a full five degrees.

Lindbergh and Carrel flushed with excitement. Might the same thing happen to a human? And, if so, did it mean there were internal changes created by that temperature drop, maybe even some with the potential to defeat death? In posing these questions about the relationship between lower body temperatures and life extension or immortality, Lindbergh and Carrel were anticipating pioneering experiments in cryobiology performed a few years later by Basile Luyet. Luyet, a scientist who was also a Jesuit priest, froze living cells in his laboratory in Madison, Wisconsin, and later was able to restore some of them to normal functioning, after rewarming. Luyet reported his results in 1940 in a monograph, *Life and Death at Low Temperatures.*

Lindbergh and Carrel continued their experiments on live primates— not just live primate cells—in 1938 by lowering the temperatures and oxygen levels inside their chamber. But a few hours later something unexpected happened: despite the colder environment and lower oxygen level, the monkey showed signs of emerging from its trance. Carrel ordered the animal removed from the chamber and checked its internal temperature. This was climbing, slowly but inexorably, finally leveling off at one or two degrees above normal. To make sure this wasn't an aberration, Lindbergh and Carrel did the test again on another monkey, and then once more, each time getting the same disappointing result.

Lindbergh was hoping that these experiments would prove as promising a potential path to immortality—or, at least, life extension—as his work on organ perfusion. But he realized that this probably wasn't going to happen. It seemed likely, he wrote in *Autobiography of Values*, that "the body of a monkey contained a heat-regulating mechanism different from that of the animals I had experimented with at Long Barn, and that the reaction of a man would probably be similar to that of a monkey."

Even so, there is evidence that Lindbergh and Carrel were investigating the possibility of trying those same temperature-lowering experiments on humans. A letter mailed by Lindbergh from Next Day Hill to Commissioner William Ellis (the same New Jersey bureaucrat who once discussed providing "feeble-minded [human] prospects" from a mental hospital to Carrel and Lindbergh) expressed Lindbergh's thanks to Commissioner Ellis for visiting Carrel and Lindbergh at the Rockefeller Institute, and for providing the two researchers with "valuable [oral] information" and an interesting "written report."

Despite the failure of the "artificial hibernation" experiments on primates, Lindbergh was thrilled to be working again at the Rockefeller Institute. Carrel, he realized once again, was right: the best place for the two men to continue their research was in a real scientific laboratory, where both could be present, working with the best equipment and support staff available anywhere. Lindbergh's motivation to continue with that work, despite the disappointments he'd experienced on this trip, was as strong as ever. So was Carrel's, as we can see from this letter he sent to Simon Flexner: "Colonel Lindbergh returned to New York, as you know," he wrote:

He continued with some highly interesting experiments he began last summer on his own account at Saint-Gildas. . . . The method of cultivating entire organs with the Lindbergh pump has developed very much. It is manifesting great potentialities. The successful preparation of artificial media has extended its applications to the organs in the anatomical regions of small animals, such as guinea pigs, rats, and mice, and also to organs removed from humans in the course of an operation, or from individuals killed by accident. In this manner, the human body will finally be brought fully into the field of experimental physiology.

THE POLITICAL TURBULENCE ABROAD kept both men motivated as well. There was no doubt in Lindbergh's mind that Europe was in danger of destroying itself in another war, and that something drastic had to be done to ensure its survival as the home of white civilization (or at least to ensure the survival of an enlightened leadership of that civilization). Even so, the excesses of the American press made Lindbergh extremely wary of returning to the Rockefeller Institute full-time.

But an intriguing alternative had surfaced. Madame Carrel, in New York on one of her rare visits, told Lindbergh that Illiec, a Channel island right next to Saint-Gildas, was for sale. The Lindberghs had seen Illiec, one of the smallest of the nearby islands, while staying with the Carrels, but had never set foot on it. Madame Carrel had, and brought photographs. The house on the island, a handsome three-story stone building with a tower, had been built in 1865 as a summer retreat for the French musician Ambroise Thomas, composer of the opera *Mignon*.

Though there was no heat, running water, or electricity (just as in the house where Lindbergh spent his boyhood summers in Minnesota), Lindbergh was eager to proceed. He loved the idea of being on an island, protected from the intrusions of the press. It would be not just a personal haven, but a scientific one, a place where he could spend the summer months furthering his research with Carrel, who'd be just a moment's walk away on Saint-Gildas when the tide was low, and just a short row away when the tide was high.

The idea of moving to France caught Anne by surprise, but Madame Carrel's photos rinsed away whatever reluctance she felt. "The pictures are beautiful," she wrote in her diary, "and the house looks quite nice. . . . C[harles] is excited. He says, 'I can see Jon digging up clams,' and that sold me." The negotiations would be complicated because of French property laws, Madame Carrel said, but she was leaving for France shortly and volunteered to serve as the Lindberghs' representative. Charles accepted her offer, then started preparing to return to England.

It wasn't easy to concentrate on that mundane task: the political situation in Europe was darkening by the hour. Hitler was talking, loudly and often, about *Lebensraum* and Germany's right to a Reich encompassing all Germans, even the 10 million living outside its borders. The French, British, and Soviets voiced their opposition to that view—and brandished their swords accordingly.

On March 11, 1938, after a final meeting between Lindbergh and Carrel, at which the aviator pledged his ongoing participation in their perfusion research, the Lindberghs walked up a ramp to their first-class cabin on *Bremen*, the same German steamship that, in 1912, had sailed past the site of the *Titanic* disaster—so closely, in fact, that passengers on the *Bremen* told the *New York Tribune* that they saw about 100 bodies still floating in the icy North Atlantic waters, some so clearly they could make out the color of their clothing. Now, twenty-six years later, *Bremen* was making the reverse crossing. At almost the same moment as it left Manhattan, Hitler's troops crossed the border into Austria.

CHAPTER 12

—

TWO MEN SITTING ON TWO ROCKS

MARCH 11, 1938, WAS ALSO AN IMPORTANT DATE FOR ALEXIS Carrel. That morning, an interoffice memorandum arrived in his mailbox at the Rockefeller Institute, sent by Herbert Gasser. Its purpose was to remind Carrel of an unfinished conversation he had with the director a few days earlier—a talk Carrel preferred to forget. This inclination was understandable, because Gasser had given him some disturbing news in that meeting: the Rockefeller Institute was implementing a new policy requiring all members to retire at age sixty-five. Carrel would reach that age in three and a half months.

Gasser had been furious with Carrel ever since Carrel's "Mystery of Death" speech at the New York Academy of Medicine in 1935. When, months later, Carrel began speaking out publicly at commencements, hospital dedications, and similar events on the need to establish an "Institute of Man," Gasser became even angrier. Some of this antipathy was rooted in Gasser's own history: he had been raised as the son of a Jewish father and a Christian mother in a small farming town in Wisconsin, a circumstance that branded him as something of an outsider. It was not in Gasser's nature to call attention to himself, or to admire others who called attention to themselves. He was particularly wary of scientists who philosophized about a subject outside their area of expertise, especially if that subject was politics.

The new retirement rule, which would take effect on July 1 of the following year (so that Carrel would actually be retired at sixty-six), allowed the institute's board—at its discretion—to continue to fund the personal research of retirees on a year-to-year basis, if those researchers so requested. The key words were "at its discretion" and "personal." The board wasn't guaranteeing ongoing financial support, and any money that was forthcoming would not be sufficient to cover a full laboratory or support staff. This is why Gasser sent his memo of March 11. "Our conversation ended without a clear understanding of what your wishes might be about the period after July 1, 1939," he told Carrel, "specifically whether you would want to continue with a laboratory on a personal basis, as I have outlined to you. I have to make a report about the retiring Division, so if you have come to any decision on the matter, I should be grateful if you would let me know."

The words "retiring Division" pained Carrel as much as the prospect of his own retirement. Gasser was ousting not merely Carrel but the entire division of experimental surgery. The Rockefeller Institute's other divisions—microbiology, pathology, chemistry, physics, and physiology—would continue as before: fully funded and fully staffed. Carrel was infuriated by these changes, and convinced they were designed specifically to thwart him. "[This] is the first time that a man who has received the Nobel Prize is being compelled to give up medical research when the research is most promising," he wrote a good friend, the surgeon Harvey Cushing, at Johns Hopkins:

> Lindbergh is being stopped at the time when he is developing apparatuses of the greatest importance. My technicians are being dispersed. In this manner, the Rockefeller Institute is suppressing techniques that are probably the most powerful tools for [medical] research so far developed.

It's no accident that Carrel mentioned Lindbergh in this letter: he knew that Gasser resented the celebrity which came Carrel's way because of his relationship with Lindbergh. Two months before Gasser informed Carrel of the new retirement policy, the director saw with his own eyes the wave of excitement that spread through the institute when word got around that Lindbergh had briefly returned to Carrel's laboratory. What others found exciting, Gasser considered a distraction from the institute's core mission: dogged, quiet research, the kind in which the experiments—not the people performing them—are the "stars." Carrel and Lindbergh's entire enterprise, inside the institute and outside it, Gasser felt, was an exercise in personal hubris and political naïveté, one that must be stopped. As the institute's director, Gasser had the power to stop half of it, so he would.

Carrel's disdain for the director was every bit as intense as the director's for him; what made Carrel even angrier was his unshakable conviction that Lindbergh and he had significantly advanced science. Clear evidence of this, he believed, was about to be published in their book, *The Culture of Organs*, scheduled for release that summer. Thinking a publicity campaign might create demand for Carrel and Lindbergh to continue their work—and force Gasser to rescind his new retirement policy—Carrel allowed a reporter from *Time* magazine into his laboratory in the spring of 1938.

This decision alarmed Lindbergh. *Time* was one of the many publications he despised for its overheated coverage of the murder of his first son. Lindbergh also thought *Time*'s original report on his perfusion pump, published in July 1935 and titled "Glass Heart," had created unrealistic expectations among the magazine's millions of readers. "I am constantly getting letters from people who wish to have 'artificial hearts' installed in themselves, or their children," Lindbergh wrote to Paul B. Hoeber Jr., the publisher of *The Culture of Organs*. "Some of these are quite pitiful, and it is unfortunate that [the work Carrel and I have done] has been distorted so greatly in the press that people are given false encouragement along these lines."

Even so, Lindbergh let Carrel have his way. The result was Carrel's second appearance on the cover of *Time*, now sharing it with Lindbergh and the perfusion device, with the caption, "LINDBERGH, CARREL & PUMP / *They are looking for the fountain of age*." The cover illustration (a painting based on a photograph) showed Carrel and Lindbergh standing on either side of the elegant glass machine, looking every bit its proud parents. Lindbergh—tall, youthful, and handsome—was depicted in a suit and tie; the shorter, older Carrel, wearing a black robe, white cap, and pince-nez, exuded the brilliance only a Nobel laureate can possess. Lindbergh was looking worshipfully at Carrel, while Carrel gazed at the machine.

"More than any other man," *Time* wrote, "Carrel has made it possible to study tissue and organs outside of their organisms, but alive." After praising the engineering skill of Lindbergh, who ensured that Carrel's organ perfusion studies could be done aseptically and with pulsatile pressure mimicking the heart, *Time* predicted two medical miracles their work might soon make possible: the regeneration of diseased organs within the body, by using the pump to ascertain the specific substances required by those organs to heal themselves; and the healing of those same organs outside the body, by removing the diseased organ, then perfusing it in the pump where it could be healed, and then reimplanting it back into the body. Thanks to the brilliance of the two men on the cover, *Time* gushed, a barrier thought impenetrable—mortality—suddenly looked porous.

Time made no mention of Carrel's retirement, though it did say he would be turning sixty-five at the end of June 1938. It also reported that the Lindberghs had just purchased the tiny French island of Illiec, just a few hundred yards away from Saint-Gildas, the larger island owned by Dr. and Madame Carrel. "It makes an arresting picture," the article said, "one that French, Roman Catholic Dr. Carrel is romantic and mystic enough to appreciate: two men, one an ageless seer, the other a young and devoted inventor, sitting on two rocks in the middle of the sea, planning ways to prolong the life and end the ills of mankind."

In another phase of his publicity campaign, Carrel spoke at the annual meeting of the American Philosophical Society, where he told the elite of the American intellectual establishment—and, by way of press accounts, the entire world—about the unprecedented results his work with Lindbergh had achieved, and might still achieve, if their research continued without interruption. Nine hundred perfusion experiments had been performed in Lindbergh's pump at the Rockefeller Institute, Carrel announced, on hearts, kidneys, livers, thyroids, ovaries, testicles, lymph glands, mammary glands, spleens, pancreases, and, in one ghoulish case, the entire limb of a human fetus obtained from a spontaneous abortion that occurred in a nearby hospital in Manhattan. Collectively, these organs had been kept alive in Lindbergh's machine for 100,000 hours—the equivalent of eleven and a half years.

Carrel's speech was covered in the *New York Times*, twice. In an editorial, the *Times* described Carrel and Lindbergh's perfusion work as "a feat of the laboratory which dwarfs anything Poe ever conceived. . . . Here, too," said the *Times*, "is material for the poet":

> That plasma-like liquid which is both blood and food to the organ in its sterile glass body, what is it but a new Elixir of Life? And the pump itself, what is it but the Fountain of Youth?

Science editor Waldemar Kaempffert rose to Carrel's defense in the *Times*'s Sunday edition. "History is being made" in Carrel's laboratory, he wrote. "The work must be finished. Carrel is at the height of his intellectual powers and technical skill. Retire just when he has broken down some of the barriers that have concealed the mystery of living and dying from the inquisitiveness of man? Impossible."

THE RUSTIC PLEASURES OF Saint-Gildas comforted Carrel when he arrived there in late June 1938. So did knowing that Lindbergh, along with his wife and two young sons, had moved to Illiec, less than half a mile away. When Carrel disembarked in Cherbourg from *Queen Mary*, the press had bombarded him with questions about his new neighbor; most of these questions were about their experiments, and nearly all were misinformed. Carrel took special pains to deny the report that Lindbergh would soon become a human guinea pig, by having his own heart replaced with a perfusion pump. "The machine was never intended as a substitute for the heart," Carrel told a reporter:

> It is an entirely experimental instrument which, by maintaining the flow of blood through the cells, keeps them alive, thus rendering possible experimentation in cures for illnesses affecting the human body. . . . It was developed gradually during my studies of living cells, and the reason I associated myself with Colonel Lindbergh was that the colonel is very skillful with his hands and is a passionate scientific researcher.

Carrel relieved his frustration at his inability to control the press—or events at the Rockefeller Institute—by controlling the Lindberghs. Anne's diary reveals her surprise at how willingly her husband, normally so protective of his independence, handed over the reins. Though Carrel had raised no children of his own, he seized the role of parenting guru to the Lindberghs and brooked no dissent. "Dr. Carrel looked at the baby in his carriage. He says he should not be so much on his back," Anne wrote on June 28. The next day she wrote: "C[harles] says Dr. Carrel will hit the ceiling if he sees Land strapped into his carriage. . . . I undo the strap." On July 5 she wrote: "Dr. Carrel asks why should the baby sleep outside in a tiny carriage when he could sleep inside in a comfortable bed. . . . He also does not think the veal broth Land is eating has any particular value."

In walks along the rocky coast of Saint-Gildas, Carrel told Charles the disturbing news from New York: sinister forces were evicting them from their scientific home just when their project was at a crucial stage. A few weeks before the Rockefeller Institute closed for its summer recess, Carrel had perfused portions of cancerous human organs excised from patients at a nearby New York Hospital. There were some technical problems—the pH level inside the perfusion medium dropped after a week or so, and the fit inside the organ chamber was tight. Still, Carrel was optimistic: the human organs had definitely survived inside Lindbergh's device. Even better, there were signs that a drop in incubation temperature decreased the cancer's growth rate. The possibilities raised by these experiments in Lindbergh's perfusion pump, Carrel said, were very significant indeed. This important work had to continue—and that meant finding another laboratory to serve as their home base. Until then, Carrel said, Lindbergh must build a small research station on Saint-Gildas. The lack of electricity and plumbing would surely limit its usefulness, Carrel knew, but it would be a start. (The Carrels did have a gas-powered electric generator.)

That night Lindbergh digested Carrel's news in his journal. "How silly it is to retire a genius because he is sixty-five [yet] keep fools of forty," Lindbergh wrote. "A retirement age is a stupid regulation." But Lindbergh knew his mentor well enough to know that the new rule was less about Carrel's productivity than his personality. "Many of the people at the Institute dislike him," Lindbergh wrote. "Carrel has never gone out of his way to make friends and, on the other hand, has made many enemies. It is often that way with a man of strong character." Many of Carrel's enemies at the institute, Lindbergh knew, were Jews. He knew this because Carrel told him so.

Carrel's attitude toward Jews had always been ambivalent. Collectively, he saw them as a coarse "race" prone to clannishness, hidden agendas, and acquisitiveness, though often adept in intellectually demanding fields such as medicine. Even so, Carrel supported the applications of several Jews, most of them physicians, seeking membership in the Century Association, where anti-Semitism was not unknown. In fact, Jews who excelled in science or other areas close to Carrel's heart often received extravagant praise from him, frequently in public. In 1932 Carrel said these words at a dinner marking the sixtieth birthday of his friend Emanuel Libman, a Jewish physician at Mount Sinai Hospital in New York:

> How strangely this strong, lone figure resembles other figures, majestic shadows in the remoteness of ages! Those inspired men whom the people of Jehovah have consistently produced since the time of Samuel. In our days prophets are born from Israel, but they do not preach to the Twelve Tribes only. They lead the human soul to the heights of mysticism . . . [and] bring to men better philosophy, more justice, and mathematical equations as splendid as the Biblical poems. They are called Bergson, Cardozo, Einstein, and now, Libman . . . who is medicine, itself.

Carrel elaborated on this theme, but with fewer rhetorical flourishes, in a letter to Stephen S. Wise of the Free Synagogue in Manhattan, the same rabbi who extolled *Man, the Unknown* as an "oasis of beauty" in 1936. Carrel's letter was in response to a new sermon from Rabbi Wise calling for Jewish assertiveness in the face of Nazi persecution. "You brought a great deal of light to the Jewish problem," Carrel wrote, "when you said, 'I am not an American of Jewish birth. I am a Jew. I have been a Jew for four thousand years.' I hope these noble and courageous words will be understood by all those of your race. We Christians will always respect the Jews who are proud of being Jews, who recognize that they differ profoundly from us."

One of the ways Jews "differ profoundly" from Christians, Carrel and

Lindbergh believed, was in their support of communism. The Lindberghs had visited the Soviet Union briefly in 1931 and again in 1933. Now Colonel Raymond Lee, an American military attaché in London, was urging Lindbergh to fly there again, to inspect Stalin's air force. This created a dilemma for Lindbergh. His commitment to Carrel and their research was as strong as ever; Lindbergh built a primitive lab in the courtyard of Carrel's home on Saint-Gildas, where, as Carrel marveled in a letter to Simon Flexner, "he has started some interesting [hypothermic] experiments on local rodents [using] his gift for being able to work with almost nothing." But Lindbergh found it nearly impossible to say no to a request from his government's military. This was especially true if honoring that request might enable him to convince the civilian leadership in America, Britain, and France that—as Lindbergh believed—the true enemy of western civilization wasn't Nazi Germany but semi-Asiatic Communist Russia. So on August 7, 1938, Charles and Anne, after leaving their sons with their nanny on Illiec and saying good-bye to the Carrels on Saint-Gildas, walked across the sea bottom to the mainland of France. They then drove to an airfield, where they boarded their black-and-orange monoplane, eventually landing, after short stops in England, Germany, and Poland, in the Soviet Union.

Lindbergh was repulsed by nearly everything he saw there. A collective farm he toured was "overgrown with weeds," he wrote in his journal. An airplane factory was "not well laid out. The machines were crowded, there was a general lack of order," and the finished planes were poor imitations of those produced in the United States or Germany. Nor was Lindbergh impressed that the Soviet air force was training female pilots. "They command men and are commanded by men [and] are given the same missions, etc. I don't see how it can work very well." In fact, the Soviets' commitment to egalitarianism struck Lindbergh as a hoax: "They preach the doctrine of dividing between people according to their need. [But] the people [here] who have do not seem greatly concerned about those who have not. . . . There is already a great difference in the salaries and privileges of different people." The one activity seemingly embraced by all Soviets—drinking huge quantities of vodka—disgusted him.

After all he'd seen (and despised) in the Soviet Union, Lindbergh would have preferred to fly directly back to Illiec, but he'd promised the American

military that he would make additional aviation inspections before returning home. The first of these was in Czechoslovakia, where Lindbergh met
with the president, Edvard Beneš, then the most nervous leader in all Europe. Hitler, who'd already annexed Austria, was now threatening to invade
Czechoslovakia because of its alleged mistreatment (alleged, that is, by
Hitler) of the Sudetenlanders, a population of ethnic Germans living in
concentrated areas on the northern, western, and southern edges of Bohe-
mia, in Czechoslovakia. The only peaceful way out of this situation, Hitler
said, was to cede the territory to him. Beneš was talking about armed resistance, and the Soviets (and others) were supporting him. But Lindbergh
warned that any fight against Hitler would be suicidal. The Czechs' pursuit
planes were far too slow to stop Germany's bombers. Prague, he said, would
be flattened.

ON SEPTEMBER 10 LINDBERGH finally arrived at Illiec, wet from a
choppy Channel crossing in a small motorboat—and more than a little worried about the future. The Carrels rowed over that night for dinner. Carrel
wanted to speak to Lindbergh about new perfusion experiments. The idea
of continuing that work—and thwarting Herbert Gasser's attempt to thwart
him—was paramount in Carrel's mind. But Lindbergh's mind was elsewhere. What he saw in the Soviet Union and Czechoslovakia, and in England, France, and Germany before that, had left a deep mark. There might
be another war soon, even more horrific than the last, he told Carrel, who
agreed, with a blend of sadness and anger. The best way to avoid that catastrophe, Lindbergh said, was for Britain and France to reach a political accommodation with Germany, even if it meant sentencing several million
Czechs to foreign domination and giving the Nazis a free hand in dealing
with their "Jewish problem."

After all, Lindbergh argued, the Czechs had been occupied by Austrians
and other foreigners for most of their history, and Lindbergh's own "research" convinced him that the "Jewish problem" in Germany had been created by Jews. Every German he spoke to, Lindbergh wrote in his journal,
wanted Jews out of the country. This was because the Germans held Jews
"responsible for the internal collapse and revolution [in Germany] after the

[First World] war." During the hyperinflation that took hold in that chaotic period, "Jews," Lindbergh wrote, "are said to have obtained the ownership of a large percentage of property in Berlin and other cities—lived in the best houses, drove the best automobiles, and mixed with the prettiest German girls." Apparently, Lindbergh accepted this vile propaganda—something that might have appeared in *Der Stürmer*, the Nazi party's most notorious anti-Semitic newspaper—as the truth.*

Carrel had trouble accepting the idea that Germany was a force for good anywhere, but he despised communism as much as Lindbergh did. Carrel also agreed that the Germans' expansion into Czechoslovakia wasn't worth fighting over. In his view only Communists and Jews—"the salaried agents of Moscow," Carrel called them—were agitating for that. Carrel's biggest worry was not politics but rather that his protégé was getting distracted from their scientific research.

In the days that followed, Carrel spent hours on Saint-Gildas and Illiec reminding Lindbergh of their achievements in organ perfusion and urging him to rejoin that project full-time. They were close to achieving the fantastic goal they'd set for themselves after their mutually life-changing introduction in New York in November 1930. The proof, Carrel reminded Lindbergh as they sat on sun-warmed boulders overlooking the sea, lay in the ovaries, kidneys, and other organs living and functioning outside the bodies that created them, in the perfusion pumps Lindbergh designed at the Rockefeller Institute. Those organs—not just a few cells, but a whole community of cells, performing complex biological functions—were the most compelling evidence yet that human immortality, a dream at least as old as the ancient Greeks, was potentially a scientific reality.

That potential, Carrel was certain, had as many political ramifications as biological. In his mind the two were linked: the survival of the white race, surrounded by faster-breeding inferiors, might depend on experimental medicine. It was imperative, Carrel argued, that select researchers, men of uncommon skill and dedication, increase the biological understanding of

---

* Typical of *Der Stürmer*'s anti-Semitic ranting was an article, published before an arrest was made in the Lindbergh kidnapping case, asserting that Lindbergh's son was taken by Jews, who drank the dead boy's blood in a religious rite.

the human body, using bold new techniques to unlock the body's secrets, and then mold the body so as to ensure the survival of western civilization by giving some men access to a life beyond normal mortal limits. There already were two men of uncommon skill and dedication working on that project, Carrel reminded Lindbergh: themselves.

Continuing their experiments in organ perfusion—with all the political possibilities those experiments raised—was a far better path to a better world than the one offered by Hitler in Germany, Carrel said. That Lindbergh was fascinated by the supposedly civilization-saving qualities of Hitler's "pagan cult," as Carrel referred to Nazism, was certainly evident to Carrel. As an alternative, he urged Lindbergh to study an authentically "pagan" political ideology: Plato's *Republic*.

It's not surprising that Carrel would enlist Plato in his cause. Many of the ideas in *Man, the Unknown* echoed those presented more than two millennia earlier in the *Republic*. Each book was utopian, presenting a blueprint for the remaking of society. Each expressed contempt for democracy as an agent of weakness and social chaos. Each disdained conventional politicians as self-interested incompetents. Each called instead for political power to be exercised by a eugenically engineered elite of experts.

Lindbergh listened to Carrel's pleas with respect. But he saw the situation differently. As depressed as Lindbergh was about the possibility of war, he saw hope for the future in Hitler's Germany. The nearly constant background of his conversations with Carrel—the sound of the waves breaking on the shore—struck Lindbergh as a metaphor: just as the surf pounded Illiec and Saint-Gildas, so were there powerful cleansing waves in history. One of those waves, he believed, had formed in Germany, where Hitler removed decades of dangerous debris from his country's cultural and ethnic landscape; reinstilled national pride in a country shattered by defeat in World War I; created prosperity; crushed the left; replaced hedonism and a burgeoning feminist movement with a culture of *Kinder, Kirche, Küche* ("children, church, kitchen"); and committed his country's best minds to the advancement of science, aviation, and technology. All this by the sheer force of Hitler's will—and all in less than five years.

Yes, Lindbergh conceded, there was opposition to Hitler's rise—not just from Jews, but from "ordinary" citizens living in neighboring nations, as well as from several political leaders in those countries. This resistance, Lindbergh believed, was a natural but misguided example of cultural inertia. "As I gain experience in life I come more and more to the conclusion that great changes come without the previous approval of the masses," he told Carrel. "In fact, I believe that any major change invariably triggers resistance. But is not civilization always guided" to higher levels "by elite minorities? Is that not one of our little recognized natural laws?"

Lindbergh also conceded there were ways in which Nazism seemed dangerous—its brutality and restriction of free thought, to name the most obvious. But these, he argued, were the correctable excesses of a bold new order, one brimming with energy and purpose that was still struggling to find its path. Nazi Germany must be allowed to become even more powerful, Lindbergh said, because it manned the farthest outpost of western civilization. It was Germany's destiny to serve as the bulwark against the two barbarisms emanating from the East: Soviet communism and the nonwhite hordes of Asia. Other western nations weren't up to the task. In England Lindbergh saw organization without spirit; in France, spirit without organization. Only Hitler's Germany had both.

Yes, there was fanaticism in Germany, but also a decency Lindbergh said he couldn't find in other European nations, a deficit he was reminded of each time he saw the headlines about murder, rape, and divorce in the English newspapers he read at Long Barn. Providing an even greater contrast was the atmosphere of youth and hope Lindbergh sensed in Germany, one that, in his view, was lacking everywhere else. Hitler's plans for the rebuilding of Germany, Lindbergh said, were proceeding with the same commitment to excellence as Germany's aviation program, which now reigned supreme in Europe. These huge public-works programs, with their simple, elegant designs, aided by factories unmatched in efficiency, were in Lindbergh's mind a shining model for the rest of the western world.

It was for all these reasons, Lindbergh told Carrel, that Nazi Germany, despite its occasional violence, was the best alternative—maybe even the only one—to another world war and to the death of western civilization. To see if he was right about that, Lindbergh was prepared to become an autodidact once again. Years earlier, as an academic washout he had transformed

Charles Lindbergh,
age ten, on a raft in
the Mississippi River
near his home in
Little Falls, Minnesota.

Lindbergh and his friend
Bud Gurney at an air show
in St. Louis, circa 1925.

Lindbergh and his wife,
Anne Morrow Lindbergh,
after a training flight on
Long Island, preparing for their
1931 trip to China and Japan.

Alexis Carrel's Nobel
Prize–winning technique
for vascular anastomosis.

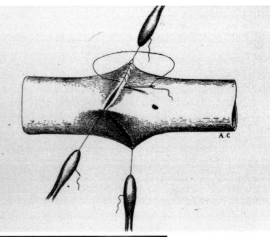

Carrel at the
Rockefeller Institute,
circa 1912, performing
an early experiment in
tissue culture.

Carrel and his wife, Anne Marie Carrel, in 1915, in front of the hospital he ran in
Compiègne, France, during World War I. It was here that Carrel codeveloped the
Carrel-Dakin method for treating war wounds. © l'harmattan.

Two of Carrel's technicians in his black-walled operating room, circa 1935.

FROM *THE CULTURE OF ORGANS*, BY ALEXIS CARREL AND CHARLES A. LINDBERGH,
ORIGINALLY PUBLISHED BY PAUL B. HOEBER, INC., IN 1938.

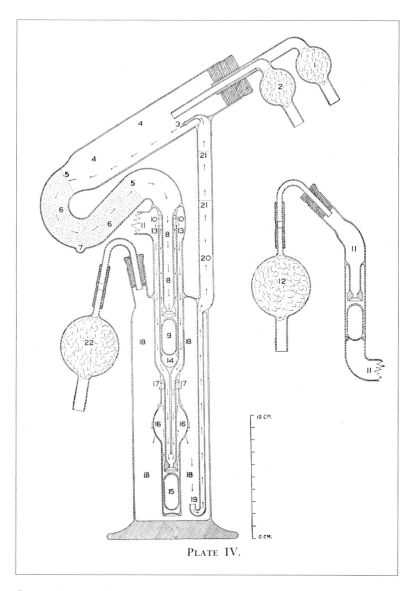

PLATE IV.

CROSS-SECTION OF LINDBERGH AND CARREL'S PERFUSION PUMP.
1: Organ chamber outer filter bulb. 2: Organ chamber inner filter bulb.
3: Organ chamber cannula. 4: Organ chamber. 5: Platinum screens. 6: Silica sand
filter. 7: End of sealed glass tube where silica sand was inserted. 8: Tube leading
to upper floating valve. 9: Upper floating valve. 10: Outlet holes at top of upper
floating valve reservoir. 11: Equalization chamber neck. 12: Equalization chamber
filter bulb. 13: Equalization chamber. 14: Upper floating valve reservoir.
15: Lower floating valve. 16: Lower floating valve reservoir. 17: Outlet holes at
top of lower floating valve reservoir. 18: Reservoir chamber. 19: Inlet to feed tube.
20: Feed tube. 21: Platinum screens. 22: Reservoir chamber filter bulb.

FROM *THE CULTURE OF ORGANS*, BY ALEXIS CARREL AND CHARLES A. LINDBERGH,
ORIGINALLY PUBLISHED BY PAUL B. HOEBER, INC., IN 1938.

Lindbergh and Carrel,
with their perfusion pump,
on the cover of
Time magazine, issue
dated June 13, 1938.

GETTY IMAGES.

Lindbergh, with wife, in
Hermann Göring's office
in Berlin, July 1936.

BAVARIAN STATE LIBRARY,
MUNICH.

Lindbergh, Carrel, and
Dr. Albert Fischer at the
International Congress of
Experimental Cytology
in Copenhagen, where
Lindbergh and Carrel
demonstrated their perfusion
pump in August 1936.

The Verdienstkreuz
Deutscher Adler ("Service
Cross of the German Eagle"),
presented by Göring
to Lindbergh on
October 18, 1938, at the
American Embassy in Berlin.

Lindbergh in the high-altitude test chamber at the Mayo Clinic in Rochester, Minnesota, September 1942.

Lindbergh with a squad of American pilots in the South Pacific, May 1944.

Lindbergh,
Carrel, and
Madame Carrel
sitting on the
rocky shore of
Saint-Gildas,
circa 1938.

© L'HARMATTAN.

Carrel in
his house on
Saint-Gildas,
shortly before his
death in 1944.

© LAURE ALBIN-
GUILLOT / ROGER-
VIOLLET / THE
IMAGE WORKS.

himself into a top aerodynamics student, then into a bioengineer, and then into a primatologist; now Lindbergh would reinvent himself once more—this time as a political scientist—by immersing himself in German culture and politics. "As time passes," Lindbergh wrote to Major Truman Smith, in Berlin, "I become more and more convinced that the future trend of world events lies in Germany's hands. . . . Whether or not one agrees with all of the German policies, there is much about the German people which commands respect and admiration":

> I have become so greatly interested in Germany, and I regard her as being of such great importance in our lives, and in our children's lives, that I am willing to do anything I can to learn more about Germany, her people and her government.

Willing, even, to disappoint the man he respected more than any other in the world.

The future of western civilization depends not on "a balance of power through the cultivation of weakness, but on strength," Lindbergh wrote in his journal, after one conversation with Carrel on Saint-Gildas. This notion surely would have sounded familiar to Carrel; he'd been saying such things to Lindbergh for years: "Weakness is the unpardonable sin." "Life loves the strong." "Pacifists never create peace; peace can only be created by the powerful." To ignore these facts, Carrel had repeatedly said, "is to ignore the biologic laws of the universe." Carrel the philosopher was proud to see how well Lindbergh had learned these lessons. But Carrel the scientist was alarmed that his student's political activism had created an unanticipated problem: the cultivation of German strength had become more important to Lindbergh than the cultivation of organs.

It was extremely difficult for Lindbergh to turn down a request from Carrel. But he did. Lindbergh could no longer commit himself to working full-time on their immortality research—not now, as Europe teetered on the brink of war. He hadn't lost confidence in Carrel, or in their work. But their research, no matter how promising, offered no guarantee of success in the short term—and the crisis facing western civilization, Lindbergh believed, had to be faced immediately. Not only must Nazi Germany be politically accommodated; it must play the leading political role in saving that civilization.

The choice he was reluctantly making, Lindbergh told Carrel in the summer of 1938, was a matter of priorities. "[How could I] spend time on biological experiments when our very civilization was at stake?" he wrote in *Autobiography of Values*. "I decided to play whatever part I could in preventing a war in Europe."

CHAPTER 13

BY ORDER OF THE FÜHRER

LINDBERGH'S NEW GOAL—NOT DEFEATING DEATH, BUT DE-feating those who would deny Germany the leading role in saving civilization—required him to remake himself once again, this time as a diplomat. In the months that followed his emotional conversations with Carrel on Saint-Gildas, Lindbergh, who had no training in statecraft whatsoever, spoke with, corresponded with, or had his views on political and military matters presented to the British prime minister, Neville Chamberlain; two former British prime ministers, Stanley Baldwin and David Lloyd George; the premier of France, Édouard Daladier; the French foreign minister, Georges Bonnet; the president of the United States, Franklin D. Roosevelt; and the American secretary of state, Cordell Hull—not to mention dozens of generals, admirals, intelligence officers, and lesser diplomats on both sides of the Atlantic. Such access was unavailable to any other private citizen in the world.

The first diplomat Lindbergh met with was Joseph P. Kennedy, the American ambassador to Great Britain (and father of the future United States president). Ambassador Kennedy cabled Lindbergh on Illiec on September 19, 1938, not even twenty-four hours after Lindbergh returned there. The situation in Czechoslovakia had reached a crisis, Kennedy wrote, and Lindbergh's views on the aviation capabilities of the potential combatants were required by officials in London and Washington. After a farewell dinner with the Carrels on Saint-Gildas, the Lindberghs took off in their plane for England the next day. One of the few personal items Charles took with him on this flight was a copy of Plato's *Republic*.

But Lindbergh's mind was less on ancient Greece than on the current crisis in Czechoslovakia. At his meeting with Kennedy at the U.S. embassy on September 21, he argued against any military response from the western democracies should Hitler invade Czechoslovakia. At the ambassador's request Lindbergh summarized his reasons for not preventing Hitler's march into Prague, in a letter that Kennedy passed on to Secretary of State Hull in Washington. Two days later, Hull confirmed to Kennedy that he had sent Lindbergh's memo to President Roosevelt.

"I feel certain that German air strength is greater than that of all other European countries combined," Lindbergh wrote in the letter:

Germany now has the means of destroying London, Paris, and Prague. . . . In the air, France's condition is pitiful. Although better off, the British air fleet cannot be compared to their German counterparts. . . . I believe that . . . Germany, on account of her military strength, is now inseparable from the welfare of our civilization, for either to preserve or destroy it is in her power.

Lindbergh was equally certain that it would be "wiser to permit Germany's eastward expansion than to throw England and France, unprepared, into war." On September 22, Kennedy handed a copy of Lindbergh's letter to Prime Minister Chamberlain, as Chamberlain boarded a plane in London to fly to Bad Godesberg, in Germany, to hear Hitler's latest demands in the ongoing negotiations regarding the crisis in Czechoslovakia.

On the same day that Lindbergh and Kennedy met in London, Hermann Göring read a report in his office—the same room where he had been photographed next to Lindbergh in June 1936—that contradicted Lindbergh's assessment. The Luftwaffe's bomber force, General Helmuth Felmy informed the air minister in that report, could not "operate meaningfully" over Britain, flying the hundreds of miles separating the two nations and then destroying British cities and factories with sustained bombardments. Luftwaffe bombers were designed for a different purpose: flying shorter missions to give cover for the German army's blitzkrieg strikes. The best Germany could hope for from its bombers over England, Felmy said, was a "nuisance effect." A German air "war of annihilation against the British"— the catastrophe Lindbergh was predicting if Britain dared to stand up to the Nazis—was "out of the question."

Unaware of that reality, Prime Minister Chamberlain, after having returned from Bad Godesberg, departed from London on September 29, this time for Munich, where he would meet with Hitler, Mussolini, and Daladier to work out a final resolution to the Czech crisis. That a humiliating outcome was certain for the Czechs was made clear when Hitler insisted that President Beneš of Czechoslovakia not participate in the negotiations. The resulting pact called for the German army to occupy the Sudetenland immediately, thus transferring de facto sovereignty over a territory of 10,000 square miles—and the 3.5 million people who lived there—from Czechoslovakia to the Third Reich.

Many Europeans, and at least one American journalist, believed that Lindbergh had helped save the world from catastrophe. "When the definitive history of the 'second World War' crisis of September, 1938, is written," Joseph Driscoll wrote in the *Washington Post*—using quotation marks to indicate his belief that the Munich treaty had made sure that this "war" would never happen—"emphasis [will] be laid on the important role played by a nonstatesman, Col. Charles A. Lindbergh."

Chamberlain, however, was happy to take the credit himself. Returning to London on September 30, he waved a signed agreement in his hand as he stood before a jubilant crowd outside his official residence, at 10 Downing Street.

"My good friends," Chamberlain said, "for the second time in our history, a British Prime Minister has returned from Germany bringing peace with honor. I believe it is peace for our time. . . . Go home and get a nice, quiet sleep."

The crowd responded with a thunderous rendition of "For He's a Jolly Good Fellow." There was one Briton, however, who couldn't understand what all the singing was about: "We had to make a decision between the shame and the war," said Winston Churchill, then a Conservative member of Parliament. "We have chosen the shame and, as a reward, we shall receive the war."

More than thirty years later Lindbergh justified his pro-German efforts toward appeasement this way. "My greatest hope," he wrote in *Autobiography of Values*, "lay in the possibility that," if war did come, it "would be confined to fighting between Hitler and Stalin. It seemed probable that Germany would be victorious in such a conflict; and by that time France and England would be stronger":

> Under any circumstances, I believed a victory by Germany's European peoples would be preferable to one by Russia's semi-Asiatic Soviet Union. Hitler would not live forever, and I felt sure the Germans would eventually moderate the excesses of his Nazi regime.

The use of the words "eventually moderate" in a sentence published more than two decades after the liberation of Auschwitz speaks volumes about Lindbergh's moral acumen.

THE CHIEF BENEFICIARY OF Lindbergh's activities on behalf of Germany in the prewar period was eager to show his gratitude. An opportunity to do so arose shortly after the signing of the Munich pact, when the Lindberghs flew to Templehof airfield, outside Berlin, on October 11 to attend, for the second consecutive year, the meeting of the Lilienthal Society for Aeronautical Research.

Less than a week before the Lindberghs' arrival, the Nazis issued a new decree making it clear just who was the "debris" they were intent on sweeping out of Europe. Beginning on October 5, all Jews in Germany were required to obtain and carry special identity cards marked with a large red "J" to make it easier for the German police to identify them. This was the latest in a series of insults that had been escalating all year: a program of "Aryanization" forced the sale of nearly all Jewish-owned businesses in Germany to Christians; and Jewish physicians, surgeons, and nurses had been forced to resign, en masse, from German hospitals in late summer. All over Germany large billboards were erected by Nazi officials and members of the Hitler Youth declaring, "Jewry Is Criminal" and "Jews Not Wanted," sentiments typically illustrated with mocking caricatures of bearded, hook-nosed men wearing caftans or large black hats. It's a virtual certainty that Lindbergh saw several of these signs through the window of the Mercedes-Benz limousine that drove Anne and him into Berlin from Templehof airfield. But if he did, he never mentioned it. Lindbergh's focus was elsewhere.

Hitler focused on Lindbergh on October 18. It had been a long day for the American: after visiting two Junkers factories several hours southwest of Berlin—one, in Magdeburg, for engines; the other, in Dessau, for fuselages—Lindbergh flew back to the capital, where he washed up and changed, before leaving with Colonel Truman Smith—who had recently been promoted from major—for a "stag dinner" at the residence of the new United States ambassador to Germany, Hugh R. Wilson.

Along with several ambassadors from nearby European nations, Wilson had invited the elite of German aviation to this event: the manufacturers Willy E. Messerschmitt and Ernst Heinkel; Adolf Bäumker, the chief of research for the German Air Ministry; and generals Ernst Udet and Erhard Milch of the Luftwaffe. These officers and diplomats, along with lower-ranking military men, all thirty of them in formal dress, milled about, talking and sipping champagne from crystal stemware, in a flower-filled room where a long dining table was set up, lit by dozens of candles in a row of silver candelabras.

The highest-ranking guest at the dinner, as was his custom, was the last to arrive. When Hermann Göring finally entered the room, he was wearing a crisply pressed cobalt blue uniform his fellow Luftwaffe officers had never seen before, and he was accompanied by an aide carrying a red box and a rolled sheaf of paper. General Göring shook hands and made small talk with his Luftwaffe colleagues until he noticed Lindbergh standing against the room's back wall. Göring immediately walked across the room, his gleaming black boots clip-clopping on the polished floor, his rouged face glowing in the candlelight. When Göring reached Lindbergh, he shook the American's hand, then turned to accept the red box from his aide and handed it to Lindbergh, unopened.

Inside was a golden cross with four small swastikas finished in white enamel, on a red ribbon with black-and-white borders—the Verdienstkreuz Deutscher Adler ("Service Cross of the German Eagle"), the second-highest medal conferred on foreigners for "service to the Reich." After Lindbergh opened the box, Göring, Hitler's most trusted confidant, looked the American in the eye, smiled broadly, and said, *"Im Nahmen des Führer."* The American consul-general in Berlin, Raymond Geist, who was standing at Lindbergh's side, translated Göring's remark: "By order of the Führer."

Lindbergh accepted the golden Nazi cross with "unqualified pride," Ambassador Wilson later wrote in his memoir, and then everyone in the room burst into applause. Soon the guests took their seats—Göring and Lindbergh at the places of honor, at either end of the table—and the banquet began. The main topics of conversation, Wilson wrote, were aviation, Hitler's architectural plans for Berlin, and the Munich pact. The contents of the red box in front of Lindbergh prompted no further comment.

Until, that is, the dinner was over, the cigars were smoked, and the brandy snifters were emptied—Lindbergh politely declined to participate in those last two activities—at which point the newest holder of the Verdienstkreuz Deutscher Adler, noticeably exhausted, returned to the rooms he was sharing with his wife at Colonel Smith's apartment, not far from the American embassy. Without speaking, Charles handed his wife the rolled-up proclamation signed by Hitler, and then the little red box. Anne opened the elegant container, then removed the ribboned medal, designed to be worn around the neck, and placed it in the palm of her hand, where she examined it. Then there was another silence, the silence of comprehension.

"The albatross," Anne finally said, sensing the criticism this swastika-covered decoration would bring her husband, once the British and American press learned of its existence. She put the medal back in the box, then closed it.

CHAPTER 14

I'LL TAKE A RAIN CHECK

LINDBERGH WENT ON AS IF NOTHING CONTROVERSIAL HAP-
pened that night at the embassy. As was his custom, he sent a handwritten
thank-you note to General Göring. "I want to express my appreciation for
the courtesy which I have always received in your country," he wrote:

> I want to thank you especially for the honor which you conferred
> upon me at the dinner given by Ambassador Wilson. I hope that
> when the opportunity presents itself, you will convey my thanks to
> the Reichs Chancellor [i.e., Hitler]. . . . It is an honor that I will al-
> ways prize highly.

He went on, "I cannot end without telling you how impressed my wife and
I were with the plans for the reconstruction of Berlin. . . . We are looking
forward to learning more about Germany and its plans for the future."

Lindbergh planned to do that learning on-site. At his instruction, Anne
had begun looking for a home to rent for their family in or near the German
capital. Shortly after accepting his Nazi medal, Lindbergh accompanied her
to see a large furnished house in the upscale suburb of Wannsee. "The most
attractive feature was the garden—a large one, well planted with trees and
shrubs, and running downhill to a river with swans," he wrote in his journal.

"We plan on taking it if we can get a satisfactory lease." The German air ministry was so eager to help Lindbergh in this transaction that it assigned a senior Luftwaffe officer to assist him. The officer phoned the owners of the Wannsee house that evening, but his report back to Lindbergh was sharply negative. The owners were asking too much rent, he said, and, even worse, were insisting that it be paid in foreign currency. There are no documents to support this account of the demands supposedly made by the property owners. What we do know is that the owners were non-Aryan. Clearly, it was unacceptable to the Nazis that the heroic, blue-eyed aviator who had been given a medal "by order of the Führer" would live with his wife and two sons in a house owned by Jews.

That same night Lindbergh composed a letter to Carrel, in New York. "There is nothing I would rather do than be working in your laboratory," he wrote. Even so, Lindbergh announced that he was moving to Berlin. Something historic was going on there, he said, something he wanted to be a part of, or at least witness firsthand: "The attention [given by the Nazis] to art and architecture is . . . comparable [to that given] their aviation program. Germany has underway the greatest building program the modern world has ever seen. . . . The designs [I] have been shown are simple and beautiful." This massive construction program, Lindbergh was certain, "necessitates peace for its completion. I feel sure that Germany wishes to avoid a major European war."

With that certainty in mind, Lindbergh decided to wait for the Luftwaffe officer to report back with a new list of rental properties in or around the German capital. Until then, Anne and he would leave their plane in Berlin, take the night train to Paris, shop for provisions, and then take another train to Brittany. From there they would row or, if the tide was low, walk, to Illiec, where their sons Jon and Land had been left behind with the domestic staff.

Shortly after Lindbergh arrived on his island hideaway, he began to realize there were some people who didn't share his enthusiasm for the Nazis or their medals. On November 13, 1938, several pages of the London *Times* were devoted to the outrage expressed throughout the civilized world at Kristallnacht, the anti-Semitic riots that took place in Germany four days earlier. This brutal mayhem, a hint of worse horrors to come, left almost 100 German Jews dead; more than 1,000 synagogues burned to the ground; and nearly 8,000 businesses vandalized or destroyed—and it resulted in more than 25,000 German Jews' being arrested and sent to concentration camps. President Roosevelt showed America's revulsion at these events by recalling Ambassador Wilson to Washington; no American ambassador returned to Germany until after the Germans surrendered in World War II.

According to the Nazis, Kristallnacht (the "night of broken glass") was a spontaneous response to a crime committed by a Jew against the German nation. The Jew was Herschel Grynzspan, a seventeen-year-old living in Paris, who shot Ernst vom Rath, a junior official at the German embassy there, on November 7. Vom Rath died two days later, which is when the violence began. Grynzspan was thought to have been seeking revenge for the fate of his parents, small business owners from Hanover, Germany, who were among the 17,000 German Jews of Polish origin arrested and forcibly relocated back to Poland by the German police in October.

The Nazis' claim that Kristallnacht was spontaneous was a lie. As the historian Peter Gay, an eyewitness to the event, later wrote, "Local authorities across [Germany] ... had the names and addresses of Jewish men on record. In Berlin ... Jewish store owners had already been compelled to paint their names on their front windows in large white letters. If ever there was a thoroughly organized pogrom, it was Kristallnacht."

Lindbergh wasn't sure what to make of the violence, organized *or* spontaneous. It seemed so "contrary to [the Germans'] sense of order and their intelligence in other ways," he wrote in his journal. He did, however, know whom to blame:

[The Germans] undoubtedly [have] a difficult Jewish problem, but why is it necessary to handle it so unreasonably? ... Do the Germans feel that in this way they can frighten all Jews sufficiently to prevent incidents such as the Herr vom Rath shooting? Or is this a countermove to the Jewish pressure on Germany?

The outrage expressed in foreign newspapers against Kristallnacht turned against Lindbergh when the Associated Press reported that he was looking for a home in Berlin. Lindbergh planned to ignore the criticism, but was warned not to by Carrel. "There is a great deal of ill feeling against you," Carrel wrote, referring to coverage of the Nazi medal in the American press.

"The situation . . . is serious." Just like Lindbergh, Carrel knew whom to blame for this turn of events. People here "have stopped thinking," he wrote. Instead, "they are influenced by the Jewish and Russian propaganda."

One example of that "propaganda" accused Lindbergh of becoming a propaganda tool himself. An official of the Institute of Propaganda Analysis in New York told the *New York Times* that, "by accepting the Nazi decoration," Lindbergh "expressed approval of the Nazi government, whether he knew it or not." Either way, the consequences for Lindbergh were clear: TWA dropped the slogan "The Lindbergh Line" from its advertising literature; and the U.S. secretary of the interior, Harold Ickes, declared in a highly publicized speech that anyone who accepts an honor from Hitler "automatically forswears his American birthright." Comments like these led Carrel to write Lindbergh again, urging him not to move to Berlin. "You [must] not lose the moral authority you have [in America,"] he wrote. "I am enclosing a clipping from the *New Yorker*. This vicious attack was written by a Gentile!"

The gentile was E. B. White, whose tone, in reality, was less vicious than ruefully ironic. "With confused emotions," he wrote in the lead "Talk of the Town" item in the issue of November 26, 1938, "we say goodbye to Colonel Charles A. Lindbergh, who wants to go and live in Berlin, presumably occupying a house that once belonged to Jews":

> He has, if any man ever had, a reason to hate democracy and admire a system that can protect privacy just as efficiently as it can destroy life and hope. We say goodbye and wish him luck. If he wants to experiment further with the artificial heart, his surroundings there should be ideal.

In the end, Lindbergh heeded Carrel and moved his family for the winter not to Berlin, but to a rented apartment in Paris, on avenue Maréchal Manoury, in the sixteenth arrondissement. The house on Illiec wasn't heated; spending the winter there wasn't a realistic option. It was very important to Lindbergh, however, that his mentor not think his decision was in any way influenced by criticism from the press. To the contrary, Lindbergh's contempt for journalists was as strong as ever: "I have found that it would be necessary to sell my character if I wished to maintain the friendship of modern journalism," he wrote Carrel. "I prefer its enmity."

HE'D GET HIS PREFERENCE. In the spring of 1939 Lindbergh came to a decision he had put off for more than three years. As Europe edged closer to war, Lindbergh heard duty call him home. This was not an easy decision to make. In America, "the country which is supposed to exemplify freedom, I have found little freedom," he wrote in his journal. "I found the most personal freedom in [Nazi] Germany."

But Lindbergh's motivation wasn't personal; it was political. If hostilities did start in Europe, someone had to lead the fight in the United States against American involvement in that war, no matter how loudly the Roosevelt administration and the press beat the drum for involvement, and no matter how vociferously those two institutions attacked anyone who spoke out against it. The someone who must lead that anti-interventionist fight, Charles Lindbergh decided, was himself.

The mechanically gifted pilot who worked so hard and secretly in Carrel's laboratory to achieve immortality in individual men would now work with the same fervor in the public arena to extend the life of western civilization. This was the duty calling him home. If war was required to save western civilization from the threat posed to it by the East—China, Japan, and, above all, the Soviet Union—that fighting must be led by the strongest western nation. "The strength of [Hitler's] Germany is the strength of youth, of spirit, of life itself," Lindbergh wrote to William C. Bullitt, the American ambassador to France. "The welfare of our civilization rests on finding some means of working with, and not against, this country."

But there was something else calling Lindbergh home, as well: Carrel had lost his own political struggle with Herbert Gasser and would be forcibly retired from the Rockefeller Institute on July 1, 1939. Carrel, angry and dispirited, had asked Lindbergh to help him find a new professional base. This was not a request Lindbergh could refuse. So, in late March, Lindbergh bought a one-way ticket to New York on the grand Cunard liner *Aquitania*. He left on April 8 from Cherbourg, a short drive from Illiec. This trip across the Atlantic, like the flight in the *Spirit of St. Louis* that made

Lindbergh world-famous, would be in a cabin for one. Anne and the boys would follow in another ship a few weeks later.

Perhaps to prepare for his new mission, Lindbergh spent much of his weeklong crossing reading J. C. Stobart's *The Glory That Was Greece*. When he was not studying Plato, Pericles, and other architects of western civilization, Lindbergh noticed that many of his fellow passengers on *Aquitania* were wealthy Jewish refugees. One of them, a Romanian in her twenties, shared a table with him on the first night out at sea in *Aquitania*'s dining room, a luxurious space with a ceiling painting done in homage to Giambattista Tiepolo's eighteenth-century masterpiece, *The Triumph of Flora*. The other places at the dinner table were unoccupied.

After an awkward silence, Lindbergh decided that he liked his attractive young companion; her English, though accented, was more than passable. Even so, he decided to change his seating arrangements at subsequent meals. "She is at least partly Jewish," Lindbergh wrote in his journal, "and she will think it is on that account. But if I don't change tables, the newspapers in America will grab her, photograph her, interview her, and then throw her in the gutter according to their usual procedure."

This was the only reference to the presence of Jews on board *Aquitania* in *The Wartime Journals of Charles A. Lindbergh*, a book that, when published in 1970, included a preface from the publisher William Jovanovich asserting that none of its entries had been rewritten, and that the only omissions from the original handwritten 3,000-page text were personal references to people still alive, passages deemed repetitious, and material "finally adjudged not important enough to warrant adding to the length of the work as a whole."

But as A. Scott Berg pointed out in *Lindbergh*, his Pulitzer Prize–winning biography published twenty-eight years later, there were "several omissions in the published texts [of *Wartime Journals*] that were substantive in nature"—nearly all of them about Jews. One of those deleted entries, dated April 10, 1939 (and written after several hours of rough seas), is particularly revealing.

"The steward tells me that most of the Jewish passengers are sick," Lindbergh wrote:

Imagine the United States taking these Jews in addition to those we already have. There are too many places like New York already. A few

Jews add strength and character to a country, but too many create chaos. And we are getting too many.

These words could have come from the mouth of Alexis Carrel—and, in fact, there's a possibility they did, a few days later, inside Lindbergh's first-class cabin. Carrel had written to Henry Morgenthau Jr., the U.S. secretary of the treasury, to ask that Madame Carrel and he be allowed to accompany U.S. customs agents as they boarded *Aquitania* when it was anchored in New York harbor, before the ship reached its dock on the West Side of Manhattan. Carrel had important matters to discuss in private with Lindbergh, he wrote in his letter. Though Carrel didn't specify what they were, the secretary gave his permission.

And so it was that the first two people Charles Lindbergh encountered on his return to the United States on April 14, 1939, were two of the people he admired most in the world: the Carrels, who knocked on his cabin door as he was sitting inside, next to his luggage. Lindbergh was thrilled at this turn of events—at least for a while. The three friends conversed happily in his cabin when *Aquitania* pulled up anchor and cruised to its dock. Once there, they continued to talk while other passengers began to disembark.

What the trio had no way of knowing was that a photographer from the *New York Daily Mirror* had managed to sneak aboard and had then bribed a steward into revealing the location of Lindbergh's cabin. This photographer entered an empty suite adjoining Lindbergh's and opened a door linking them; he then jumped in and snapped several photos with his Speed Graphic, each with the loud and intrusive flash that were that camera's unofficial trademark.

One of those photos appeared on the front page of the *Mirror* the next morning. Caught completely unawares, Lindbergh—his once lean face now fleshier and his blond hair noticeably darker and thinner than they were when he became the most photographed person on earth in 1927—was beaming as he sat next to an older woman in a dark jacket and feathered hat, while looking across the room at a thickset bald man in a wool suit whose back was to the camera. These two people, identified as "fellow passengers" beneath the photo, were Anne and Alexis Carrel. "Lindy Comes Home Smiling," the caption in the *Mirror* began.

But he did not keep on smiling. Minutes later, all the other passengers on *Aquitania* had disembarked, to be replaced by more press photographers. Several New York police officers entered Lindbergh's cabin and offered to aid him and his friends as they moved through this mob. Lindbergh accepted. "I went first, Mme. Carrel next, and last, Dr. Carrel," he wrote in his journal:

> Both sides of the corridor and stairs were lined with cameramen and flashing, blinding lights. They started shoving and blocking the way in front of us. The police immediately formed a wedge and pushed them out of the way. . . . All the way along the deck the photographers ran in front of us and behind us, jamming the way, being pushed aside by the police, yelling and falling over each other on the deck. . . . It was a barbaric entry into a civilized country.

AFTER SPENDING A QUIET night at Next Day Hill, Lindbergh went to work on the two projects that had brought him home: keeping America out of Germany's way in Europe; and finding a new scientific base for Carrel. The first mission led him to drive the next morning in the Morrow family's De Soto to West Point, where he met with General Henry H. "Hap" Arnold, chief of the army air corps. After listening to Lindbergh describe the Luftwaffe's amazing growth—a conversation that took place in the grandstand next to a baseball field, where the West Point Cadets were playing Syracuse University—General Arnold urged Lindbergh to briefly go on active duty in the air corps, so that he might tour American air bases and assess America's air capabilities. Lindbergh instantly agreed. He also acceded to Arnold's request that he travel to Washington, where he could give his report on the Luftwaffe to the secretary of war, Harry Hines Woodring; and, after that, to the commander in chief, President Franklin Delano Roosevelt.

The two famous Americans met in the Oval Office at the White House on April 20. The president "leaned from his chair to meet me," Lindbergh wrote that night, "and it is only now that I stop to think about it [that I remember] . . . he is crippled. I did not notice it . . . during our meeting." Roosevelt, whose pro-British, anti-German position was hardly a secret, listened to Lindbergh's report, then gave his own impressions of Europe and its leaders. Unlike Lindbergh, Roosevelt had read *Mein Kampf.*

The President is a "suave, interesting conversationalist," Lindbergh wrote. "But there is something about him I did not trust, something a little too suave. . . . Still, he is our President. . . . I [want] to work together as long as we can; yet somehow I have a feeling that it may not be for long."

The next day Lindbergh drove to Bolling Field in the capital to go on active but unsalaried duty in the air corps. He was assigned the most advanced fighter plane in the service, the Curtiss P-36A, which he took up on a test flight. He left in that same plane the following morning on the inspection tour he'd talked about with General Arnold, making twenty-three stops in thirty days.

At every base he visited on this hectic itinerary, Lindbergh—invariably met with awe—told American pilots that the Germans were superior in the air and probably unstoppable on the ground, that the Nazis had infused Germany with a vibrant new spirit, that Britain and France were weak and directionless, and that it was imperative for the sake of western civilization that America stay out of Germany's way as the Germans guarded the frontier against the West's true enemies: the Soviet Union, China, and Japan. Lindbergh also suggested that the air corps employ eugenic measures to screen candidates for its ranks, a step that he said should also be taken for potential spouses for those pilots. Lindbergh even asserted that the final say in those marriages be given to the air corps. Alexis Carrel couldn't have said it better himself. In fact, he'd made a similar point about society at large in *Man, the Unknown.*

Carrel would talk about his own future with Lindbergh when Lindbergh returned. Perhaps because of his imminent forced retirement, Carrel was showing signs of increasing paranoia. Before Lindbergh left on his tour of air bases, Carrel sent him this warning: "I am glad you have decided to go on active military duty. But you have dangerous enemies. I have reason to believe they may attempt to harm you. Sabotage of your plane and of your parachute can be feared. Anne"—Carrel's pendulum-swinging, future-seeing wife—"asks that you be extremely watchful."

After Lindbergh came back, safely, to Next Day Hill, where his own wife, Anne, and their two sons had arrived in late April, he drove into Manhattan to meet Carrel at the Rockefeller Institute. It was an emotional homecoming for Lindbergh. He was delighted to see Carrel in the setting where they'd once worked so long and hard together. And vice versa: Carrel, in one of the collarless surgeon's smocks he typically wore to work, nearly jumped out of his chair behind his desk to embrace Lindbergh when the latter entered his large office. After patting his lanky protégé on his back, Carrel insisted that they walk through the various workstations of the division of experimental surgery, just as they'd done when they met, in November 1930. At one of those stations, technicians continued to monitor the chick-heart cells that were still beating, twenty-seven years after Carrel "gave birth" to them in that very spot. Lindbergh happily sat down to examine the tissue through a microscope, and his own heart began to race: he felt the same awe that he felt on his first visit to Carrel's laboratory, nearly nine years earlier.

Carrel was equally eager to show Lindbergh the ovaries, kidneys, and other whole organs from cats, dogs, and similar donor animals—organs that were being kept alive and fully operational, outside the bodies that created them, in Lindbergh's perfusion pumps. The two friends then walked into Carrel's "mousery," where thousands of rodents—some resting, some fighting, the rest noisily burrowing through soil—were participating in one of the most rigorous life-extension experiments in the history of zoology. Moments later Lindbergh and Carrel climbed the staircase that linked Carrel's laboratory with the operating rooms above. There Lindbergh's mind drifted back to those thrilling mornings when he had stood silently in those rooms in his own black robe and hood, watching Carrel slice into the anesthetized bodies of animals with such precision, confidence, and grace.

Those experiences—partly medical, partly mystical—had transported him, Lindbergh wrote in *Autobiography of Values*, far beyond "the world men ordinarily live in." To be in those rooms once again, with their black operating tables and black sinks, a place where, in Lindbergh's words, "life merged with death so closely I sometimes could not tell them apart," only enhanced Lindbergh's admiration for the bald Frenchman who had welcomed him into the exhilarating world of biological research, and, by doing so, transformed a few daydreams Lindbergh had in the Utah desert into reality. Other than his father and maternal grandfather, Lindbergh had never loved a man so much or so well.

Carrel told Lindbergh about the installation of one of their perfusion pumps—with the thyroid of a dog inside it—at the Hall of Medical Science at the 1939 World's Fair, which had opened several weeks earlier in Queens. Carrel's visit to the fair to examine the installation was widely reported in the press. "It might seem purposeless to a visitor to take so much trouble to keep a small pink organ, hardly the size of a string bean, pulsating in a glass tube," a reporter for the *New York Times* wrote. "But Dr. Carrel says that in permitting the separate study of tissue growth and prolonging the action of vital organs, this apparatus becomes one of the principal instruments for the investigation of the mystery of age—the first laboratory approach to the problem of enabling man to live indefinitely." It's possible Carrel repeated those very words to Lindbergh in his office, hoping they might inspire Lindbergh to once again focus his considerable energies on the project that had brought the two men together.

Lindbergh couldn't commit himself to that, but he did devote much of June 1939 to helping Carrel find a new professional home. On June 3, a day after Lindbergh rented a house for his family in Lloyd Neck, on the North Shore of Long Island, he lunched with Carrel at the Century Association in Manhattan, where they were joined by Ralph Wyckoff, formerly a biophysicist at the Rockefeller Institute. The presence of Dr. Wyckoff, an expert on X-ray crystallography and the purification of viruses by centrifugation, was significant. He, too, had been pushed out of Rockefeller by Herbert Gasser, who believed Wyckoff too much of a physicist, and not enough of a biologist, an imbalance that allegedly made his work "medically inappropriate."

After leaving the institute in 1938, Wyckoff joined Lederle Laboratories in New Jersey, where he quickly proved Gasser wrong. Using his ultracentrifugation technique, Wyckoff isolated the virus responsible for sleeping sickness in livestock, and then created a vaccine, cultured in chick embryos, to prevent the spread of that illness. Lederle was soon shipping nearly 50,000 doses of Wyckoff's vaccine each day—resulting in a huge windfall for the company, which planned to use some of those profits to build Wyckoff a large tissue culture laboratory modeled on the one established by Carrel twenty-five years earlier at the Rockefeller Institute. Once this new lab was built, Wyckoff could explore the possibility of adapting his sleeping-sickness treatment there for use on humans.

Despite that generous offer, Wyckoff was open to other opportunities. Lindbergh suggested that he join with Carrel to launch a new super-institute, one that would be part virus research center, headed by Wyckoff; and part "Institute of Man" think tank, headed by Carrel. Wyckoff said he was interested, so Lindbergh gave him Carrel's "Memorandum as to a Proposed Center of Integrated Scientific Research," a document in which Carrel called for a "nucleus" of men of unquestioned achievement and "universalist minds" to direct a small, elite group of specialists who would amass data and study all the issues crucial to the creation of a "civilized" society. These researchers' findings would become the basis of reforms in the medical, legal, and educational systems—indeed, in virtually all aspects of life, from the breast-feeding of babies to the content of public entertainment. To Wyckoff's surprise, Lindbergh suggested that this new institute might be situated at High Fields, the now unoccupied estate near Hopewell, New Jersey, where Lindbergh and his wife had been living when their first child was kidnapped.

Two weeks after that lunch, Lindbergh and Wyckoff, joined by the businessman James Newton, a friend of both Lindbergh's and Carrel's, spent two hours driving from Manhattan to High Fields. Lindbergh hadn't set foot there for more than four years. But if this return was hard on him emotionally, he didn't say so in his journal. Instead, he talked about his former home as if he were a realtor.

"The place has changed considerably since I last was there," he wrote, on June 18, 1939. "The bushes have jumped upward, and there is a feeling of much heavier foliage. The house needs a new coat of whitewash, and the shutters should be painted, but everything is in good condition."

Though Lindbergh was reluctant to write about the pain he'd experienced in that house as the world's most famous man, *Time* magazine was not. In an article that was on the newsstands at the very moment Lindbergh and his friends were touring High Fields, *Time* called for an end to the "long dark years of war" between Lindbergh and the press. "For twelve years Charles Lindbergh has been a hero, and twelve years is too much," *Time* declared. It was time for the press to back off, to respect Lindbergh's privacy, and to allow him and his family to lead a normal life. Undercutting its own position, *Time* put Lindbergh's face on its cover.

Lindbergh, Carrel, Wyckoff, and Newton—who once worked as a labor-relations adviser to the tire baron Harvey Firestone—met again several days later at a Japanese restaurant in midtown Manhattan. "Everyone has agreed that the [new] Institute should be built at High Fields," Lindbergh wrote. Carrel and Wyckoff estimated that the enterprise would require $50,000 to $100,000 per year in funding. Newton volunteered to be chief fund-raiser. "We now have the men and the property," Lindbergh wrote. "I hope the financing will not prove too difficult."

Carrel, satisfied that his future was in good hands, left soon afterward for Saint-Gildas, leaving Lindbergh free to refocus his attention on the political situation in Europe and America. In Lindbergh's view, it was worsening in both places. He was distressed that Britain had begun working to create an anti-German alliance with France and the Soviet Union. He was even less happy when Britain's King George VI was received as a hero in Washington by President Roosevelt, who then called on Congress to revise America's Neutrality Act so that the United States might aid Britain, with supplies and financial assistance, in resisting German aggression in Europe.

Lindbergh was equally troubled by what he saw as pressure from American Jews to confront the Nazis, especially in the pro-interventionist drumbeat emanating from the American media, which Lindbergh saw as controlled by Jews. On August 23, the same day that Nazi Germany and the Soviet Union signed a nonaggression pact, Lindbergh talked about the Jewish pro-war lobby at a dinner at the Washington home of William R. Castle, a former State Department official in the Hoover administration. Castle's other guest was Fulton Lewis, a political commentator on the Mutual radio network.

"[We] discussed the European situation and the action this country should take if war breaks out over there," Lindbergh wrote in his journal that night. "We are disturbed about the effect of the Jewish influence on our press, radio and motion pictures."

Once again, however, as A. Scott Berg was the first to point out, Lindbergh's published entry on this dinner was abridged. Left out of *Wartime Journals* was this:

Whenever the Jewish percentage of total population becomes too high, a reaction seems to invariably occur. It is too bad because a few Jews of the right type are, I believe, an asset to any country. . . . If an anti-Semitic movement starts in the United States, it may go far. It will certainly affect the good Jews along with the others.

Lewis was so delighted that Lindbergh shared his concerns regarding this subject that he made an offer: "Colonel, I'm going on vacation soon. I'm asking a number of prominent people to fill in as guest commentators on my show. Why don't you be one of them?" Lindbergh was flattered, but said his active-duty status in the air corps made that impossible for now. "But," he said, "I'll take a rain check."

Five days later, Lindbergh invited Dr. Wyckoff to his home for a swim, a meal, and a chance to talk further about their plans for a new institute to be jointly led by Wyckoff and Carrel. If Lindbergh seemed distracted when Wyckoff got there, he had reason: for several months, Colonel Truman Smith, now stationed in Washington, had been sending Lindbergh terse telegrams regarding the likelihood of war in Europe. The latest one had reached Lloyd Neck just before Wyckoff arrived. "YES, 80," Smith wrote, meaning that he now estimated the probability of war at 80 percent. After Lindbergh mentioned this prediction to Wyckoff, both men agreed, Lindbergh noted in his journal, "that nothing could be done" regarding the new institute "until the crisis has turned one way or the other." Four days later, on September 1, 1939, it turned. Germany invaded Poland and World War II was under way.

—

CHAPTER 15

NOT MERELY A SCHOOLBOY HERO, BUT A SCHOOLBOY

F OR LINDBERGH, HITLER'S BLITZKRIEG STRIKE INTO POLAND was the beginning of the war he had returned home to fight—or, to be more accurate, to fight against. After Wyckoff left, Lindbergh sat down at his desk and started writing down his thoughts about the situation in which America now found itself. For the next week or so he did little else, other than listen to news reports on the radio. Then, on September 10, he phoned Fulton Lewis at his weekend home in rural Maryland. "Mr. Lewis, this is Charles Lindbergh," he said. "I want to talk to you about the radio address you suggested."

The two men decided that Lindbergh would give his address on September 15, in Washington, after he secured his release from active duty in the army air corps. Lindbergh left New York on a train, alone, just before midnight on September 12, and went to the capital, where he'd rented a pied-à-terre several weeks earlier. Anne would join him there in two days. By the time she arrived, it was arranged that her husband's speech would be carried not just by the Mutual radio network, but by the CBS and NBC networks as well—blanket coverage usually reserved for the president of the United States.

News of Lindbergh's impending speech, and its contents, soon reached the White House, which responded with a fascinating bit of politics. Using Colonel Smith as its reluctant messenger, the Roosevelt administration offered Lindbergh a deal: if he didn't give a radio speech urging America to remain neutral in the war, the president would create a new cabinet post—secretary of air—and give it to him. Lindbergh wasn't surprised by the offer, but he was surprised that Roosevelt thought he might accept.

Shortly after six p.m. on September 15, Lindbergh and his wife took a taxi from their apartment to Fulton Lewis's home in the capital, where they had supper. Then they drove in Lewis's car to the Carlton Hotel, the neutral site selected by the three competing radio networks for Lindbergh's address. Just before 9:45 p.m., Lindbergh cleared his throat and walked to the six microphones, two for each radio network, that had been set up for him in a small room above the hotel lobby. He spoke standing.

"In times of great emergency, men of the same belief must gather together for mutual counsel and action. If they fail to do this, all that they stand for will be lost," he began, in his high-pitched, slightly nasal voice—which hadn't been heard in public in America since he testified at the trial of Bruno Richard Hauptmann in 1935. "I speak tonight to those people in the United States of America who feel that the destiny of this country does not call for our involvement in European wars." But Lindbergh knew there were many listening who felt otherwise; to those listeners he made an argument that could just as easily have been made by Carrel.

The current war in Europe is not one "in which our civilization is defending itself against some Asiatic intruder," Lindbergh said. "There is no Genghis Khan nor Xerxes marching against our Western nations. This is not a question of banding together to defend the white race against foreign invasion."

To remain aloof from the new European war wouldn't be easy, Lindbergh conceded, because Americans will be "deluged with [interventionist] propaganda, both foreign and domestic—some obvious, some insidious. Much of our news is already colored," he said. "We must learn to look behind every article we read and every speech we hear. We must not only inquire about the writer and the speaker—about his personal interests and his nationality—but we must ask who owns and who influences the newspaper, the newsreel and the radio station."

Lindbergh soon made similar arguments elsewhere. In November 1939, an article he wrote, "Aviation, Geography and Race," appeared in America's best-selling magazine, *Reader's Digest.* The racial views of Carrel—particularly those on the differing intellectual capabilities of the races—suffused Lindbergh's text. Aviation, he wrote, "is a tool especially shaped for Western hands, a scientific art which others copy in a mediocre fashion, another barrier between the teeming millions of Asia and the Grecian inheritance of Europe—one of the priceless possessions which permit the White race to live at all in a pressing sea of Yellow, Black and Brown."

"We, the heirs of European culture," Lindbergh wrote, "are on the verge of a disastrous war, a war within our own family of nations, a war which will reduce the strength and destroy the treasures of the White race. . . . It is time to turn from our quarrels and to build our White ramparts again. . . . Our civilization depends on a united strength among ourselves, . . . on a Western Wall of race and arms which can hold back either a Genghis Khan or the infiltration of inferior blood. . . . Let us not commit racial suicide by internal conflict."

There were many Americans who considered Lindbergh's words heroic. The Republican senator William E. Borah of Idaho was so inspired that he urged Lindbergh to run for the Republican nomination for president in 1940. Herbert Hoover, whom Lindbergh had visited in his apartment at the Waldorf Towers in Manhattan, was also warm to this idea. (Lindbergh politely declined.) Thousands of letters and telegrams arrived at the Lindbergh home in Lloyd Neck, and at the radio stations that carried his speeches, the newspapers that reprinted them, and the magazines that published his other articles. "No one . . . exert[s] a deeper influence on public opinion than yourself," DeWitt Wallace, the founder of *Reader's Digest*, told Lindbergh. Ninety percent of the mail arriving at Lloyd Neck was positive, but Lindbergh was savvy enough to know that "the people who like what you say are more likely to write than those who don't—at least," he wrote in his journal, "that's true in the intelligent classes."

But there was criticism, too, some of it from unexpected sources. One American who attacked Lindbergh publicly was the former boxing champion Gene Tunney, who'd met Lindbergh several times after Lindbergh's flight to Paris. (Tunney held the world heavyweight title in 1927, having vanquished Jack Dempsey.) Those meetings were friendly, so Lindbergh was surprised to learn of a speech Tunney gave in Boston, denouncing as "shocking impertinence" Lindbergh's suggestion that America "desert" Britain in its time of need. Tunney, who'd served overseas in the U.S. Marine Corps in World War I, also heaped scorn on Lindbergh for accepting a medal from "those gangsters" in Germany.

As for expected sources, a veritable onslaught of scorn was directed at Lindbergh in Britain, where one member of Parliament stood up to say that Lindbergh's Nazi medal—the "Service Cross of the German Eagle"— should have been shaped in the form of a "double cross." The Briton Harold Nicolson, who wrote the authorized biography of Anne Lindbergh's father, Dwight Morrow (a project that brought Nicolson to stay for a time at Next Day Hill when the Lindberghs were living there), caught Lindbergh by surprise with an unflattering public appraisal of his psychological makeup. (As mentioned before, Nicolson was also co-owner, with his wife, Vita Sackville-West, of Long Barn, the house in the English countryside where the Lindberghs lived when they left America in December 1935.)

In the British magazine the *Spectator*, Nicolson wrote that during Lindbergh's struggle to remain himself after becoming the world's most famous man, and again, later, after the horrific crime that robbed him of his firstborn son, "his virility and ideas became not merely inflexible but actually rigid; his self-confidence thickened into arrogance and his convictions hardened into granite. He became impervious to anything outside his own legend—the legend of the young lad from Minnesota whose head could not be turned."

But something had turned Lindbergh's head. "He liked [the Nazis'] grim efficiency," Nicolson wrote, "and he was not at all deterred by the[ir] suppression of free thought and free discussion." In Lindbergh's mind that suppression was an intelligent alternative to the liberty and license of America, the nation where Lindbergh had been hounded like an animal by a "free" press, then forced to bury his infant son, who was stolen for ransom from Lindbergh's secluded mansion. Lindbergh "admired the [Nazis'] conditioning of a whole generation to the ideals of harsh self-sacrifice; the rush and rattle of it all impressed him immensely," Nicolson wrote. In the end, this didn't surprise the Englishman at all. Lindbergh, he said, speaking of a man he knew well, is "not merely a schoolboy hero, but a schoolboy."

ANNE LINDBERGH WAS STUNNED by these remarks. "Bitter criticism," she wrote in her diary. "Personal attacks. C. is a 'Nazi.' He will be punished. Our other two children will be taken. . . . I feel angry and bitter and trapped again." Anne was especially angry with Nicolson; she'd considered him a friend. "If only he had 'attacked' C. fairly [and] intellectually on issues," she wrote. "But no—he has to hit him with . . . 'The lad from Minnesota,' and all that. I try to remember that [Nicolson] is a disillusioned idealist."

What Anne tried to forget—but couldn't—was that there were members of her own family who were contemptuous of her husband's views. Anne's mother, Betty, then acting president of Smith College, her (and Anne's) alma mater, made a public show of this rift by joining the Nonpartisan Committee for Peace through Revision of the Neutrality Law, a revision Lindbergh had specifically argued against in his second radio address. Anne's sister Constance, now married to Aubrey Morgan, the widower of Elisabeth Morrow Morgan, was even more vocal in arguing against Lindbergh's position. This wasn't just a political issue for Constance. It was personal. Her husband was assistant to the director general of the British government's information service in New York.

Anne felt pulled in two directions. Charles demanded that she choose sides: was she a Morrow or a Lindbergh? After a period of intense anguish, Anne chose the latter, even if it meant, as she wrote in her diary, making herself over into "someone else." That someone else was Charles; the public proof of her transformation was an article she wrote for the Christmas 1939 issue of *Reader's Digest*, "A Prayer for Peace." That Anne felt less fortified than debilitated by this makeover seems clear in her first sentence:

I am speaking as a woman, a weak woman, if you will—emotional, impulsive, illogical, conservative, dreaming, impractical, pacific, unadventurous, any of the feminine vices you care to pin on me.

Even so, Anne found the strength to restate the arguments her husband had been making for months: that aggression was a fact of political life; that the current war was less about preserving freedom than redrawing borders; that Germany's restlessness was the product of unfair restraints placed on it by the Treaty of Versailles; that democracy was a flawed institution inspiring mediocrity; and that China, Japan, and especially the Soviet Union, with its false vision of equality—not Nazi Germany—were the true enemies of western civilization. This last point was true, Anne declared, no matter how extreme the Nazis might seem at times. "Hitlerism is a spirit and you cannot kill or incarcerate a spirit," she wrote. "It can only . . . be exorcized. To exorcize this spirit you must offer Germany not war but peace."

Now that he'd brought his wife on board, Lindbergh was free to confront or ignore his critics outside his family. Just as he'd convinced himself as a young flier that he "lived on a higher plane than the skeptics on the ground," Lindbergh was now certain that his political opponents were on a lower intellectual level than he: they were ill-informed, ignorant of human nature, and, worst of all, intimidated by strength. Lindbergh didn't need a college diploma to know he was right: as a pilot he'd "tasted a wine of the gods of which [others]"—especially elected politicians and newspaper columnists—"could know nothing." If Lindbergh found himself attacked by his inferiors, so be it. Lindbergh took pride in his isolation. Hadn't his mentor, Carrel, taught him life's most incontrovertible lesson? The path of the heroic truth-teller is never easy, for the weak multitudes are afraid of his truth.

Nearly all of Lindbergh's assertions—about the precarious state of western civilization, the notion that western culture means "white" culture, the morality of the strong dominating the weak, and the different intellectual capabilities of the different races—had their origins in Carrel's countless early-morning monologues, which Lindbergh had eagerly listened to in the surgeon's office atop the Rockefeller Institute. That Lindbergh now faced loud opposition to his isolationist views only proved, in his own mind, that he was doing as Carrel would have him do: walking down the bumpy road of the truth-teller, criticized by those too ignorant to accept his righteous message.

This is why it was such a shock to Lindbergh to see that one person who loudly disputed his "truth" about Germany—especially the idea that Hitler's armies should be free to sweep across western Europe—was Alexis Carrel.

CHAPTER 16

FOR WE ALL KNOW WHAT AWAITS US

CARREL WAS WEEDING HIS GARDEN ON SAINT-GILDAS WHEN World War II started, just as he had been when World War I began twenty-five years earlier. But now, unlike the last time, Carrel was a man of divided loyalties: he had a French passport yet an American spirit. He also had an American wallet and, even more relevant, access to fatter American wallets whose owners might provide the funds required to make his dream—establishing the "Institute of Man"—into a reality on Lindbergh's estate in New Jersey. For that reason Carrel almost returned to the United States after France declared war on Germany on September 3, 1939. But the patriotic fervor that swept France after the declaration caught Carrel in its emotional pull: he donned the scratchy French army uniform he had worn between 1915 and 1918 at the hospital he ran at the front and rushed on a packed train to Paris to offer his services to the French ministry of health.

There was so much to do, he told the health minister, Marc Rucart, and so little time to do it: a state-of-the-art laboratory had to be built—with Carrel in charge, of course—where he could study new treatments for shock and poison gas, assess the latest gangrene-inhibiting technology, experiment with anticoagulants, and organize a system for the preservation and transportation of blood for battlefield transfusions. Instead, Rucart asked Carrel to investigate the state of French preparedness in those areas and report back to him. The idea of a laboratory was tabled.

Carrel went right to work but, to his horror, found that it was nearly impossible to achieve his appointed task without paying bribes. Instead of uniting to fight the Germans, the French were fighting among themselves, usually over money. As was his custom, Carrel expressed his contempt for this state of affairs to anyone, no matter how high-ranking, within earshot. The officials weren't pleased to hear themselves called "stupid" and "corrupt." Few would forget the insults, deserved or not.

Carrel's angry report was written, filed, and ignored. He was so frustrated that he instructed his secretary at the Rockefeller Institute, Katherine Crutcher, to renew the lease of his apartment on East Eighty-Ninth Street, in case he decided to return to New York on short notice. That this was a real possibility is shown in a letter Carrel mailed to his brother Joseph, in Lyon. Working conditions in Paris, he told Joseph, were "abominable. . . . The pettiness, moral laxness, cowardice, jealousy and baseness of everyone is truly inconceivable. I am in a world that is as alien to me as the world of the Greeks and the Syrians!"

Carrel became even more anxious when he learned, while reading a medical journal, that the Germans were not only more unified than the French, but bigger. Apparently, they'd done experiments, even before Hitler, similar to those that Carrel performed on rodents in his mousery at the Rockefeller Institute. "The Germans have trained their youth and made athletes of them by exercise and changing to a proper diet," Carrel wrote a friend. "They have added two inches to the average stature in one generation. France could have achieved the same result," he said, but didn't, and now it would "have to pay the price on the battlefield."

To escape the frustrations of Paris, Carrel returned to Saint-Gildas as often as possible. One day there, his wife sat at the kitchen table and threw runes, a group of twenty-four tiny stone tablets, each marked with a letter from the ancient futhark alphabet, that when tossed and interpreted were said to predict the future. Anne Carrel wept when she was finished: unless France found a benefactor, the tablets told her, the road ahead would be marked by chaos and humiliation.

Her husband tried to assume that benefactor's role in a radio address delivered to the French nation on December 6, 1939. "We are like cells in

the living body that is France," said the world's foremost expert in cell culture. "We must act toward the country as cells do in the body: first, we must
do our own work, and, secondly, we must participate in the larger effort to
achieve the overall health of the nation":

> For we all know what awaits us if we do not: forced labor for life for
> our workers and peasants, deportation to Africa for large masses of
> the populations in our richest provinces, and mass executions.

This was intended to scare France out of its torpor and turpitude, but
Carrel was convinced that no one had heard him. The infighting among
Frenchmen continued; so did the bribery and inefficiency. Even before the
Germans he so despised invaded France, Carrel was sure his homeland was
defeated. "I am disgusted. We shall be beaten. I am leaving," he wrote in his
journal. "I do not want to be present at the final collapse." After informing
his wife, now serving as a nursing officer in the French army, Carrel purchased a one-way first-class ticket on the French liner *Champlain,* leaving
on May 10, 1940, from Saint-Nazaire, for New York. Before boarding, he
wrote another letter to his brother. "Our only hope at this point is America,"
Carrel said. "I believe we are capable of resisting *les boches*, but we are not
strong enough to defeat them. If the Americans could give us two or three
thousand airplanes immediately, that would change everything."

Carrel's call for American aid was precisely the course Lindbergh was
arguing against. His third radio address, "The Air Defense of America," was
delivered on May 19, while Carrel was still at sea. Eight days later Carrel
arrived in Manhattan. To the reporters who met him at the dock and asked
him about the situation abroad, Carrel said, "I went to my farm in Brittany
[before my departure] and found it as peaceful as it always has been. The
men were all away with the army or the navy, fighting for the France they
love so well." Less than three weeks later, the swastika was flying atop the
Eiffel Tower.

LINDBERGH LEARNED OF HIS mentor's arrival in the *New York Times.*
He immediately phoned Carrel at his Upper East Side apartment and said
he'd come over that afternoon. This was more than fine with Carrel; there

was much he wished to discuss with Lindbergh. While crossing on *Champ-lain*, Carrel had read an article Lindbergh wrote for *Atlantic Monthly*, "What Substitute for War?" Though Carrel recognized many of his own ideas in the piece, he was unhappy with Lindbergh's conclusions, and most of all with his assertion that Germany had a "natural right" to expand—not just eastward into Poland, but westward into France.

The "substitute for war" that Lindbergh called for in his article was a negotiated peace with Germany *after* it exercised that right to expand. Those "countries which, like England and France, are well satisfied with their position and possessions follow the type of political ideology that comes with luxury, stable times, and the desire to enjoy rather than ac-quire," Lindbergh wrote:

> The countries which, like Germany, have recently gone through great hardships and chaotic times have the political system that springs from such times and which involves rigid discipline and the subordination of individual freedom to the strength of a recuperating state—a state whose people must acquire *before* they can enjoy.

The new war wasn't a battle between right and wrong, Lindbergh said; it was a battle between "differing conceptions of right." The Allies champi-oned "the static, legal 'right' of man," while the Germans pursued "the dy-namic, forceful 'right' of nature."

Carrel was every bit as admiring of the aggressive aspects of nature as Lindbergh. He was also willing to admit that much of the weakness and moral decline in France and Britain originated from within, and that the ultimate enemies of the West were "Asiatic barbarians" and the godless Communists of the Soviet Union.

What he could not accept was Lindbergh's vision of Nazi Germany as the West's great protector. Carrel paced back and forth across his bare-walled living room, gesticulating with his hands, his voice getting higher and louder, as he reminded his protégé of the summary executions of inno-cents and similarly murderous or maiming cruelties committed against sol-diers and civilians alike by the Germans in World War I—in fact, for much of modern history. He'd personally seen the gruesome results of German handiwork in the hospital he ran at Compiègne, fifty miles northeast of

Paris. Of course war is hell everywhere and for everyone, Carrel said, but only the Germans take such hellish delight in it.

He also reminded Lindbergh that Nazi Germany embodied a dangerous regression to Europe's pre-Christian past. Hitlerism, Carrel said, was a heartless, pagan death cult with no respect for the dignity of life or the humanity of man. Yes, the antireligious Soviet Union was worse, but there was an insidious parallel between the false equality preached in the Soviet Union, Carrel warned, and the Nazis' commitment to finding *Lebensraum* for the German nation. If the Nazis were successful in that quest, there would only be two very unequal classes in the world: Germans and everyone else. This was not a future Carrel wanted to be a part of. It would mark not the saving of western civilization, but its death.

But Lindbergh wasn't persuaded. "Carrel is able to discuss the war objectively and sees the causes clearly," he wrote in his journal. "On only one major point do I disagree with him. Carrel feels that if Germany wins, Western civilization will fall."

Though passions were ignited in their discussion, the two men embraced and parted as friends, agreeing to meet for dinner a week later at the art-filled apartment of Boris Bakhmeteff, one of the core members of the Philosophers Club. The debate would continue there, among the Orthodox icons and depictions of Holy Mother Russia. They'd also discuss what was happening in the search for funding for Carrel's "Institute of Man."

The news was bad with regard to fund-raising. In fact, philanthropic sources in the United States had dried up so dramatically because of the possibility that America might soon be at war that Lindbergh withdrew his offer of his estate in New Jersey as the site of Carrel's new institute. Instead, he would be deeding High Fields over to the state of New Jersey, which would use it as an orphanage or a similar institution for needy children. Carrel was disappointed to hear this but accepted Lindbergh's decision with equanimity.

Their friendship would be tested by their conflicting views on the situation in Europe, and what America should do about it. That the Roosevelt administration was moving toward increasing its support for Britain and France was made clear in a speech the president delivered on June 11, 1940, at the University of Virginia. "Let us not hesitate to proclaim certain truths," Roosevelt said. "We are convinced that military victory for the gods of force and hate would endanger the institutions of democracy in the Western World—and, therefore, the whole of our sympathies lie with those nations that are giving their life blood in combat against those forces." There are others, Roosevelt said, who "hold to the delusion that we can safely permit the United States to become a lone island in a world dominated by the philosophy of force." But such an island, said the president, "represents the nightmare of a people without freedom, of a people lodged in prison, handcuffed, hungry and fed through the bars from day to day by the contemptuous, unpitying masters of other continents."

Lindbergh listened to Roosevelt on the radio with his own brand of contempt. "[Roosevelt's] speeches never seem to me to be those of a normal man," he wrote in his journal. To Lindbergh's ears they were "dramatic and demagogic," usually at the same time.

But Carrel thought Roosevelt's speech was magnificent. And when he learned that Lindbergh was going on the radio again to speak against the president's policies—on June 15, one day after the Nazis entered Paris— Carrel telephoned Lindbergh at home. He'd been following Lindbergh's new career as a radio commentator with growing alarm. When a friend in the French army wrote to Carrel in New York and asked him to use his influence to get Lindbergh to make a public display of support for the French, Carrel said he would try, but he wasn't optimistic. "Most people [in America] have begun to understand what this war means," Carrel wrote back, "but one of the few who hasn't is the unfortunate Lindbergh. He remains absurdly isolationist. What stupidities he is uttering!"

Carrel's conversation with Lindbergh started calmly enough, with talk of family and pleasant memories of Illiec and Saint-Gildas. Then Carrel asked Lindbergh to include in his next radio address a "friendly reference to France"—in particular, one that promised American support for the anti-Nazi resistance.

"I'd like to," Lindbergh said, "but I don't see how I can, since [my speech] is primarily an argument against our entry into the war." Besides, Lindbergh added, such aid from America would only prolong the fighting in Europe, causing more deaths in Britain and France, and thus further contribute to the destruction of western civilization, by making Germany's apparently inevitable victory more costly to both the losers and the winners.

"But it's the Nazis who are destroying western civilization!" Carrel insisted.

Lindbergh told Carrel he was being irrational, then got off the phone.

Carrel and Lindbergh spoke again, at Carrel's apartment, on June 22. This was just over a week after the Wehrmacht marched into Paris, and less than twenty-four hours before Adolf Hitler arrived there to tour his most prized war trophy so far. (Hitler stayed in Paris for only three hours, rushing through the Paris Opera, the Arc de Triomphe, the Eiffel Tower, Sacré Coeur basilica, and Napoleon's tomb, before returning on a train to Berlin.) In truth, Carrel had reason to be depressed even before the fall of Paris: he hadn't heard from his wife in weeks. He thought she was serving somewhere in Brittany, but he wasn't sure. The combination of that uncertainty with the humiliation of his homeland by the despised Germans left Carrel a changed man. Gone was the impatient vitality that had made him such a productive, even Napoleon-like, figure. Gone, too, was the boundless confidence that had given him the ruthless intellectual certainty others took for arrogance. In their place was a new and heavy self-doubt—so heavy, in fact, that a heretofore inconceivable role reversal occurred between the onetime mentor and his protégé. Now it was Carrel asking Lindbergh for guidance.

"What do you think the Germans will do in France?" Carrel asked.

"I don't know," said Lindbergh. "But it probably won't be as bad as people in this country think."

"What do you think I should do?"

Lindbergh urged patience. Until the situation in France was sorted out, it would be impossible to make an intelligent decision, especially not in New York, 3,600 miles away. "It might be possible for you to work with the new French government," Lindbergh said. German influence would create a new sense of efficiency there, the kind of order that prewar France had lacked.

Lindbergh wasn't immune to the charms of prewar France: after the Germans marched into Paris, his wife and he sat down in their living room on Long Island and listened to *Mignon*, the opera composed on Illiec by Ambroise Thomas. But Lindbergh left art and music out of his discussion with Carrel, choosing instead to remind the Frenchman of how long he had yearned for order in his native country. That was true, of course, but what

Lindbergh failed to realize was how despondent Carrel would be to see that discipline established by the hated Germans.

Lindbergh didn't see it, because he was mesmerized by something he thought he did see: the birth of a new Europe, one that, led by Germany, promised exciting and historic changes. Lindbergh had just made this point in a letter to Truman Smith: "Germany has demonstrated an ability in war that has astonished the rest of the world. There will soon be an opportunity for her to show an even more amazing [political] leadership," he wrote.

The choice every European had to make, Lindbergh believed, was whether or not to be a part of Europe's political and cultural rebirth. Lindbergh would never presume to tell Carrel what to do—he had far too much respect for the scientist to do that—but he did urge Carrel to seriously consider being a part of the new, German-dominated French leadership, then still forming. "But if [returning to France] isn't practical for you," he said, "I hope you will stay in America and write."

Lindbergh's advice was fascinating for two reasons. For starters, an article had just appeared in the *New York Times* in which a senior Nazi official, gloating over the fall of Paris, declared what the "new France"—indeed, all of the "new Europe"—would look like. The official was Alfred Rosenberg, the Nazi Party's official racial ideologue and the author of *The Myth of the Twentieth Century*, a book, published in 1930, which argued that history's driving force was the war between two races: Aryans, the creators of all noble values and civilization; and Jews, agents of all cultural corruption and disease. The Nazis' capture of Paris, Rosenberg said in the *Times*, signaled the start of the "disinfection" of western Europe, which would be accomplished by scrubbing away from its body all "international speculators, abstract philosophies and Jews." (The "disinfection" of eastern Europe had already begun.) It is unlikely that this article, headlined NAZI SEES "NEW ERA" REFORMING EUROPE, would have escaped Lindbergh's notice, appearing, as it did, in a newspaper he checked daily for war reports.

The alternative offered by Lindbergh—that Carrel remain in America and continue writing—is equally revealing, because it shows the paradoxical situation Carrel now found himself in. Just as Carrel was losing confidence in himself, the American public was turning to him more and more for guidance. This was because Carrel had begun writing articles for *Reader's Digest* in the summer of 1939, and the articles had become something of a sensation, helping to raise the magazine's circulation to record levels. An article titled "Married Love" appeared under Carrel's byline in the July issue, almost at the precise moment Carrel was forced into retirement at the Rockefeller Institute. "Do You Know How to Live?" followed in August; and "Work in the Laboratory of Your Private Life" was about to come out just as Lindbergh and Carrel were speaking in Carrel's apartment. As the titles indicate, these weren't scholarly articles on tissue culture or surgical methods. They were the kind of instructive ruminations on happiness, sexual fulfillment, and personal growth that might be written today by people like Phil McGraw, Ruth Westheimer, and Deepak Chopra.

The "author bio" that accompanied these pieces in *Reader's Digest* shows the respect, even awe, accorded to America's new "self-help guru":

> In 1939 Dr. Alexis Carrel concluded 33 years of brilliant biological research at the Rockefeller Institute. He had reached the Institute's retirement age, 65. But the man who had perfected the surgical technique which made blood transfusion easy; who had kept a section of chicken heart alive for a quarter-century; who had developed, with Colonel Lindbergh, an amazing artificial heart; who had won the Nordhoff-Jung medal for cancer research and the Nobel Prize for success in suturing blood vessels—this man could not retire. He went to his native France to volunteer his surgical experience to the government; and he has returned [to America] with firsthand knowledge of what can happen to an easygoing people lacking in self-discipline.

In the summer of 1940, the ironic situation for Alexis Carrel was this: though the public saw him as the "answer man," he had huge questions to face in his own life—and no answers.

—

CHAPTER 17

THERE IS MUCH I DO NOT LIKE THAT
IS HAPPENING IN THE WORLD

LINDBERGH HAD ANSWERS—NOT JUST FOR HIMSELF, FOR ALL
of America. He was as sure of his ability as a political analyst as he was of
his skill as a flier, so he decided to take his speaking campaign to the next
level. Rather than talking to a few broadcast engineers in a hotel room, he
would speak to thousands, in person, in stadiums and convention halls.

He gave the first of these speeches on August 4, 1940, at Soldier Field
in Chicago, a huge horseshoe-shaped stadium built with mock Doric colon-
nades to give it the appearance of an ancient Greek arena. Squinting into
the bright sun, Lindbergh called for a treaty between the United States and
Germany before any hostilities started between them.

"A war between us could bring all civilization tumbling down," he said.
"An agreement between us could maintain civilization and peace as far into
the future as we can see." As a sign of his new relationship with Carrel,
Lindbergh repeated something in Chicago that he'd said on Saint-Gildas,
when Carrel was pleading with him to return to their immortality research
on primates and humans, and Lindbergh was explaining why he had to
leave that work to begin championing Hitler's Germany as the guarantor
of western civilization's immortality. Previously, Lindbergh's public comments were often restatements of ideas he'd heard from Carrel. At Soldier

Field the ideas Lindbergh expressed were assertions Carrel had heard first from him.

"From 1936 on, as I traveled through Europe, I saw the phenomenal strength of Germany growing like a giant," Lindbergh told his audience. "In England there was organization without spirit; in France there was spirit without organization; only in Germany was there both."

The *Chicago Tribune* praised Lindbergh's speech as well argued; his personal mail ran twenty-to-one in favor; and a college student in upstate New York named Kurt Vonnegut wrote a piece in the Cornell *Sun* calling Lindbergh "one helluva swell egg" and a "sincere and loyal patriot." But there were critics. William Yandell Elliott, a professor at Harvard (where he later mentored Henry Kissinger), mocked Lindbergh's foreign-policy credentials, noting that "sustained thought in [that] difficult field" demands "more training than preceding 'Wrong Way Corrigan' across the Atlantic." Walter Winchell started calling the Lone Eagle the "Lone Ostrich." Ralph Ingersoll, publisher of New York's new left-leaning tabloid, *PM*, unleashed his editorial cartoonist on Lindbergh, a task the cartoonist performed with great gusto in the months to come. In one cartoon Lindbergh was depicted patting the contented head of a giant swastika-covered monster that had just swallowed up most of Europe. Another had a caption that read:

The Lone Eagle had flown,
The Atlantic alone,
With fortitude and a ham sandwich.
Great courage that took,
But he shivered and shook,
At the sound of the gruff German landgwich.

This cartoonist would be heard from again. He signed his work "Dr. Seuss."

Lindbergh ignored the attacks and moved on, joining forces with several young men who also would be heard from again. They were students at Yale, most of them at the law school, who had contacted Lindbergh after his first radio address, in September 1939, to say they supported his position. Now these men, who included Gerald Ford, Sargent Shriver, Kingman Brewster, and Potter Stewart, had formed an organization on the

Yale campus they called the Committee to Defend America First. The director of the committee was another Yale student, R. Douglas "Bob" Stuart Jr., who postponed his final year of law school to lead an effort to transform their tiny student group into a national organization that would speak for the anti-interventionist cause with one loud voice, on campus and off.

Stuart succeeded, using a combination of family connections—his father was a senior executive at Quaker Oats—and personal charm. He persuaded the retired U.S. Army general Robert E. Wood, now chief executive of Sears-Roebuck, to serve as national chairman of America First. Soon the organization's executive committee included Eddie Rickenbacker, America's most celebrated flying ace during World War I, and now the president of Eastern Airlines; the writer John P. Marquand (author of *The Late George Apley* and the creator of "Mr. Moto"); the perennial Socialist candidate for the U.S. presidency, Norman Thomas; Alice Roosevelt Longworth (Theodore Roosevelt's daughter); the actress Lillian Gish; the liberal journalist John T. Flynn, of the *New Republic*; and (briefly) Henry Ford.

But the man most desired by Stuart for a spot on that committee was Charles Lindbergh. The courting began in the fall of 1940, when the original chapter of America First invited Lindbergh to speak at Yale. Lindbergh hesitated to accept, perhaps because the invitation brought back memories of his own academic failures. While mulling it over, he reread his old journal entries. Suddenly his self-confidence returned. "All the logic I used in concluding that war would probably not come before 1940 turned out to be correct," he wrote on October 10, 1940, after making that review (and despite the fact that the war began in September 1939.) "I thought Germany would expand eastward—she did":

I thought the French Army could not break through the [German] West Wall—they failed. I felt there was no way to give effective aid to Poland—none was given. France has been defeated, and England is on the defensive, fighting desperately for her life. I knew the German Air Force was vastly superior to all other air forces in Europe combined—that is now a well-established fact. In every technical estimate I was right.

"Nevertheless," Lindbergh wrote, "war was declared—declared by the weak and not the strong, by those who would lose, not by those who would win."

This, he decided, was a warning that must be delivered to the American public, so he phoned Kingman Brewster at the Yale *News* (two decades later, Brewster became Yale's president) and said he would accept the invitation.

The thirty-minute address Lindbergh gave on October 30 was by far the longest he'd ever given. One sentence showed the enduring influence of Carrel, who'd often told Lindbergh that history follows a biological course: "[Your] generation," Lindbergh said, "is taking over the problems of life during the greatest period of mutation that man has ever known." Those tumultuous changes, he argued, demand a choice: Americans "must either keep out of European wars entirely or participate in European politics permanently."

Lindbergh was received with rapt attention, followed by a standing ovation. This response was especially thrilling to him because a few days earlier he had slipped unrecognized into a movie theater in Manhattan where, when his face appeared on the screen in a newsreel, nearly everyone in the audience hissed.

Less than three months later, in Washington, Lindbergh walked into a brightly lit room filled with newsreel cameras and 1,000 spectators. Inside, the Foreign Affairs Committee of the House of Representatives was holding hearings on the Lend-Lease bill just proposed by President Roosevelt. If passed, Lend-Lease would give the president the power to transfer war matériel to any nation deemed vital to American interests. This would be the most interventionist action yet taken by the U.S. government.

Lindbergh was ready to testify against the bill, but he wasn't ready for the questions asked by Representative Luther A. Johnson of Texas.

JOHNSON: You are not in sympathy with England's efforts to defeat Hitler?

LINDBERGH: I am in sympathy with the people on both sides, but I think it would be disadvantageous for England herself, if a conclusive victory is sought.

JOHNSON: ... Are you not in sympathy with England's defense against Hitler?

LINDBERGH: I am in sympathy with the people but not with their aims.

JOHNSON: You do not think it is in the best interests of the United States . . . for England to win?

LINDBERGH: No, sir. I think that a complete victory [by England] would mean prostration in Europe and would be one of the worst things that could happen there and here. . . . I believe we have an interest in the outcome of the war.

JOHNSON: On which side?

LINDBERGH: In a negotiated peace [between the sides.]

JOHNSON: Again: which side would it be in our interest to win?

LINDBERGH: Neither.

JOHNSON: Have you ever expressed any opposition to Hitler?

LINDBERGH: Yes, but not publicly. I believe we should maintain neutrality publicly. There is much I do not like that is happening in the world, on both sides. . . . However, there is not as much difference in philosophy [between the Allies and the Nazis] as we have been led to believe.

After Lindbergh repeated these views before the Senate's Foreign Affairs Committee, where he declined an invitation from Senator Claude Pepper of Florida to criticize the "outrageous wrongs which have been perpetrated and are still being perpetrated by the German government," the Lend-Lease bill was passed and signed into law. This disturbed Lindbergh, but what really upset him was reading Senator Pepper's "explanation" of why his famous witness had refused to criticize the brutal excesses of Nazism: "The man who built a mechanical heart," the senator told a reporter, "knows more about an artificial heart than he does a human one."

PEPPER'S REMARK WASN'T THE only disparaging one. The journalist Cornelius Vanderbilt Jr., a great-great-grandson of the railroad tycoon, cabled to Lindbergh after his testimony: "WHAT AN UNPATRIOTIC DUMB-BELL YOU ARE." The *Richmond News Leader* editorialized, "Millions today would vote to hang Lindbergh or to exile him as enthusiastically as they

[once] cheered and extolled him." The lesson Lindbergh took from this public relations fiasco was simple: he would continue to speak against American involvement in the new war, but not in Congress. Instead, he'd speak behind lecterns sponsored by America First, and not merely as a friend of that organization but as its most famous board member.

With that decision began the most heated portion of what historians now call the "great debate" of 1941, a battle of ideas focusing on one question: What should America's posture be in a world where war had started in Europe, but not here? The rivals in this debate, Charles A. Lindbergh and Franklin D. Roosevelt, were as different in background and temperament as they were in political ideology. Roosevelt was a fun-loving, patrician Democrat, educated at Groton and Harvard, whose family, wealthy from real-estate holdings, had been in North America since the seventeenth-century Dutch colony, Nieuw Nederland. Lindbergh, the son of an aloof school teacher from Michigan and a gloomy U.S. Congressman born (out of wedlock) in Sweden, was a self-educated engineer less comfortable with people than with machines. Roosevelt was a smoker and social drinker; Lindbergh loathed both activities. Roosevelt enjoyed the limelight; Lindbergh hated it. Roosevelt was witty and (when he needed to be) ambiguous; Lindbergh was humorless and (by choice and nature) blunt.

The two never debated face-to-face; Roosevelt positioned himself above the fray. When he thought it useful, the president sent out his secretary of the interior, Harold Ickes, who was delighted to play attack dog with Lindbergh, a man Ickes believed to be a "ruthless and conscious fascist motivated by hatred for . . . democracy." But even Ickes never debated Lindbergh in person; he merely mocked him in speeches and in comments to the press. And Lindbergh was an easy target. Between April and October 1941, he spoke at thirteen rallies sponsored by America First, in virtually every section of the nation. At each of these events, which received enormous press attention, Lindbergh was the unquestioned star of the evening, though others also spoke. On every occasion he delivered a speech he wrote himself.

On April 23, 1941, Lindbergh spoke in New York at the Manhattan Center on West Thirty-Fourth Street, built in 1906 by Oscar Hammerstein I as an opera house. Thirty-five thousand protesters chanted anti-isolationist slogans outside, more than three times as many as were seated in the auditorium. Many of these picketers belonged to a group called the Friends of Democracy, whose founder, Reverend Leon M. Birkhead, described the America First meeting to the *New York Times* as "the largest gathering of pro-Nazis [in America] since the German American Bund rally in Madison Square Garden," in February 1939.

Birkhead's opinion surely would have been bolstered had he heard the cheers in the hall when Lindbergh declared, "England is losing this war." Lindbergh's wife, who was inside, found those cheers made her "curl inside with shock." Even so, Anne was awed by her husband's courage in stating his opinions so forthrightly. The crowd was "silenced" by his mere presence, she wrote in her diary:

> And when he [spoke,] slowly, with emphasis, I felt his great strength and power and I watched that crowd looking at him, with faith, with undivided attention, with trust—leaning on that strength. . . .
> He was not using his power *intentionally*. No—it was simply there—in him.

After a columnist for the *Hamburger Fremdenblatt*, in Germany, wrote that Lindbergh's speech showed him to be "a real American"—a statement transmitted across America by the Associated Press—Roosevelt struck back at his rival. The president knew Anne was naive: her husband *was* using his power intentionally. So when, at a press conference the next day, a reporter asked the president about Lindbergh's speech, Roosevelt responded with a folksy history lesson.

"Once upon a time there was a place called Valley Forge," he said, "and there were an awful lot of appeasers there who pleaded with Washington to quit, because he 'couldn't win.'"

"Are you [thinking] of Lindbergh, sir?" another reporter asked.

As the *New York Times* reported the next morning: "A simple and emphatic affirmative was the answer."

Lindbergh was so angry at having his patriotism challenged by the president of the United States that he resigned from the army air corps reserves—and, in so doing, fell into Roosevelt's trap: to the public Lindbergh appeared to be shirking his patriotic duty at the very moment when other Americans were, for the first time in the nation's history, registering for a peacetime draft under the provisions of the Selective Service Act, which had become law in October 1940.

Now Lindbergh's speeches became even more strident on the subject of Roosevelt. On May 23, speaking before 20,000 people at Madison Square Garden in New York, Lindbergh took his attacks to a new level. After shouts of "Lindy!" "Our next president!" and a commotion so intense that Anne feared a deranged spectator might shoot her husband, Lindbergh lamented that Americans, supposedly living in the land of representative government, had been given no real choice in the last presidential election, as both major candidates, Roosevelt and Wendell Willkie, supported aid to the Allies. That so-called election, Lindbergh said, gave American voters as much of a choice about the direction of their nation's foreign policy "as the Germans would have been given if Hitler had run against Göring."

This bizarre assertion came back to burn him. Harold Ickes, speaking weeks later in New York, reminded his audience that Lindbergh had kept a Nazi medal given to him by Göring, with a proclamation signed by Hitler, even after resigning from the military of the United States. One must wonder not just about the loyalty of such a man, Ickes said, but about his very humanity.

"This Knight of the German Eagle has never denounced Hitler or Nazism or Mussolini or Fascism," Ickes said. "I have never heard Lindbergh utter a word of pity for Belgium or Holland or Norway or England":

> I have never heard him express a word of pity for the Poles or the Jews who have been slaughtered by the hundreds of thousands by Hitler's savages. I have never heard Lindbergh say a word of encouragement to the English for the fight they are so bravely making for Lindbergh's right to live his own life in his own way, as well as for their own right to do so.

Then, borrowing an image from Senator Pepper, Ickes mocked Lindbergh's scientific interests. If Lindbergh was still looking for an artificial heart, the interior secretary said, "he could readily locate one," inside Lindbergh's own chest, "with an X-ray machine."

The press coverage of Ickes's speech was devastating. More than 200

American libraries pulled the Lindberghs' books—Charles's and Anne's—from their shelves. Lindbergh was badly wounded by the publicity campaign directed against him by President Roosevelt. But the biggest wound was yet to come. And it would be self-inflicted.

THIS HAPPENED ON SEPTEMBER 11, 1941, when Lindbergh gave a speech in Des Moines, Iowa, he called "Who Are the War Agitators?" The title had been brewing in his mind for some time. In a radio address delivered in May 1940—almost a year before he joined America First—Lindbergh had referred to "powerful elements" in American society who were agitating for war. Now he was ready to name them.

It was a long trip from Martha's Vineyard, where the Lindberghs had rented a house for the summer, to Des Moines: ferry and train to Boston; TWA flight to Chicago; bus to Chicago for supper with Bob Stuart; cab back to the airport; then a United flight (with a stop in Moline, Illinois) to Des Moines. After a short rest in a hotel, Lindbergh arrived at the Des Moines Coliseum, a hall typically used for livestock shows and professional wrestling. The arena was already full, with nearly 8,000 people in attendance, but there was an unexpected problem. President Roosevelt had arranged to give a speech of his own that night, announcing a new "shoot on sight" policy for the U.S. Navy against any German or Italian warships in the "American defense zone," which pretty much included all of the Atlantic Ocean. The local America First chapter decided that its speakers would address the audience after the president's speech, which was broadcast into the hall over the public address system. The America First speakers would wait behind a curtain.

Truth is, they were expecting a tough crowd. Iowa was not a hotbed of isolationism. Roosevelt's vice president, the Iowan Henry A. Wallace, was an enthusiastic supporter of aid to the Allies. So was the state's most important newspaper, the *Des Moines Register*, which, on that very morning, ran a cartoon on its front page showing Hitler applauding an address by Lindbergh, over the caption: "His Most Appreciative Audience."

When Roosevelt's speech was finished and Lindbergh took the podium, he was "met by a mixture of applause and boos—the most unfriendly crowd of any meeting to date, by far," he later wrote. "The opposition [was] organized and there were groups of hecklers strategically located to be effective for the microphones." But Lindbergh had become a better, and certainly more experienced, speaker in the past few months. He was determined to get his message across clearly. He did.

"It is now two years since this latest European war began," he said. "From that day in September 1939, until the present moment, there has been an ever-increasing effort to force the United States into the conflict. . . . Who is responsible for changing our national policy from one of neutrality and independence to one of entanglement in European affairs? . . .

"The three most important groups who have been pressing this country are the British, the Jewish, and the Roosevelt administration. . . . I am speaking here of agitators, not of those sincere but misguided men and women who, confused by misinformation and frightened by propaganda, follow the lead of the war agitators. As I have said, these war agitators comprise only a small minority of our people, but they control a tremendous influence. Against the determination of the American people to stay out of war, they have marshaled the power of their propaganda, their money and their patronage."

After dealing with the British, Lindbergh spoke of the "enemy" within: "It is not difficult to understand why Jewish people desire the overthrow of Nazi Germany," he said. "The persecution they have suffered in Germany would be sufficient to make bitter enemies of any race. No person with a sense of the dignity of mankind can condone the persecution of the Jewish race in Germany":

> But no person of honesty and vision can look on their pro-war policy here today without seeing the dangers involved in such a policy both for us and for them. Instead of agitating for war, the Jewish groups in this country should be opposing it in every possible way for they will be among the first to feel its consequences.
>
> Tolerance is a virtue that depends on peace and strength. History shows it cannot survive war and devastation. A few far-sighted Jewish people realize this and stand opposed to intervention. But the majority do not. Their greatest danger to this country lies in their large ownership and influence in our motion pictures, our press, our radio and our government.

I am not attacking either the Jewish or the British people. Both races I admire. But I am saying that the leaders of both the British and the Jewish races, for reasons which are as understandable from their viewpoint as they are inadvisable from ours, for reasons which are not American, wish to involve us in the war.

We cannot blame them for looking out for what they believe to be in their own interests, but we must also look out for ours. We cannot allow the natural passions and prejudices of other peoples to lead our country to destruction.

Writing in his journal later that night, Lindbergh noted, "When I mentioned the three major groups agitating for war . . . the entire audience seemed to cheer." He wouldn't grasp the size of the furor he'd created for several days.

He traveled east in a private compartment on the Broadway Limited and didn't see the New York papers until September 13, when he arrived at Pennsylvania Station. He had expected to be criticized in the *Times*, which was owned by a Jewish family, and in the left-leaning *PM*. What surprised him was the attack against him in the *Herald Tribune*, the newspaper of the city's WASP elite. "We have sustained Mr. Lindbergh against attacks upon his patriotism, which we believe to be unwarranted by the facts," the *Herald Tribune* said on its editorial page:

This was done in the conviction that the discussion of America's relation to the great world conflict was an American debate conducted by Americans who are resolved to maintain the American system, and differing only in their concept of how that is to be accomplished. But the Des Moines speech, marking the climax of a series of innuendoes and covert allusions by Isolationist leaders, opens new and ugly vistas, and seeks to inject into open debate subjects which all good Americans pray might be confined to the pages of the *Völkischer Beobachter* and the addresses of Adolf Hitler.*

---

* The *Völkischer Beobachter* ("People's Observer") was the official newspaper of the National Socialist Party.

Other papers equally free of the "large [Jewish] influence [over] our press" expressed identical sentiments. WE ARE NOT NAZIS, said a headline in the Gannett-owned *Rochester* (New York) *Times-Union.* IMITATOR OF HITLER, said a headline in the *St. Paul* (Minnesota) *Pioneer Press;* GANGRENED "AMERICANISM," said the *Charleston* (South Carolina) *News Courier.*

Many of these editorialists pointed out that Lindbergh's charges about Jewish "influence" on America's media and government were false: A study just completed at the University of Notre Dame concluded that Jews "controlled" just 3 percent of the American press. (A survey by *Editor and Publisher* put the figure at around 1 percent.) The government agencies responsible for foreign policy were almost entirely in the hands of non-Jews, and President Roosevelt had only one Jewish American in his cabinet, the secretary of the treasury, Henry Morgenthau Jr., the same man who, in 1939, gave Alexis Carrel permission to board the ship bringing Charles Lindbergh back to America, when that ship was still anchored in New York harbor.

President Roosevelt never commented on Lindbergh's address in Des Moines. He didn't have to; Republicans did it for him. Wendell Willkie, Roosevelt's opponent in the election of 1940, described Lindbergh's speech as "the most un-American talk made in my time by any person of national reputation." Thomas E. Dewey, who had been the district attorney of New York County and would become the governor of New York state—and the Republican presidential candidate in 1944 and 1948—called it "an inexcusable abuse of the right of freedom of speech." Walter Winchell gloated that with just a few hundred words in Iowa, "Lindbergh's halo became his noose."

Back on Martha's Vineyard, Lindbergh was mystified by the uproar. In his view, he'd merely said in public what people said in private: the truth. "It seems that almost anything can be discussed today in America except the Jewish problem," he wrote in his diary. On September 15, Lindbergh was summoned to Chicago by General Wood for an emergency meeting of America First's executive committee. After a contentious debate, America First issued its only public statement on the matter, accusing the Roosevelt administration and its supporters of trying "to hide the real issue by flinging false charges at the America First Committee":

Colonel Lindbergh and his fellow members of the America First Committee are not anti-Semitic. We deplore the injection of the race issue into the discussion of war or peace. It is the interventionists who have done this.

America First was so committed to having Lindbergh as its public face that he continued to speak for it, in Fort Wayne, Indiana, on October 3; and again in New York, on October 30. (Lindbergh accepted General Wood's suggestion that he not mention Jews in either speech.) At Madison Square Garden, Lindbergh studied the faces of the people sitting below him (just as Carrel had studied his face eleven years earlier). "They were *far* above the average for New York," he wrote in his journal. "These people are worth fighting for." When Lindbergh took the podium moments later, his speech was delayed by a standing ovation lasting six minutes.

Lindbergh's next speaking engagement, in Boston, wasn't for several weeks. So he returned to Martha's Vineyard, where he helped his sons, Jon and Land, build a tree house. As he hammered nails with his boys, Lindbergh couldn't help noticing the American ships—military and merchant—passing by. All of them had a fresh coat of dark paint: the color of war.

In the first week of December, Lindbergh began working on his Boston address, scheduled for December 10. America First, much to his pleasure, had weathered the Des Moines storm. In fact, a chapter in upstate New York had just demonstrated its renewed fighting spirit by sending a forceful letter to President Roosevelt: "What's all this saber rattling in connection with Japan?" the members demanded to know. The letter was dated December 6, 1941.

REPORTERS CALLED LINDBERGH ALL the next day. He ignored them. "All that I feared would happen has happened," he wrote in his journal. "We are at war all over the world, and we are unprepared for it from either a spiritual or a material standpoint."

Lindbergh wanted to do his patriotic duty, but how? A simple, direct man would have marched to the nearest recruiting office and signed up. But Lindbergh didn't; he wrote to his former commanding officer in the army air corps, General Henry H. "Hap" Arnold, on December 20, requesting a meeting. When he hadn't heard from Arnold after several weeks, Lindbergh followed up with a phone call. General Arnold's aide, Major Eugene Beebe, spoke for his boss and informed Lindbergh that his future with the air corps would be decided by the secretary of war, Henry L. Stimson. Lindbergh and Stimson met in Stimson's office in Washington on Monday, January 12, 1942. Lindbergh said his preference, now that the country was at war, was to offer his services to the army air corps. Stimson, a lifelong Republican who had served as secretary of state in the Hoover administration, said he would respond with complete frankness.

"From my reading of [your] speeches," he said, "it is clear to me that you take a very different view of our friends and enemies in the present war— not only from [the administration,] but from the great majority of our countrymen, and that [you] evidently lack faith in the righteousness of our cause." Stimson, who had been an artillery officer in World War I, added, "We are going to have a very difficult war on our hands, and I am unwilling to place in command of our troops as a commissioned officer any man who has such a lack of faith in our cause."

Lindbergh couldn't believe his ears. Stimson—no doubt speaking for the president—had blackballed him from service in the United States military. Equally crushing, Stimson made it clear that the president would have to approve any job related to the military that Lindbergh might seek in the private sector.

Lindbergh returned to Martha's Vineyard to consider his limited options. He was humiliated once again when Stimson's decision became public. Hundreds of miles away from Martha's Vineyard, the legendary antifascist troubadour Woody Guthrie wrote a song called "Mister Charlie Lindbergh," which he began to perform in concerts and soon recorded:

Mister Charlie Lindbergh, he flew to old Berlin,
Got 'im a big Iron Cross, and he flew right back again,
To Washington, Washington. . . .

Hitler wrote to Lindy, said "Do your very worst,"
So Lindy started an outfit that he called America First.
In Washington, Washington. . . .

Lindy tried to join the Army, but they wouldn't let him in.
'Fraid he'd sell to Hitler a few more million men,
In Washington, Washington.

Lindbergh had crashed to earth before, as a gypsy flier, as a mail pilot, and several times as a cadet in the army air corps. But never like this.

CHAPTER 18

THE TISSUES, THE BLOOD, AND THE MIND OF MAN

O F ALL LINDBERGH'S CONFIDANTS, THE ONE HE RESPECTED most had no inkling of the predicament the fallen hero now found himself in, even though that same confidant had predicted Lindbergh's tumble from grace a year and a half earlier. "He is committing suicide with the stupidities he is uttering!" Carrel said of Lindbergh in the spring of 1940. The reason Carrel was unaware in the winter of 1942 that his friend was disdained by the U.S. military and despised by the American press was a simple matter of geography: Carrel was no longer in America. He was in Paris, which was now occupied by the Germans and subjected to a censorship so stifling that reports from America were nearly impossible to get.

Carrel had made his prediction about Lindbergh's fall in a fit of pique: France surrendered to Hitler in June 1940 and, much to Carrel's anger, Lindbergh refused his request to say something positive in public about the French resistance. This triggered perhaps the only argument the two friends ever had. Soon after that spat Carrel sought the aid of a New York therapist, Dr. David Schorr, for insights into Lindbergh's psyche. In truth, it was Carrel who might have benefited from psychiatric treatment at this point—not because he was deluded, but because he was depressed.

The fall of France, coming so soon after his forced retirement from the

Rockefeller Institute and his failure to find a new home for his organ perfusion research or funding to create an "Institute of Man" in the United States, had robbed Carrel of his self-confidence. When Lindbergh suggested, in June 1940, that Carrel offer his services to the French collaborationist government then forming in Vichy, the surgeon was too numb to respond. The two men didn't see each other for a while after that conversation. Lindbergh was busy crossing the United States speaking for America First, while Carrel was at home in his Manhattan apartment, feeling betrayed and useless, his mind filled with worry not merely for himself, but for his wife and brother in defeated France.

That worry turned to high anxiety when German soldiers came ashore on Saint-Gildas, where Madame Carrel had returned after the French defeat. These men barged into the Carrels' home on August 5, hoping to intimidate her. But as a descendant of one of Napoleon's most decorated generals, Madame Carrel had been raised not to show fear to anyone—least of all to Germans. After searching the Carrels' property, the soldiers appropriated their motorboat, all their fuel, and most of their livestock, as Madame Carrel watched, muttering curses under her breath.

The same soldiers rowed over to Illiec, where they pillaged much of the Lindberghs' possessions. According to an article later published in the *New York Times*—an article that probably originated in Joseph Goebbels's propaganda office in Berlin—these soldiers found a framed photo of Hermann Göring on a small table in the Lindberghs' living room, inscribed, above the air minister's ornate signature, "To Colonel Lindbergh, in friendly remembrance of a visit to Berlin in 1936."

What we know for sure is that, after the looters left, Madame Carrel rode over to Illiec from Saint-Gildas on a mule, at low tide, under cover of darkness. Once there, she used a flashlight to find and rescue the one item she knew the Lindberghs were anxious to keep, which, through a combination of good luck and simple philistinism, the soldiers had left behind: a hand-carved wooden Madonna. A lone Wehrmacht sentry, seeing Madame Carrel make her escape, fired rifle shots in her direction. She lowered her head next to the mule's neck, hugged the Madonna close to her chest, and kept riding.

Her husband learned of these harrowing events more than a month after they occurred, when letters written by Madame Carrel were smuggled out of occupied France into the unoccupied zone—the zone ruled from Vichy by Marshal Pétain—where transatlantic mail service was still in operation. Getting news from Saint-Gildas shook Carrel out of his self-pitying stupor. He began working, with Frederic Coudert and others, to finance a volunteer ambulance corps to aid the underfunded hospital system in unoccupied France. Carrel also called on another friend from the Philosophers Club, the architect Charles Butler, to draw plans for a temporary field hospital equipped with 100 beds, an X-ray department, and several laboratories, which would also be shipped abroad. But even while Carrel was working on these relief projects, his commitment to eugenics remained inviolate. "The idea of bringing children from France to the United States is an excellent one," Carrel wrote to an agency looking into that possibility. "But I feel strongly that only children of the best stock should be sought."

In early 1941 Carrel's concerns about the eugenic disaster looming abroad led him to leave the safety of America and return to Europe to assess the situation firsthand. A friend, James Wood Johnson, a member of the family that grew rich selling Band-Aids, told Carrel one night at dinner that he was leaving shortly for Spain and France, as a private citizen, to deliver more than a ton of medicines and vitamins to civilians in need.

"I will go with you," Carrel said, as the waiter at the Century Association cleared their plates.

Before leaving America Carrel asked Lindbergh to meet him at the Rockefeller Institute, where Carrel retained an office but no staff other than his secretary, Katherine Crutcher. It's likely Carrel chose this setting, where he hadn't set foot in months, to send a message to Herbert Gasser: the director had succeeded in stopping the research done by Carrel and Lindbergh at the institute, but he hadn't succeeded in ending their friendship.

Carrel and Lindbergh hadn't seen each other for weeks. Lindbergh was thrilled to notice that the impending trip had replenished Carrel's self-esteem. On and on the surgeon went, just as he used to, when the two immortalists, sitting in the same room in identical black robes, discussed the decadent state of the world, and what they would do to fix it, if only they had the opportunity. The current disaster in Europe was all the more depressing, Carrel told Lindbergh, because it was so predictable. Carrel himself had predicted it in *Man, the Unknown* and in countless speeches, all of them, he remembered with great satisfaction, met with standing ovations. As disheartening as the European crisis surely was, it presented a once-in-a-lifetime opportunity for rebirth—but only if the healing process was directed wisely. And who was better qualified to ensure that it was, Alexis Carrel asked his friend, than Alexis Carrel?

Before they parted, Lindbergh handed Carrel a letter he had written the night before to Carrel's wife. Its contents show that Carrel had already begun expressing his eagerness to direct the rebirth of France to Lindbergh in their telephone conversations before their meeting.

The French government in Vichy, Lindbergh wrote to Madame Carrel, "must be reminded that [Dr. Carrel] foresaw clearly what was happening in France many years ago. He told people what it was leading to, but they would not listen. Now he has been demonstrated right, and now is the time to make use of his services in reconstructing the nation along sound and enduring lines."

But writing in his journal after his meeting with Carrel, Lindbergh worried that his friend's mission to France—a mission that had no backing from a government agency in France or any other nation—would be sabotaged by Carrel's most enduring nemesis: himself. "I believe Dr. Carrel can be of great value to France," Lindbergh wrote:

> [But] Carrel makes enemies easily and makes little attempt to "sell himself." I admire him for this, but I am afraid it may result in wasting the powers he could otherwise contribute to the reconstruction of France. He antagonizes so many of the people he comes in contact with, and few of them are able to see beneath his surface shell to the great mind it contains.

CARREL AND JOHNSON LEFT for Europe on February 1, 1941. Each had made numerous transatlantic crossings in the past, always on luxury liners such as *Queen Mary* or *Île-de-France*; but on this occasion they traveled on *Siboney*, a mere freighter. According to the *New York Times*, which ran a story headlined DR. CARREL SAILS FOR SURVEY IN EUROPE; WILL STUDY EFFECTS OF COLD AND MALNUTRITION, *Siboney* set sail that morning with twenty windows on its promenade deck boarded up because they'd been smashed during a storm at sea on its last voyage. A letter from Coudert arrived just before Carrel embarked. In it, one member of the Philosophers Club reminded another of the eugenic crisis facing western civilization.

"It is with mixed emotions that I watch you leave for Europe, so filled with conflict and suffering," Coudert wrote to his good friend. "Doubtless you will be obliged to witness many unpleasant situations. But I am glad you are going, because I am certain you will end the suffering and begin the rebirth of the ancient races that built our civilization."

Carrel and Johnson arrived in Madrid on February 19. They saw poverty and hunger in Portugal, where *Siboney* docked, but nothing prepared them for what they found in Spain, a nation still reeling from a civil war that ended in 1939 with victory for the fascist general Francisco Franco. That the visitors were staying at Madrid's poshest hotel—the Ritz—surely brought the deprivation outside into sharper focus. Leaving that environment of velvet sofas and white-gloved desk clerks each morning, Carrel and Johnson were accosted by skinny, shabbily dressed children, many with open wounds or large skin rashes, all of them begging for handouts.

Johnson chronicled what they saw in two articles published in the *Saturday Evening Post*: "We Saw Spain Starving" and "Ah, Madrid! Rumors, Suspicion, Fear!" Though Johnson painted a flattering portrait of Carrel's activities, even Johnson couldn't help noticing that the scientist viewed his encounters with Spanish children through the eyes of a committed eugenicist. Among the dozens of dying youngsters the two men saw daily, Carrel would often find one or two who were robust, despite the disease all around them. Nature, Carrel told his friend, had selected those young people to build Spain's future. Spain's doctors, Carrel said, should be tending to *them*.

When, on March 14, it was time to cross the border into France, Carrel and Johnson faced an unexpected problem. They were hoping to ship their remaining supplies into unoccupied France, but were warned not to. The medicines would be stolen by corrupt French customs officials, they were told, and then sold on the black market, perhaps even to Germans. After hearing this warning, Carrel and Johnson left the supplies behind in Spain with aid agencies they trusted. This decision, made quickly and with the best of intentions, would be called into question by leaders of the resistance when France was eventually liberated.

But liberation was only a dim hope in the spring of 1941. Arriving by plane in Vichy, Carrel was surrounded by reporters at the airport. "I was not sent here by anyone," he said. "I have not been entrusted with any mission. I am just a doctor who has come on his own to see how things are." On March 16 Carrel secured an audience with Vichy's head of state, Marshal Pétain. Soon to be eighty-five years old, Pétain was still known to all Frenchmen as the "hero of Verdun," the place where, in the spring of 1915, the French army, under his leadership, had stopped the Germans' westward advance in World War I.

We have no transcript of the conversation between Carrel and Pétain, as they walked through the Parc de Sources, in Vichy's spa district, but Carrel's biographer Robert Soupault is certain it was a harmonious meeting. After all, the marshal had issued a series of "messages" to the French people after their defeat in 1940—nearly all of these on themes Carrel had been sounding for years. France fell to the Germans, Pétain wrote, because the "spirit of pleasure won out over the spirit of sacrifice," causing the "body of state" to be infected with "immorality," "laziness," and "incompetence," which spread "like gangrene." For France to recover, it must "suppress alcoholism," "encourage the family," and "reform public education." This rebirth, Pétain warned, would require self-denial, and even pain. Hadn't Carrel made the same point in *Man, the Unknown* when he wrote, "Man cannot remake himself without suffering"?

Carrel surely mentioned his idea of creating an "Institute of Man" at this meeting with Pétain. But the marshal, perhaps needing time to think, gave Carrel a different assignment—touring the provinces of "Free France" to assess the health of French children, as Carrel had just done in Spain. In the next several weeks, Carrel, accompanied by Johnson, visited orphanages, hospitals, and private homes in Lyon, Avignon, Aix, Marseille, and many villages in between. The trip left Carrel shocked. Life in the provinces of southern France was a depressing echo of the corruption and depravity he had encountered in Paris in 1939, when he was sent by the ministry of health to assess the readiness of France's military hospital system—only to learn he'd have to pay bribes to get the data he needed. The situation in "Free France," Carrel wrote in his journal, was one of "total selfishness." Every citizen "is concerned with food and nothing else." "Too much wine" is being drunk. No one "is making the slightest effort to help anyone else." There is "total amorality."

After this tour, Carrel was so pessimistic about France that he thought of abandoning it forever and returning to America. He couldn't make that decision without consulting his wife, however, so he left Johnson in Marseilles and crossed into occupied France, reaching Saint-Gildas on April 14.

For Madame Carrel, the idea that her husband would turn his back on France in the darkest moment of its history was unthinkable. She knew there were only a few Frenchmen Carrel would listen to. She herself was in that tiny group; she made sure that two other members were on Saint-Gildas when it came time to discuss her husband's future. These two men were Dom Alexis Presse, a Benedictine monk from the nearby Breton mainland; and the engineer André Missenard, a graduate of France's famous Polytechnic.

Madame Carrel, who attended morning Mass daily, had introduced Dom Alexis to her husband several years earlier. The surgeon was immediately taken with the sparsely bearded, robed monk, who, in long walks in the Breton countryside, spoke to Carrel about the enduring value of the monastic life—especially as a source of inspiration and progress for mankind. In fact, the passages in *Man, the Unknown* that called for Carrel's "high council of experts" to work in monastic isolation were a direct result of those conversations. Carrel was equally impressed by Dom Alexis's eagerness—an impatient, sometimes even angry, eagerness—to serve a big political idea.

In April 1941, while sitting in the Carrels' rustic summer home as a fire crackled in a nearby fireplace, Dom Alexis said that history was now calling Carrel to serve such an idea. This idea was saving France, a country defeated and in desperate need of cultural, political, and spiritual renewal. The agent of the renewal, the monk said, would be the "Institute of Man" that Carrel had written of so eloquently. Yes, it was an enormous challenge, but who better to face it? Did a man of Carrel's abilities even have the right to absent himself from the crisis his countrymen now found themselves in? If an individual could help, didn't that mean God had given him a sacred obligation to do so? Was it not a sin to walk away?

After Dom Alexis spoke, Missenard, a thin, dark-haired man with a prominent Gallic nose, followed. The opportunity now available to Carrel, he said, was a chance to use science to build a new civilization, one without the flawed structure that had made the current crisis in France inevitable.

France could rise from the ashes of its defeat, but it must be guided by the men of "unquestioned achievement" and "universalist minds" Carrel had described in his writings. There were such men in France, said Missenard, men ready to commit themselves to that history-making project under Carrel's leadership. Missenard was one of them, and there were others—physicians, architects, engineers, economists, statisticians, educators, eugenicists—all of them eager to work in an effort led by the great Nobelist and best-selling philosopher who'd provided a blueprint for the renewal in *Man, the Unknown*. It was after reading that book in 1936 that Missenard had made a point of introducing himself to Carrel in Paris.

Dom Alexis's "big idea" argument certainly resonated with Carrel: *Man, the Unknown* was itself a "big idea." Carrel was surely also impressed with Missenard's declaration that there were scientists in France ready and eager to follow Carrel's lead. To remove those men from the culture that had led France to the greatest humiliation in its history, and place them in a special institute where they would be decontaminated from the decadence all around them, wasn't just a good idea, as Carrel saw it; it was good science. He'd always envisaged his "Institute of Man" as a kind of "perfusion pump" for western civilization—a place where the best minds could not only be isolated from the degeneracy around them, but nourished (under his direction) and made even stronger. These fortified minds would study the sick culture around them, assess the data they collected while living in monastic isolation, and then heal that culture outside with a series of prescriptive edicts.

To attempt this undertaking in France, of which half was occupied by the Germans and the other half led by a collaborationist government, would require compromises Carrel was loath to make. He had always assumed that his institute would be established in peacetime in the United States.

But maybe Dom Alexis and Missenard were right: perhaps France—despite the obvious problems created by war and defeat—was a suitable location. The "messages" issued by Pétain, many of them echoing ideas Carrel had expressed in his own writings, showed that the hero of Verdun was interested in a national reformation project. Perhaps he could be persuaded to fund Carrel's institute.

Madame Carrel's intervention worked. Less than a month later her husband was in Vichy, meeting with Pétain to discuss the idea of creating his

"Institute of Man" in Paris. A witness to those talks, Henri du Moulin de Labarthète, Pétain's chief of staff, marveled in his memoir at Carrel's "exceptional learning and the almost athletic fluency of his mind."

When the press in Vichy noted Carrel's presence there, its reports were transmitted abroad by wire services. James Wood Johnson, now back in New York, read them and assumed the worst: he'd heard with his own ears Carrel's contempt for the "total selfishness" they had encountered in southern France—a contempt so deep Carrel said he would probably return to New York, permanently, in May. If Carrel was still in Vichy, Johnson interpreted that to mean he was being held against his will, and this is precisely what he told a reporter for the New York Times.

CARREL BELIEVED DETAINED BY NAZIS, read the Times's headline on May 16, 1941. Carrel's secretary, Katherine Crutcher, cabled from New York to ascertain the truth. Like Johnson, Crutcher knew that Carrel had talked of returning to America: she'd made a reservation for him on the Pan Am Clipper, the transatlantic airliner flying out of Lisbon. Carrel had missed that flight, but she would happily book another, if he wanted. Carrel sent this response: "THANKS VERY MUCH FOR YOUR CABLE. I AM RETURN-ING TO THE OCCUPIED ZONE. IMPORTANT WORK IN THE OFFING."

Carrel's lobbying campaign with Pétain had succeeded. He was going to Paris to form his institute. Before he arrived, Carrel sent the marshal a letter. The institute they were about to create, he wrote, would "regenerate the race" and "remake a people"—a challenge even Carrel, who rarely doubted himself, described as "daunting." In his journal Carrel expressed his hopes and anxieties in the form of a prayer: "Lord make me an instrument of peace. Where there is hatred, let me sow love. Where there is despair, hope. Where there is darkness, light." This he would do, or die trying, Carrel decided, in occupied Paris.

CARREL'S "INSTITUTE OF MAN"—RENAMED the Fondation Française pour l'Étude des Problèmes Humaines (the French Foundation for the Study of Human Problems), but known everywhere in France as the Carrel Foundation—was created by a decree signed by Marshal Pétain on November 17, 1941. "The aim of the Foundation," Carrel wrote, "is to find means of improving our physiologic, mental, and social condition, to bring to the problems of individuals and society more complete solutions than have ever been available before ... and by doing so, to create a civilization that, like science, will be infinitely perfectible. For the first time in history, science has been summoned to apply its methods to the simultaneous reformation of the tissues, the blood, and the mind of man." Has a bigger "big idea" ever been proposed?

Living with his wife in their small apartment on rue Georges-Delavanne, Carrel threw himself into achieving his two most immediate goals: finding a headquarters for his foundation, and finding the "universalist minds" to staff it. Right from the start there were problems. The Nazis took a keen interest in real estate transactions in Paris, so much so that Carrel was forced to meet with the German ambassador, Otto Abetz.

When Abetz summoned Carrel to his embassy—the palatial former home of Napoleon's stepson, Prince Eugène de Beauharnais—to discuss potential sites for the foundation, a huge party, replete with magnums of champagne and trays of caviar, was under way for Nazi dignitaries visiting from Berlin. Carrel's presence at this reception, however brief, was noted by Frenchmen already suspicious that he was too cozy with their German occupiers. Other Frenchmen—the scientists at the Pasteur Institute—were insulted when Carrel looked into the possibility of moving his foundation into their building, which would have meant displacing them from offices and laboratories they'd occupied for decades. In the end, Carrel established his headquarters on rue de la Baume, in a building formerly occupied by a philanthropic organization, the Rockefeller Foundation, which willingly turned the site over, as its American employees had already left the city. The Germans had been planning to appropriate this building for their own use. Knowingly or not, Carrel outmaneuvered them.

Either way, it was clear the Germans were hoping to use the famous Nobel laureate for their own propaganda purposes. Offers of special treatment were made: a free car and driver; unlimited access to food (when most Parisians had to hustle for anything more extravagant than bread and potatoes); and extra fuel to heat his apartment. Pierre Laval, the Nazi sympathizer chosen by Pétain to run his government as premier, summoned Carrel to Vichy and offered him—Carrel said "demanded that he take"—the post of minister of health.

Carrel refused all these offers, preferring to make the daily journey from his home to the Carrel Foundation on bicycle, to eat whatever was available, and to wear extra sweaters when he did his reading in his apartment—where, according to a letter he sent to Frederic Coudert, the temperature indoors rarely reached forty-five degrees Fahrenheit during the winter.

Unfortunately for Carrel, he shivered for naught: most Frenchmen (and quite a few Americans) assumed that he had accepted the offers. Even the *New York Times* was misinformed, reporting, in early 1942, that Carrel was about to become the minister of health in Laval's pro-Nazi government. This erroneous report led Herbert Gasser to refuse a request from the Carrel Foundation for a selection of reprints from medical journals in the Rockefeller Institute library. As Gasser explained to Carrel's secretary, Paris was occupied by America's enemy, and any help sent there—even a few medical articles—would give aid and comfort to that enemy. Days later, Gasser told the secretary to remove Carrel's personal items and papers from his office at the Rockefeller Institute, as the director was planning to give that space to someone else.

In Paris, Carrel was too busy thinking about filling the offices in his own foundation to worry about Gasser's insults. As was his wont, he went about the hiring process in unorthodox fashion. His wife joined him during most interviews, and her opinion of the applicant's handwriting carried significant weight. Similarly, more than one applicant was rejected because Carrel didn't like the look of his face. Though nearly all the men and women who entered this screening process were university graduates, none, by Carrel's fiat, were professional academics. The salaries being offered weren't large—10,000 francs a month—but Carrel said that what he was offering transcended monetary value: a chance to share in the adventure of creating a new French civilization, one that would serve as a model for the rest of the world.

By the summer of 1942, there were nearly 300 full-time employees—each hired on a six-month contact, renewable after a performance review—in nearly a dozen buildings spread throughout greater Paris. (The building on rue de la Baume housed the foundation's executive committee, accounting office, and support staff.) Most of the people Carrel hired were laboratory or clinical scientists, roughly half from medical fields; the rest were engineers, statisticians, and the like. This staff was divided into six departments, the largest of which focused on population studies (headed by

André Missenard), child development, and workplace medicine. The foundation's expenses were underwritten by a subsidy of 40 million francs from Vichy.

The pro-Vichy journalist Alfred Fabre-Luce profiled Carrel and his new foundation for a magazine in a piece titled "The Man Who Wishes to Understand Man." "The ultimate aim" of Carrel's foundation, Fabre-Luce wrote, "is the creation of a superior man by means of heredity, nutrition, and social organization. The immediate aim is quite simply the survival of man as he is. Man's future is now on the dissecting table, and Carrel is at the bedside to draw the lesson." But another French journalist, writing anonymously, was less impressed: "Is it really possible," he asked, "that every Frenchman is required to donate one franc so that Doctor Carrel can create a race of supermen?"

That question, though hyperbolic, contained some truth. In France Carrel put into operation the eugenics program he'd first suggested in 1937, in a speech at the University of Illinois. Before "civilized man" can be created by scientific experts, Carrel had said in that address, the current state of man must be assessed in a comprehensive demographic survey. Teams of researchers, all of them members of Missenard's department of population biology, were sent out into the field. The short-term goal was to measure France's birthrate, which had begun falling even before the war; the long-term goal was to suggest methods to reverse that decline. Selective methods.

"Measures to encourage higher birthrate must be in harmony with eugenic considerations," Missenard wrote in a foundation memorandum. "There is an obligation on the part of the hereditarily and biologically gifted to mate only with others of equally superior quality." Missenard's colleague Jean-Jacques Gillon, chief of the department of child biology, went even farther, ascribing a religious aspect to eugenics: It is "a method ordained by God," Gillon wrote, "based on the laws of Nature, and aimed at making our children healthy, energetic, and intellectually gifted."

One "divinely inspired" survey undertaken by the foundation involved medical examinations of 100,000 schoolchildren. The results confirmed Carrel's worst fears: "gifted" Frenchmen were not reproducing at a sufficient rate. Instead, there were thousands of children with physical defects and an equally alarming number with serious learning disabilities. When one of Carrel's more altruistic employees, who had developed a program to educate retarded children, spoke to him about it, Carrel responded with surly impatience: "We're interested in people of quality!" he barked, before throwing the proposal into the trash.

Missenard, a civil engineer turned social engineer, was even less patient regarding this subject. He urged that those diagnosed as insane, epileptic, or retarded be sterilized, incarcerated, or both. This idea was rejected by Carrel as too extreme, but others weren't. A bill drafted by Missenard, and submitted to the Vichy government, created the first eugenic law in the history of France. After December 16, 1942, no one in France could be married without a prenuptial certificate from a physician attesting that he or she was free of syphilis.

The foundation also studied immigration, urging that considerations of "biological worth" be applied to all who sought entrance into France. The presence of "biologically undesirable" foreigners, Carrel believed, posed a threat every bit as serious as the declining birthrate among France's "biologically gifted." Immigration needn't be stopped entirely, but immigrants must be "ranked according to their character and biological desirability." Researchers traveled to rural French villages inhabited almost entirely by immigrant populations, and subjected the resident aliens to extensive tests to ascertain their "biological desirability." One scientist from the foundation took dental impressions to determine if the immigrants had teeth appropriate to the consumption of French cuisine.

CARREL READ THESE REPORTS with a growing sense of frustration. One reason was that he was just about the only person in France who *was* reading them. Serge Huard, the physician who accepted the post of minister of health from Premier Laval after Carrel turned it down, considered the foundation's medical surveys an intrusion into his own area of responsibility, so he ignored them. Scientists at two other French institutions—the University of Paris and the National Center for Scientific Research—reacted with similar resentment. Carrel was criticized by these establishment academics for being "too American" (because he'd made his career there), too

German (because of his dealings with the occupation authorities), and too Bolshevik (because one of his departments focused on improving the health of factory laborers).

The public showed little interest in the foundation, pro or con, and that made Carrel even more frustrated. Though he published his foundation reports with large press runs, sales never climbed out of the hundreds. As usual, he responded with contempt. Those pitiful sales, Carrel wrote to his brother, proved only that, "There are no people more hateful than the French. Their ignorance is truly astounding!"

Carrel was equally unhappy to see that his staff hadn't turned out to be all he hoped. He had expected to find, among his nearly 300 employees, a core of "true believers": a dozen or so brilliant, hardworking men whom he would establish at a château he had purchased for the foundation outside Paris, at a cost of 3 million francs. There, in semirural isolation, those men would be relieved of fact-finding and would become synthesizers of data. They would be the "high council of experts" Carrel described in *Man, the Unknown.*

But this utopian construct fell apart under the weight of bloated expectations. Yes, there were some men dedicated to Carrel's vision—Missenard and Gillon, certainly—but the majority of those who'd joined the foundation, Carrel learned, were merely looking for a safe place to work until the war was over, no matter which side won. Over time Carrel's irritation became obvious at staff meetings, which he called with little enthusiasm even in the happiest of times. If Carrel found a presentation at one of those meetings to be "stupid" or "unscientific," he walked out of the office in midpresentation. His *own* office.

By mid-1943, Carrel's frustration was affecting his health. He experienced shortness of breath and chest pains while vacationing on Saint-Gildas—likely signs, he knew, of an impending heart attack. In August, just after his seventieth birthday, he was proved right. Madame Carrel helped him into a motorized dinghy, which they took to the mainland. There they boarded the night train to Paris, where Carrel was treated by the well-known cardiologist Édouard Donzelot. The attack wasn't a major one, but it left Carrel weakened, emotionally as well as physically. His enemies, inside and outside the foundation, sensed an opening.

The first strike occurred at the foundation's headquarters. As Carrel was recuperating on Saint-Gildas, the staffer he'd chosen to run the foundation, the economist François Perroux, organized a coup. Perroux had long been quietly unhappy with Carrel's management style: his refusal to demand "statements of objectives" or firm delivery dates from his research teams, his rudeness at staff meetings, and so on. When Carrel got word of the coup from his wife, who had stayed in Paris, he returned to the capital and fired Perroux. Just as Carrel was exulting at having "purged the Foundation of this paranoid," as he put it in a letter to his brother, other news arrived to end his euphoria. Premier Laval, still smarting from Carrel's refusal to join his cabinet, slashed the foundation's subsidy by nearly 30 percent.

Carrel was soon living full-time in Paris, but in a new apartment, at 56 avenue de Breteuil, which his wife had recently inherited from a cousin. Though his foundation was less than a mile away, he rarely visited it. Carrel gave day-to-day executive power to Missenard, Gillon, and Dr. André Gros, three men he trusted completely. In truth, as the winter of 1944 turned into spring, Carrel found himself thinking less and less about the grand reformation project he'd undertaken. And when he did think about it, he wondered if it hadn't been a huge mistake.

Another man might have accepted that the foundation was at best a personal fantasy and at worst hubris. Such a man might have belatedly understood that it was one thing to experiment on animals (as Carrel had done so successfully at the Rockefeller Institute), but quite another to experiment on humans. He might have realized that office politics in a private institution (such as the Rockefeller Institute) were child's play compared with the real politics one encounters running an agency funded by the state.

But Carrel wasn't such a man. He blamed France. Once again his homeland had let him down—just as it had when the medical faculty at the University of Lyon blackballed him years earlier. Now, decades later, he'd been undermined again by intellectual inferiors. Once again his genius had provoked only jealousy and resentment. "The most difficult thing about France is trying to make Frenchmen not detest one another," Carrel wrote his brother in March 1944. "If I had grasped the full extent of that pettiness and mean-spiritedness I would never have started this Foundation." In another letter, mailed days later, he said, "My Foundation is nothing more than a puff of smoke. It would take perhaps twenty years to find and train the right men to do this work correctly. I have not been given that time."

Or the energy. As time passed, Carrel retreated into an internal world. Before the war started he'd begun writing a sequel to *Man, the Unknown* in English, but had never finished. He picked it up again on avenue de Breteuil, this time writing in French. *Réflexions sur la conduite de la vie* was the working title. If Carrel couldn't achieve immortality for western civilization through his foundation, perhaps he could achieve it through his writings.

This was the thought that kept him going. His primary contact with the outside world was through his wife—and even that contact was tenuous. Madame Carrel was just as sickened as he was by the failure of France to embrace the foundation. After a while she too rarely left their apartment, choosing instead to flip tarot cards, throw runes, or make pendulum readings there over a map of Europe with a certain Madame Laplace, a psychic whom she revered. Laplace had predicted that the Germans' advance into the Soviet Union would falter at Stalingrad. Months before then, while visiting the Carrels on Saint-Gildas, she'd predicted an enormous aerial attack on London with bombs "that look like comets." Now, in the spring of 1944, there was talk of a massive Allied invasion to liberate France, perhaps at Normandy. Madame Laplace consulted her stones, cards, and pendulums: that invasion, she told the Carrels, will never happen.

IT DID HAPPEN, OF course, on June 6, and that is when the final chapter began for Alexis Carrel. Two months later American forces were approaching Paris. Just before the capital was liberated on August 25, Carrel, now seventy-one, suffered a second heart attack. Much to his horror, there were Parisian cardiologists who, on hearing of his condition, urged their colleagues not to treat him. "This man does not deserve any consideration," said one. "He has tricked everyone with his false modesty. In Vichy he demanded that two official motorcyclists precede his car as he passed through the city; at banquets, he insisted on a seat of honor next to Marshal Pétain." (None of this was true.) Another cardiologist was still seething from the insults Carrel had leveled at the Parisian medical community in 1939: "He left slamming doors behind him and then he came back like a conquering dictator. That is unforgivable!"

Carrel did get treatment—once again from Dr. Donzelot—but the prognosis was poor: this second attack left him virtually immobile. He had trouble sleeping and, more often than he'd like, even breathing. This physical discomfort was matched by his emotional distress. His name, he learned, was on a list of collaborationists prepared by the resistance. While being visited by an old friend from Lyon, Dr. Louis Aublant, who, unlike Carrel, had accepted a post in Laval's ministry of health, Carrel learned that Aublant had opened his mail recently to find a miniature coffin. This was an unmistakable message that the resistance had marked him for death. Carrel's reaction was nervous laughter.

No coffins arrived at 56 avenue de Breteuil, but other bad news did. Pasteur Vallery-Radot—grandson of Louis Pasteur, a leading scientist at the Pasteur Institute himself, and now the newly appointed minister of health for liberated France—issued a terse public statement: "Doctor Alexis Carrel, Regent of the French Foundation for the Study of Human Problems, is relieved of his duties." Radio programs and newspaper articles in Paris repeatedly identified Carrel as a pro-Nazi racist. A policeman was stationed outside his apartment, lest he attempt to flee the country.

One day an unimposing middle-aged man stopped at the front door of Carrel's apartment. The bored policeman on duty there rudely told him to leave, but when the visitor pointed to the band encircling his right arm—the French flag, with a blue Cross of Lorraine and the letters F.F.I. (Forces Françaises de l'Intérieur) emblazoned on the white stripe—the policeman recognized its wearer as a member of the resistance, and waved him in.

Carrel, who was sitting in his living room wearing a dressing gown, recognized his visitor immediately. It was Raymond Paumier, a member of his foundation's mother and child unit—the very man whose proposal for educating the retarded children of France Carrel had tossed into the trash.

Paumier hadn't come to taunt. To the contrary, his respect for Carrel was as great as ever. A provincial schoolmaster, he'd been hired by the foundation after he sent Carrel a letter filled with praise for *Man, the Unknown*, which Paumier read for the first time in 1942. A phone call from Madame Carrel then summoned him to the foundation's headquarters on rue de la Baume. After a brief conversation—and a close examination of Paumier's face—Carrel offered the schoolmaster a position, which Paumier accepted as a great honor.

But that was two years ago. Now Carrel could only tell Paumier the obvious: "As you can see, Paumier, I am not well. Some people say I am to be arrested; some people say I am already under house arrest. What have I done that they should hate me so much?" Paumier began to cry as he answered, "You were too big for them, *patron*, and that is difficult for some people to forgive." The resistance man and the accused collaborationist shook hands. They would never see each other again.

By this time reports of Carrel's precarious situation in Paris had reached New York. FRENCH HEALTH CHIEF DISMISSES DR. CARREL, announced the *New York Times* on August 28. That chief, Vallery-Radot, explained his action as the first in a series of "necessary purifications." A follow-up story, DR. CARREL DENIES HE AIDED GERMANS, ran three days later. "I was living tranquilly in the United States when I decided France needed me," Carrel was quoted as saying. "I came and founded my Institute and put my theories into practice. I had one aim and I achieved it. I did not do anything against France." Weeks later, the American political journal the *Nation* expressed pleasure at Carrel's fall, publishing a tiny poem about him written by Melville Cane, an attorney in Manhattan who was nearly as celebrated for his light verse as for his legal advice:

> *This man of science, with uncanny art,*
> *Contrived to keep alive a chicken heart.*
> *A sorry triumph, death thus to postpone,*
> *And in the process, atrophy his own.*

Carrel's friends in New York rallied to his defense. Frederic Coudert met with Secretary of War Stimson and asked Stimson to use his influence to make sure that Carrel was not arrested. Simon Flexner, the retired director of the Rockefeller Institute, sent Stimson a similar plea by telegram.

At almost the same instant, a letter from Carrel suddenly arrived at the Rockefeller Institute, addressed to Katherine Crutcher. The blend of fact and fantasy, hope and despair expressed in the letter surely can be attributed to Carrel's declining health.

"I hope that you are well and all my friends at the Institute are well," Carrel began. "Please remember me to them and tell them I will see them soon after this nightmare is over. We have suffered a great deal. Mme. Carrel fortunately is in good health. I have been severely ill for years. I am not yet well. We are suffering mostly from bad food and excess of work. Our work has succeeded quite well. I have organized an Institute like the one described in my 'Construction of Civilized Man' speech read in 1937 at Dartmouth College. The situation here is terrible. Please pay the rent on my apartment. I hope to see you soon."

It was not to be. As September ended and October began, Valley-Radot's subordinates at the ministry of health continued their investigation into Carrel's wartime activities. As the investigation moved forward, Carrel's health deteriorated. He was now bedridden, and his pulse weakened it seemed, by the hour. Fluid filled his abdomen, and soon his chest. Physicians from the foundation—Gillon and Gros sharing the leadership role— tended to Carrel around the clock. Most of the time there wasn't much to do: the patient was asleep. When he awoke he often showed flashes of his old self—not all of them pleasant. When one doctor asked if there was something he could do for him, Carrel snapped, "Yes, improve my health! Give me a way of standing up, of resuming the work that my enemies rejoice at seeing left half-done!" There were also moments of depression: "I wanted to do good, and they have *murdered* me!" And delusion: "When I'm better, we'll take up where we left off at the Foundation on an international scale."

On November 2 the fluid buildup in Carrel's lungs reached the point where Dr. Gillon suggested to Madame Carrel that she call Dom Alexis, in Brittany, so that he might tend to her husband in his final hours. Carrel seemed to revive in the presence of Dom Alexis, who arrived on November 4. But Carrel had no delusions: "It is in the hour of one's death that you feel the insignificance of all things," he whispered to Dom Alexis. "I achieved fame. Out in the world, people spoke of me and my works. Yet now I am nothing more than a little baby before God."

As dawn broke on November 5, Gillon called Dom Alexis and Madame Carrel to Carrel's bedside. Two candles were lit; Dom Alexis performed the last rites. Moments later Carrel's body began to quiver under his wool blanket. At 5:15 a.m., as Dom Alexis and Madame Carrel chanted the rosary, Alexis Carrel—the scientist who, as the poet Percy Mackaye once wrote in the *New York Times*, had clutched "death's frozen fingers" in a battle of strength no mortal before him had ever fought so hard or so well—breathed his last.

CHAPTER 19

—

THE WORLD WAS NEVER CLEARER

Lindbergh learned of carrel's death in the *new york times.*
The two men hadn't seen each other since the day before Carrel left to deliver medicine to the sick children of Spain and France, only to find, once he
got there, that his scientific interest was piqued most by the children who
didn't need any medicine. Lindbergh's final meeting with Carrel was nearly
four years ago, which only intensified his grief. "I cannot yet make myself
realize he is gone—I still see him in life too clearly," Lindbergh wrote to Dr.
Irene McFaul, a longtime aide to Carrel at the Rockefeller Institute. "One
of my greatest sorrows lies in the lack of communication we had in recent
years. I had not dared to write Dr. Carrel since the occupation of France,
because of political conditions and the fear that a letter from me might be
used by his enemies to injure him."

Lindbergh was pained to learn that those enemies had injured his friend
a few hours *after* Carrel's death. A radio report was broadcast twice in France
on the afternoon of November 5, 1944, declaring that "Dr. Alexis Carrel,
wanted by the police for collaboration, fled from his home in Paris this
morning. He is thought to be hiding in the city." Only after Madame Carrel
called the French ministry of information and said that her husband had
died that morning—in his own bed, with no warrant for his arrest ever
issued—was the error corrected, and then without an apology.

Lindbergh felt that he too had been smeared by false charges. The worst occurred when the secretary of war impugned Lindbergh's patriotism and said he was not welcome to serve in the U.S. military. Lindbergh then sought employment at several aircraft manufacturers fulfilling military contracts with the government, only to learn the Roosevelt administration had black-balled him from working there, too. Only one manufacturer was stubborn enough to resist the anti-Lindbergh pressure. This was the Ford Motor Company, then building the B-24 bomber at its large Willow Run plant, near Ypsilanti, Michigan. On March 24, 1942, Lindbergh took a tour of the 2.5-million-square-foot facility that American newspapers called "the largest room in the world." His guide was Henry Ford himself.

This wasn't their first meeting. In August 1927 Lindbergh had landed in Detroit while taking his victory lap around America in *Spirit of St. Louis*. He visited his mother at her home in Grosse Pointe (she was then teaching chemistry at Detroit's Cass Technical High School), and he spoke at a huge outdoor rally; then Lindbergh took Ford on his first airplane flight. The cabin of *Spirit of St. Louis*, as the world knew, was designed for only one oc-cupant, so a temporary seat was hastily installed for Ford by a team of car-penters from his auto plant. Ford, who was then sixty-four years old, had to bend awkwardly to keep his head from hitting the cabin's ceiling. He actu-ally spent most of the five-minute flight perched on the edge of Lindbergh's seat, an arrangement made possible only because he was one of the few men in America even thinner than Lindbergh.

On April 2, 1942, Lindbergh became a production executive at the Wil-low Run plant, where thousands of assembly-line workers—including ten midgets recruited from the circus to hammer rivets in tight-fitting spaces inside the B-24's wings—worked three shifts a day, trying to make good on Ford's promise to President Roosevelt that his plant would produce one thirty-ton bomber every hour. Not only was the plant nowhere near that rate of productivity, but the few bombers it did produce weren't very good. After taking one B-24 up for a test flight Lindbergh wrote these impres-sions in his journal: The controls "were the stiffest and heaviest I have ever handled. . . . I would certainly hate to be in [this] bomber if a few pursuit planes caught up to it [in combat.]"

Lindbergh had plenty of time to worry about this because he was living alone at the Dearborn Inn, a nearby hotel owned by Ford. Anne and the children had stayed behind on Martha's Vineyard, but they finally joined Lindbergh in Michigan that summer, in a house he'd rented in Bloomfield Hills, a suburb north of Detroit. Anne, once again pregnant, was very happy in Bloomfield Hills. The Lindberghs' house adjoined the famous Cranbrook Academy of Art. After giving birth to Scott Morrow Lindbergh on August 12, Anne took classes at Cranbrook and socialized with the Finnish architect Eero Saarinen and other members of its illustrious faculty.

But her husband was growing restless—and still holding grudges. On September 18 President Roosevelt visited Willow Run. Lindbergh went home early. ("He heard the devil was riding into town, so he took off for the hills," Roosevelt said.) Not long afterward, Lindbergh spoke to Ford privately and proposed a new arrangement between them: Lindbergh would forgo his modest salary—$666.66 a month, the same pay he would have received had he been serving as a colonel in the military—to become a consultant reimbursed only for expenses related to his work for Ford. This would give Lindbergh the freedom to pursue what he described to Ford as "aviation research elsewhere in the country." There was one field of research Lindbergh was especially keen to pursue. That's because it reminded him of work he'd started with Alexis Carrel.

That new research focused on high-altitude flying. A new American fighter plane, the P-47 Thunderbolt, could cruise at 400 miles per hour at 40,000 feet. This was not only four times faster than the speed of *Spirit of St. Louis,* but four times higher than its maximum possible altitude when Lindbergh flew to Paris in 1927. The increase in altitude had obvious military benefits, as no antiaircraft shells could reach that high. What was less clear to the War Department was whether a pilot could fly in the thin atmosphere up there without passing out.

To answer that question, Dr. Walter M. Boothby, chief of the aviation medicine research unit at the Mayo Clinic, in Rochester, Minnesota, built a pressurized steel chamber that could simulate the physiological conditions of pilots flying at high altitudes. When Lindbergh read about this research, he was reminded of the pressure tank a manufacturer had built for him in England, which he then shipped to New York for the "artificial hibernation" experiments he performed on monkeys in 1938 with Carrel at the Rockefeller Institute.

Lindbergh had always wanted to advance that potentially life-extending research on humans—starting with himself—but Carrel forbade it, saying it was too dangerous. Now Lindbergh saw a chance to continue that work. He contacted Boothby and volunteered to serve as his test subject. On September 22 he arrived at the Mayo Clinic.

But before Boothby would allow Lindbergh inside the chamber, he insisted that Lindbergh "desaturate" himself by walking on a treadmill for thirty minutes. The purpose was to reduce the nitrogen levels in Lindbergh's blood and tissues—and thus reduce the possibility that life-threatening nitrogen bubbles might form there, created by the extreme pressure inside the chamber. Those bubbles could injure the central nervous system, cause swelling and pain in muscles and joints, create blood clots, or even block the blood flow entirely. (When contracted by scuba divers, this condition is informally known as "the bends.")

Boothby's chamber—a steel cylinder with three portholes on one side and a door at one end—looked like a submarine Captain Nemo might have used to fight a giant squid in *Twenty Thousand Leagues under the Sea*. There were two compartments inside: the testing area was filled with vacuum pumps, fans, oxygen tanks, electrical wiring, and a solitary high-backed chair for the test subject; the second room was an air lock that permitted Boothby to observe Lindbergh, who, when an experiment was under way, was strapped into the high-backed chair with an oxygen mask covering his nose and mouth, his head stabilized by an elastic band reaching around his forehead from the back of his seat.

That headband wasn't for show. The prevailing wisdom in the small community of aeromedical experts in the fall of 1942 was that it was impossible for a pilot to train his senses to become aware that he was entering a state of hypoxia—the medical term for inadequate oxygenation of the blood—in time to take action to overcome that state before passing out. Boothby and Lindbergh were hoping to prove the contrary.

On September 24 the pressure inside Boothby's chamber was set to equal the pressure inside a plane flying at 40,000 feet. After a signal from Boothby, Lindbergh simulated the physical demands that would be required of a pilot forced to make an emergency parachute jump at that altitude. He removed his oxygen mask, stood up, and began moving his arms back and forth—not because he was cold, but to mimic the activity required to open a cockpit hatch. Then, after Lindbergh returned to his chair, Boothby altered the pressure inside the tank to simulate the changes experienced by a parachuting flier as he descended. Seconds later Lindbergh was gasping for breath; then he passed out.

When Lindbergh came to, he told Boothby—who was visibly alarmed—about the artificial hibernation experiments he'd performed on rodents in England with a vacuum pump and glass bell jar in 1936. "Dr. Boothby is much more concerned about the danger of becoming unconscious than I am," Lindbergh wrote in his journal. "In running experiments on rodents under low pressure some years ago, I found they would regain consciousness quickly, after passing out, when the air pressure was increased," without negative consequences.

But the point of those experiments on rodents (and the ones Lindbergh and Carrel later performed on monkeys) was rather different from that of the experiments Lindbergh was now participating in. The current experiments were about enabling American pilots to shorten the lives of their enemies. The earlier ones, Lindbergh remembered in his journal, were about lengthening life.

On October 1 Lindbergh finally tried an artificial hibernation study on himself. As the pressure inside the tank (and the temperature along with it) was lowered by a technician, Lindbergh experimented with the yogic breathing techniques he'd learned in India in 1937. His goal was to document a scientifically verifiable link between those breathing exercises, the lowered internal temperatures they created, and the life-extending benefits that were said to accrue to the Indian holy men who practiced them, often while meditating, wearing only a loincloth, in the snows of the high Himalayas.

Lindbergh found after experimenting with various methods of pulsated breathing that his internal temperature quickly dropped two degrees inside the tank. So far so good. But then, moments later, just as had happened with the monkeys he experimented on at the Rockefeller Institute, his temperature began to rise, until Lindbergh had a "fever" of nearly 101 degrees. "Clearly," he wrote, "a more complicated procedure would be required to overcome the human body's temperature-regulating mechanism." Lindbergh's biggest regret was that he couldn't share the details of this experiment with Alexis Carrel.

When Lindbergh resumed his experiments with Boothby, he learned to recognize three symptoms of an approaching blackout: difficulty breathing, a darkening of vision, and a steady loss of mental acuity. Even more important, he discovered that roughly fifteen seconds of consciousness remained after he experienced the first of those symptoms.

This knowledge saved Lindbergh's life one October morning after he returned to Willow Run. He was descending from 41,000 feet in a P-47 he was test-flying when, at 36,000 feet, he felt a blackout approaching. His oxygen supply seemed to have cut off, even though the gauge on his control panel showed fifty pounds still remaining. As his vision got darker and his mind began to lose focus, Lindbergh struggled to remember what he'd learned at the Mayo Clinic: when he did, he ripped off his oxygen mask, shoved his flight stick forward, aimed his plane straight down, and passed out. His P-47 went into a screaming dive of more than 20,000 feet—not that Lindbergh noticed: he was unconscious in his seat, his head bobbing forward, his arms dangling at his side.

Finally, he was revived by the thickening air in his cabin at the lower altitude. After shaking his head to become more alert, Lindbergh grabbed his flight stick and leveled his plane; if he hadn't, it would have crashed into the hard Michigan farmland below, raising a bright orange fireball and littering the countryside with debris—most of it metal, the rest of it human. When, moments later, Lindbergh taxied to a stop on one of Willow Run's runways, he climbed out of his cockpit, his body drenched in sweat, and asked a Ford mechanic to check his air gauge. It was giving false readings, the mechanic reported back, after examining it. Lindbergh's oxygen tank was completely empty.

Lindbergh's expertise in high-altitude flying eventually took him where he wanted to be: combat against the Japanese. This process began in December 1942 when the president of United Aircraft invited him to East Hartford, Connecticut, to inspect the F4U Corsair fighter plane United had just started supplying for navy and marine pilots in the Pacific. Lindbergh jumped at the chance to test this plane, one of the most recognizable American fighter planes ever built: the F4U's wings bent up and its engines created a high-pitched whine so distinctive that Japanese fighter pilots called the aircraft "Whistling Death."

Lindbergh made eight trips to United's East Hartford plant in 1943, many lasting several weeks. He worked there as a test pilot, a flight instructor, and an aeronautical engineer. The first job, Lindbergh's favorite, meant he got to fight in mock aerial battles. One afternoon in the skies over central Connecticut he engaged in a high-altitude gunnery contest against two pilots half his age who were said to be the best in the Marine Corps. According to one eyewitness, Lindbergh "outflew and outshot" both of them.

It wasn't long before word of Lindbergh's skills as a Corsair pilot reached the War Department, which had been getting conflicting reports from the Pacific theater about the relative value of single-engine versus twin-engine Corsairs. Because Lindbergh had experience test-flying both, he was summoned to the Washington office of Brigadier General Louis E. Wood of the Marine Corps to give his opinions. Sensing an opportunity, Lindbergh said he couldn't answer the general's questions with certainty unless he flew those planes himself in a combat situation. Much to Lindbergh's shock, General Wood called the next day and said Lindbergh was cleared to proceed to the South Pacific as a technical representative of United Aircraft. Though he'd be a civilian, Lindbergh would have to wear a uniform in the war zone; in fact, Wood said, he'd have to buy his own. But this uniform, Wood made clear, must be free of any service rank or insignia. If Lindbergh was captured by the enemy, he'd be a man without a country and protected by no international treaties.

That was fine with Lindbergh; he didn't plan on getting captured. On April 24, 1944, after getting inoculated against typhoid, typhus, tetanus, cholera, and smallpox, he took off for the South Pacific, from San Diego, wearing a uniform he had bought at Brooks Brothers in Manhattan. Lindbergh's only luggage was a small canvas bag holding a waterproof flashlight (purchased at Abercrombie and Fitch), batteries, a razor, soap, shoe polish, chocolate bars, a toothbrush, underwear, a spare uniform, and a pocket edition of the New Testament. "Since I can carry only one book—and a very small one—this is my choice," Lindbergh wrote in his journal. "It would not have been my choice a decade ago, but the more I read [it,] the less competition it has."

AFTER STOPS AT HAWAII and other American island bases, where he did little more than shake hands and go snorkeling with aviation officers, Lindbergh stepped out of a tent at daybreak on Green Island, one of the smallest of the Solomon Islands in the South Pacific, not far from Guadal-canal. He poured some water from a nearby can into his helmet, washed his face, and reported for duty at the pilots' tent; at forty-two, he was old enough to be the father of nearly everyone else there. Along with three other pilots, Lindbergh was issued a .45 automatic pistol, knife, parachute, and life raft. Four Corsairs, each bearing 1,600 rounds of ammunition, were being dis-patched that morning to fly 100 miles northwest to the Japanese-held town of Rabaul, on the island province of New Britain, in Dutch New Guinea. Their mission was to provide air cover for a large American bomber raid by B-17s and B-24s, slower and less agile planes that needed protection from Japanese fighter planes trying to shoot them down. Lindbergh would be flying one of the four Corsairs on that cover mission. This—not shaking hands or snorkeling—was why he'd come to the South Pacific.

If the Corsairs encountered no opposition in the skies over Rabaul, said the Marine colonel announcing the day's orders, the pilots were free to strafe Japanese targets below on the way home. The colonel thought it was under-stood that this directive applied only to the three pilots wearing genuine marine uniforms. Lindbergh somehow missed that.

So did Major Alan J. Armstrong, the head of Lindbergh's flying team. The Corsairs hadn't encountered any opposition while performing their cover mission, so Armstrong took his team down to strafe several buildings on or near Rabaul's airstrip, even though Rabaul was protected by antiair-craft gunners said to be the best in the Japanese army. Lindbergh, following Armstrong's lead, ignored the shells exploding around him and dived down toward the target area, firing a long burst of bullets into the largest building there. All were direct hits. "I hope there was no one inside except soldiers—no women or children," he later wrote in his journal. "I will never know."

When the Corsairs returned to base, the commanding officer looked tense. He wasn't happy to know Lindbergh had done more than observe.

"You didn't fire your guns, did you?" he asked.

"I did," Lindbergh said.

"You never should have done that!" the officer screamed. "You're on ci-vilian status. If you had to land and the Japs caught you, you would have been shot."

"I'm guessing they would have shot me no matter what," Lindbergh said.

The tension on the officer's face cracked. "All right," he said. "We'll wait a day or two, just to see if [the top brass] kicks up a fuss."

None was kicked. A week later, at 8 a.m. on May 29—his mother's birthday—Lindbergh learned in a briefing that he would be sent up in another Corsair. No cover duty this time: he'd be dropping a 500-pound bomb on the Japanese-held town of Kavieng, capital of New Ireland, another island province of Dutch New Guinea. He took off at 9:45, one of four pilots on the mission. Less than an hour later, Lindbergh dived down on Kavieng, the sun at his back. At 5,500 feet, with antiaircraft fire exploding all around him, he dropped his load. "You press a button and death flies down," he later wrote. "It is like listening to a radio account of a battle on the other side of the earth."

LINDBERGH LONGED FOR COMBAT with an enemy he could see. He got his wish on July 28, 1944. He was piloting one of eight P-38 Lightnings taking off from Biak, a small island off the north coast of mainland New Guinea, on a mission to bomb and strafe "targets of opportunity"—i.e., Japanese planes parked on runways—on Ceram, an island 500 miles to the west. For Lindbergh, the mission didn't start promisingly. He was slow retracting his wheels after takeoff—so slow that one of his fellow pilots sent him a radio message heard by everyone else on the mission: "Lindbergh from Doakes," the pilot said into his transmitter. "Get your wheels up! You're not flying the *Spirit of St. Louis!*" (*Spirit*'s wheels weren't retractable at all.)

When Lindbergh's team got to Ceram, they descended through cloud cover only to find the Japanese runways empty. Disappointed, they turned back. But as they circled over Elpaputih Bay, on Ceram's south coast, they learned from radio transmissions that Japanese aircraft were headed their way—chased in that direction by other American fighter planes. Those American pilots had used up all their ammunition, however, so it was up to Lindbergh's team to take the enemy on.

By now Japanese soldiers on Ceram were aware of the American planes above them, so they started firing their antiaircraft guns, loudly and often. After emerging from a thick black cloud of gunnery smoke, Lindbergh got his first sight of a Japanese plane in the air—a Mitsubishi Ki-51—flying directly at his nose at nearly 300 miles per hour, the same speed Lindbergh was flying at it. This head-on pass, Lindbergh knew, was the most dangerous encounter one could have with an enemy fighter plane. Like it or not, he was in a deadly game of "chicken," contested thousands of feet up in the air with twenty-millimeter cannons.

Lindbergh and his opponent started firing. "[He] flies directly at me. I hold the trigger down and my sight on his engine as we approach head on," Lindbergh later wrote. "My tracers and my 20's splatter on his plane." The smell of gunpowder filled his nostrils; the sound of his mechanized cannons was deafening. Both planes continued on their collision course. "We are close—too close—hurtling at each other at more than 500 miles an hour." Then, a split second before impact, Lindbergh pulled back on his controls. So did his opponent. "Will we hit? His plane, before a slender toy in my sight, looms large in size. A second passes—two—three—I can see the finning on his engine cylinders. There is a rough jolt of air as he shoots past behind me. By how much did we miss? Ten feet?"

He would never know. "There was no time to [think] or even be afraid." Lindbergh scanned the sky for aircraft. "There are only other P-38s and the plane I just shot down." That plane, with its pilot still inside, "turn[ed] slightly as it pick[ed] up speed—down—down—down toward the sea." Then "a fountain of spray—white foam in the water—waves circling outward as from a stone tossed in a pool." Seconds later "the foam disappeared" and the ocean, having swallowed its prey, was "just as it was before." That night, after the eight P-38s returned to base, word spread quickly from one lantern-lit tent to another on Biak: "Lindbergh got a Jap."

Three days later one almost got him. Ironically, the day started with the cancellation of the scheduled mission—another strike against Ceram— because of bad weather. The only Japanese-held area with clear skies overhead, Colonel Charles MacDonald learned from the military's weatherforecasting service, was over the Palau Islands, 600 miles to the north. According to air force intelligence, those islands were protected by at least 150 Japanese Zeros, the same type of plane that had destroyed the U.S. battleship fleet at Pearl Harbor.

"Want to go to Palau?" MacDonald asked Lindbergh and two other pilots, at breakfast. They took off in their P-38s just before 9:30 a.m. MacDonald's plan was to pick off Japanese aircraft in isolation there, but when none crossed their path he decided to "ring the doorbell." This he did by leading a dive over Koror, an island about halfway up the Palau chain, where their P-38s drew ack-ack fire from gunnery installations below before zooming off to the largest island in the group, Babelthuap, where the Japanese had their main airstrip.

Japanese fighter planes would be waiting for them. "Bandits, two o'clock high!" Major Meryl Smith, one of Lindbergh's team members, shouted into the radio. Lindbergh looked to his right: there they were, three of them, in Zeros. Smith, MacDonald, and the fourth American pilot, Captain Danforth Miller, took off in hot pursuit. Then, seemingly out of nowhere, a fourth Zero appeared.

At first it headed toward Smith's plane; then it suddenly changed direction. An ill-advised turn by Lindbergh had placed his own P-38 directly in the fourth Zero's sights. Worst of all, the Zero was above him. *"Zero on your tail!"* MacDonald yelled into Lindbergh's receiver. Lindbergh twisted his head back to look. MacDonald hadn't exaggerated. "I bank right," Lindbergh later wrote. "I push the r.p.m. full forward and the throttles to the fire wall. [My] propellers surge up past 3,000. . . . But he has the altitude advantage."

Lindbergh knew he was in mortal danger. The Zero "is closing in on my tail. I am not high enough to dive." All Lindbergh can do is "nose down a little and keep turning to avoid giving him a direct shot." Seconds pass. "He must have his guns on me now—in perfect position, on my tail." The forty-two-year-old man who, just a few years earlier, had searched for immortality, "commended [his] soul to God" and prepared to die:

> I hunch down . . . and wait for the bullets to hit. I think of Anne— of the children. My body is braced and tense. There is an eternity of time. The world was never clearer.

But, amazingly, "there is no sputtering of [my] engine, no fragments flying off [my] wing, no shattering of glass on the instrument board in front of me." Somehow, the Zero pilot missed the helpless target right in front of him, and was chased away by MacDonald, Smith, and Miller, then returning from destroying the other three Zeros. At 4 p.m., the four exhausted Americans landed on Biak.

—

LINDBERGH HAD OTHER CLOSE brushes with death in the South Pacific, though none as personal. Biak was held by the Japanese for two years before the Americans won it back in the spring of 1944, just weeks before Lindbergh's arrival, after a fierce struggle. One morning an air force officer took Lindbergh to see the former Japanese fortifications on Biak, many of them in caves on the western side of the island. Japanese bodies were everywhere, sprawled in positions that only bodies mangled by an air strike can take. Ants had eaten much of the flesh, leaving behind sun-bleached bones and the overwhelming stench of death. Many of the heads on those corpses had been smashed in, apparently by rifle butts.

"I see that our infantry have been up to their favorite occupation," the air force officer said. That "favorite occupation" was knocking out Japanese soldiers' gold-filled teeth for souvenirs. Another Japanese infantryman nearby hadn't been so lucky. Lindbergh found his body lashed upright to a post, outside a cave—the soldier's head had been cut clean off. A few feet away a bomb crater was filled with garbage. Mixed in with the refuse were other Japanese corpses. American soldiers had dumped trash on those dead men. "I have never felt more ashamed of my people," Lindbergh wrote in his journal that night.

Even so, there's no denying that Lindbergh enjoyed his time in the South Pacific—the camaraderie with his fellow pilots most of all. They were inextricably linked, he realized, by their mortality. "We are . . . constantly in the presence of death," Lindbergh wrote. "We live to pounce upon our enemies and to be pounced upon by them. . . . Like rabbits playing in a clearing, . . . we are unaware of the circling hawk until it swoops. . . . We feel only the companionship of each other."

In a letter to Anne, Lindbergh enclosed a photograph of himself, standing next to four young American fliers. The smile on her husband's face brought a smile to Anne's as well. Since the murder of their first child in 1932 she hadn't seen that smile very often. Looking at the photo she knew that, in the South Pacific, Lindbergh had finally regained the happiness he'd experienced in the old days—the really old days, long before the flight to Paris that had changed Lindbergh's life forever: the days of barnstorming flights through the South and Midwest, of training as an air cadet in Texas, of flying the mail from St. Louis to Chicago. Now—decades after he was sure he'd never experience those feelings again—her husband had finally rejoined the wild and free fraternity of the skies. Anne was happy for him.

She was even happier to learn that he was coming home. But she had news, too: the owners of the house in Bloomfield Hills couldn't renew the Lindberghs' lease, so the family had to move by September 1. Lindbergh expressed regret about leaving this burden on Anne alone but, as he wrote in a return letter, "there is no one I trust as much" to make the right decision. The family could stay in Michigan, if Anne liked, but Lindbergh told her that he thought his work at Ford was over. "I have done about all I can there," he wrote. "The [issues] there are now primarily connected with mass production, . . . which they understand far better than I." One possibility on his mind was rejoining United Aircraft, which was continuing to develop high-altitude planes in Connecticut. But he made it clear to Anne that she should choose a home in whatever part of the country she liked, and he would happily join the family there.

She chose Westport, Connecticut, where she found a large house nestled among trees, set far back from the road, and close to Long Island Sound. It was a choice that pleased her husband. United Aircraft was just a short drive away. After Anne took possession of the house, she parked the children at Next Day Hill and set about furnishing it.

On September 20, 1944—after fifty combat missions in the South Pacific; a flight from Hawaii to San Diego; another flight from San Diego to Pittsburgh, followed by a night train from Pittsburgh to New York and a commuter train from Grand Central to Westport; and, last, a ride from the Westport train station with a taxi driver who had to stop twice to get directions to the house—Charles Lindbergh was home from the war he'd fought so hard to keep America out of, and then fought so bravely once he was in it.

CHAPTER 20

THE GRANDEUR OF HIS LIFE

SEVEN MONTHS LATER LINDBERGH WAS IN A PLACE HE NEVER expected to be again: the War Department, in Washington. Even more to his surprise, he was being invited to serve his country. President Roosevelt was dead; the new commander in chief, Harry Truman, had no feud with Lindbergh. With Allied forces closing in on Berlin, it was clear the European conflict was nearly over. Though the Germans were on the edge of ruin, their skills in jet propulsion and rocketry were undeniable. This expertise wasn't merely respected by the U.S. military; it was coveted.

The hot war in Europe was about to end, but it appeared a cold war was about to start. A technical mission was being organized at this meeting, held on April 19, 1945, to address that likelihood. Its objective was to go to Europe to study the Germans' aviation expertise and recruit a few German scientists to share their knowledge with the United States after Germany's inevitable defeat—and not with the Soviet Union. There was a place on this mission for Lindbergh if he wanted it. He did, but not just to serve his country. Lindbergh had a personal mission: he wanted to learn all he could about the final days of Alexis Carrel.

The Germans surrendered by the time Lindbergh and eleven other mission personnel took off from Washington on May 11, 1945. After crew changes in Newfoundland and the Azores, their navy transport plane moved on toward Paris. Just after sunrise on May 13, Lindbergh—asleep on the plane's floor—was awakened by a series of shouts. He saw that his fellow passengers were standing, their heads pressed against windows, talking excitedly about what they were seeing below in the soft morning light. Lindbergh wiped the sleep from his eyes and joined them. It was the abbey of Mont-Saint-Michel they were admiring—rising, in all its architectural glory, from the sea. Lindbergh's still groggy mind tried to process what he was seeing behind the abbey.

"The lush green of spring crops, the deeper green patches of forest, the brown of lately worked fields. It was the coast of Brittany!" he wrote in his journal. "If only I had woken sooner, I might have seen Illiec and Saint-Gildas." But as his head cleared, Lindbergh realized the transport plane was too far south.

Hours later Lindbergh was in a hotel room in Paris. He tried to phone a friend of Carrel's, John B. Robinson, an American attorney in the Paris office of Frederic Coudert's law firm, but there was no directory in his room. So he walked down to the lobby, where he was handed one dating from before the war. Instinctively, he flipped first to the "C" pages. There it was: Alexis Carrel, 5 rue Geo.-Delavanne—Invalides 73 05. "I stand looking at it," Lindbergh wrote. "The name, the number, just as it had been before the war. I wanted to pick up the phone and call, to hear his voice at the other end, the precise French accent, the dignity, the warmth of welcome":

> But he is dead, killed by the unfairness of war, by the false accusations of men who never made a fraction of the sacrifice he did for his country. Possibly Mme. Carrel is there. But I did not dare to call her. With the extreme political antagonisms that exist in France, I must inquire about her situation before I phone. A call from me might cause her great trouble.

Lindbergh finally reached Robinson, who said he'd see what he could learn about Madame Carrel's situation; hours later he had disappointing news: she had sailed for the United States that morning. "Six years away," Lindbergh wrote, "and I miss her by a few hours!"

As Lindbergh passed the days in Paris reading intelligence reports, he began to fear he'd fail in his quest to learn about Carrel's end. Then, less than forty-eight hours before the mission team was to leave for Germany, he heard from Robinson. The attorney had found a woman who was with Carrel almost up to the moment of his death. She was the duchess de Chaulnes, a member of one of France's oldest families, as evidenced by her splendid home on avenue d'Orsay, where an eighteenth-century ancestor had once entertained King Louis XVI.

"*He was innocent!*" the duchess said of Carrel, seconds after Lindbergh entered her parlor. "We must clear his name of all the charges that were made." The duchess, a rapid-speaking woman in her sixties, handed Lindbergh a sheet of paper. "Here they are, translated into English." Lindbergh read the page and was shocked: the most serious accusation referred to Carrel's visit to the German embassy in Paris when a party was under way. A party? Surely a great man's reputation wasn't ruined by something as trivial as that. But so it was, the duchess said—an injustice all the more deplorable because Carrel's visit to the embassy wasn't even social: it was about finding a home in Paris for Carrel's foundation, and the visit occurred only because the German ambassador had summoned Carrel to report on his progress. Carrel hadn't known a party would be going on, the duchess insisted, and he had left immediately after speaking with the ambassador.

After wiping her eyes with a handkerchief, the duchess described Carrel's final days. He was tended around the clock by doctors from his foundation, who showed ceaseless devotion to their leader. But it was hard to find proper food, and Carrel became weaker and weaker. This was a horrible irony, she pointed out, because Carrel's enemies had accused him of accepting extra food from the hated Germans. Though his body deteriorated, Carrel's mind remained clear, making the situation even worse: Carrel knew of the baseless charges being made against him. The injustice of it all sapped his will to live. Carrel died less of a heart attack, the duchess said, than of heartbreak. His body had been interred in a temporary vault at Père Lachaise cemetery in Paris, then transported to Saint-Gildas.

AFTER MEETING THE DUCHESS, Lindbergh borrowed an old Renault and took a lonely drive from Paris to the Côtes-du-Nord to pay his last respects to the man who had meant more to him than any other except his father. Shell-pocked walls lined the narrow roads through the provincial towns Lindbergh passed on his way to Port-Blanc, the village on the Breton coast across from Saint-Gildas. Parking his car there, Lindbergh's mind, he wrote in *Autobiography of Values*, went back to the night nearly ten years earlier when, at that same spot, he pulled a rubber raft out of the trunk of a French taxi, inflated it, and then rowed, in the darkness, to Saint-Gildas, where the Carrels were asleep. This time there was daylight and the tide was out, so he walked. The tidelands, he was delighted to see, hadn't changed: small fish still got trapped in tiny pools of water, and the local fishermen's boats, looking comically useless, sat on the sea bottom.

He walked first to his own former home, on Illiec. There was no structural damage to his house, but the surrounding land had suffered: most of the island's trees had been cut down for firewood. Not a soul was there. Lindbergh expected Saint-Gildas to be equally deserted, but found, once he walked over, that Madame Carrel had turned the island over to an order of monks who now lived in the old farm buildings. The Carrels' house had been reserved as a retreat for scientists and scholars looking for a place to study and think, away from the bustle of Paris or Lyon. It had been left, at Madame Carrel's instruction, exactly as it was when the Carrels lived there. The Carrels' Breton maid, Antoinette, had been retained to make sure of that. She was surprised and delighted when she opened the front door after Lindbergh knocked.

After a halting conversation—Lindbergh had never learned French—Antoinette took her old friend to see Carrel's final resting place, inside the dirt-floored, ivy-covered stone chapel where Lindbergh had seen Madame Carrel go off to pray so often in years past. The tiny sanctuary was barely large enough to contain Carrel's grave, which lay in front of the wooden altar, covered by an oblong patch of sand, on which Antoinette placed fresh-cut flowers every morning.

Standing there, alone, Lindbergh felt his mind flooding with memories. He thought of his final encounter with Carrel, in January 1941, at Carrel's office at the Rockefeller Institute.

"It was a solemn meeting," Lindbergh wrote. "We, like other men, felt channeled into destinies beyond our understanding. Carrel, at the age of sixty-six, . . . decided to give up his [medical] research [in the United States] and return to his native France. 'My country is in trouble,' he told me." For Lindbergh, Carrel's decision showed courage and patriotism. Others, he knew, saw it differently. The two friends embraced when they parted that winter day in New York. They knew they were living in a time of war, uncertainty, and loss. What they didn't know was that they would never see each other, speak to each other, or even exchange letters ever again.

Lindbergh hadn't merely lost a mentor; the world, he believed, had lost a great man. The monumental scope of Carrel's life, Lindbergh wrote of this moment, filled his consciousness. He thought of Lyon, where Carrel mastered the vascular-suturing technique that changed surgery forever; of Stockholm, where he accepted his Nobel Prize; of Compiègne, in northern France, where he devised a method to stop infection in the wounds of soldiers during World I, one of the most important breakthroughs in the history of battlefield medicine; of Paris, where Carrel wrote *Man, the Unknown*; of New York, where Carrel and Lindbergh, working in black robes in a series of black rooms, used the insights of tissue culture and mechanical engineering to keep whole organs alive and functioning, without infection, outside the bodies that created them, for weeks at a time. This was a feat no one before them had ever achieved—one that, had their work not been interrupted, might have led to the realization of the ultimate human dream: a life without end, at least for some.

But of all the locales in Carrel's career the one Lindbergh thought of most as he stood at his mentor's grave was the windswept island where he was standing that very moment: Saint-Gildas, where he had so often walked alongside Carrel. The surgeon's windbreaker would be buttoned tightly, his beret canted down, his mismatched eyes twinkling beneath his pince-nez, as they talked about life, death, and ending the latter's dominion over the former. This picture in Lindbergh's mind was as sharp as any photograph. As waves crashed onto nearby boulders and receded back into the sea, there, in his mind's eye, they still were: two men, two friends, two partners in science, two seekers of truth. Those had been the happiest days of Lindbergh's life.

Looking at the sandy grave before him, and the wildflowers that lay atop it, Lindbergh felt that loss as intensely as he'd ever felt anything.

"Carrel's body lay a few feet beneath the surface," he wrote. "The dead body of the man whose life's work had been concerned with living bodies. The man who had preserved life in isolated organs and given immortality to isolated cells. His simple grave contrasted with the grandeur of his life."

Minutes later Lindbergh returned to the main house and said good-bye to Antoinette. If he didn't leave soon, the tide would be back, and he wouldn't be able to walk to his car. The ride back to Paris was even lonelier than the ride out.

A FEW DAYS LATER Lindbergh was in Germany for the first time in seven years. He'd written a letter to Carrel in 1938, marveling at the Nazis' "simple but beautiful" plans for the rebuilding of Berlin and Munich. Now the need to rebuild those cities was no longer a matter of aesthetics. Landing at the same Munich airport where he'd touched down in his black-and-orange monoplane in 1938, Lindbergh was shocked by the devastation he saw. Virtually no building was left standing in Munich's city center; automobile traffic was possible only because piles of rubble had been pushed off the roads by American soldiers. "It is a city destroyed," Lindbergh wrote in his journal.

After being assigned an interpreter (the marine colonel George Gifford), Lindbergh set out to find some of the leading figures in German aviation and then recruit them to the American side. The reason Lindbergh himself had been recruited for this mission was that he knew many of those men. In fact, two of his targets—Dr. Willy E. Messerschmitt, the world-renowned aircraft designer and manufacturer; and Adolf Bäumker, former chief of research for the German air ministry—had been present at the U.S. embassy in Berlin on October 18, 1938, when Hermann Göring, resplendent in a cobalt-blue uniform and a rouged face, presented Lindbergh with the Verdienstkreuz Deutscher Adler "by order of the Führer."

Messerschmitt was in formal dress that night at the embassy, but not so on May 21, 1945, when Lindbergh found him living in a shack, not far from his large country home, now liberated by the U.S. Army, in Oberammergau, some fifty miles south of Munich. Messerschmitt had recently returned to Germany after briefly being held in England as a prisoner of war. Lindbergh was surprised that the British had let such a prize slip through their fingers. In the fall of 1944 the Luftwaffe introduced the Me-262—the first turbojet fighter plane in the history of air combat, a product of six years of research and development at the Messerschmitt factory in the Bavarian city of Augsburg, much of it directed by Messerschmitt personally.

Now the father of the military jet looked like a broken man, which only intensified Lindbergh's determination to treat him with respect. After all, Lindbergh considered Messerschmitt one of the greatest minds aviation had ever produced. Lindbergh spoke at length with Messerschmitt about the German jet program. He told Messerschmitt that Wernher von Braun, the German rocket scientist, had agreed to work for the United States. This was no small upgrade in know-how for the War Department: von Braun was the primary developer of the liquid-fueled V-2 rocket, the first un-manned guided ballistic missile to be used in combat. Lindbergh urged Messerschmitt to join von Braun in coming over to the Americans, but Messerschmitt said he needed time to think. In the end, he stayed in Ger-many, where, after being prosecuted in the postwar denazification program, he was imprisoned for two years.

Lindbergh was more successful with Adolf Bäumker, whom he located in Worthsee, about twenty miles west of Munich. Working with a new translator, the army lieutenant E. H. Uellendahl, Lindbergh found Bäumker to be even more anxious than Messerschmitt about his future. A Soviet in-terrogator had called on him several days earlier, brandishing a pistol and demanding information. Bäumker was equally worried about retaliation from renegade Nazi SS troops he thought were hiding in the nearby Bavar-ian woods, unwilling to surrender and ready to kill anyone who spoke to either the Soviets or the Americans.

Even so, Bäumker was willing to help, in hopes that his cooperation would enable him to get his family to safety in the United States. Bäumker gave Lindbergh a list of scientists and engineers who had worked with him in the German rocket program. He also volunteered to write a detailed his-tory of that program, once he arrived in America. In return for Bäumker's information and promises, Lindbergh gave Bäumker all he had with him: a can of fruit juice, two cigars, two packages of cigarettes for his wife, and a candy bar for their daughter.

These meetings reinforced Lindbergh's view that Germany was just as much a part of the western family of nations as Britain or the United States. This seemed even more obvious to him when, as Uellendahl and he were driving on a rural Bavarian road, the lieutenant asked directions from a small group of German soldiers. They were "disarmed and apparently plod-ding along on their way home," Lindbergh wrote. "A half-dozen young men, courteous, giving us directions as best they could, showing no trace of hatred or resentment, or of being whipped in battle. They looked like farmers' sons." At least they did until Lindbergh was forced to confront the facts: first about Nazi Germany; then, ever so slowly, about Alexis Carrel and himself.

THAT PROCESS BEGAN ON June 10, when Lindbergh arrived in Nordhausen, a town in the Hartz Mountains 160 miles southwest of Berlin, where the Germans had built their V-2 rocket factory. In truth, the words "rocket factory" are grotesquely insufficient to describe what took place in Nordhausen between 1943 and 1945. It was a death factory for slave laborers who were trucked into the complex from the nearby Buchenwald concentration camp. Those prisoners—some 20,000 at any given moment (and more than 60,000 in the nearly three years the factory was in existence)— were housed in fifty-eight unheated barracks, surrounded by barbed wire, snarling dogs, and SS men, in a concentration camp called Dora, located just outside a tunnel leading into the V-2 factory.

That factory was built underground from a series of mining tunnels that were expanded by pickaxes wielded by Dora's prisoners. While that backbreaking work was under way, the prisoners lived in the tunnels without running water or latrines, in lice-infested bunks stacked four-high, breathing air thick with gypsum dust and fumes from the blasting work, which continued twenty-four hours a day. The SS troops beat, whipped, and shot "stragglers" who collapsed from overexertion. When construction was finished, there was a complex of some fifty tunnels, several large enough to be served by railroad tracks that carried the completed V-2 rockets out of the factory. (The laborers carried the rocket parts in with their bare hands.) More than 4,000 rockets were produced in the plant; 26,000 laborers died in the process.

Lindbergh had to drive through Camp Dora to reach the factory's entrance. "I felt I had woken in another planet," he wrote. Hundreds of emaciated men were there, all in dirty striped uniforms, many with hollowed-out cheeks smeared with caked-in grime, their eye sockets encircled with black lines left by overwork and lack of sleep. None of them, it seemed, had the strength to speak. Some were still living in Dora's foul-smelling barracks; others were unwilling to sleep there, so they'd improvised. "Dozens of [V-2] tail sections were being used by the [former prisoners] for houses," Lindbergh wrote. The sections were "laid flat on the ground, fins sticking out behind, shiny metal all over, a shelf to sleep on built across the interior like a floor." Even there, in those makeshift metal shelters, yards away from the filthy barracks, Lindbergh was overwhelmed by a stench. It was a smell, he wrote, that seemed less human than rodentlike—an unclean smell: the odor of urine, feces, sweat, disease, degradation, hunger, fear, and death.

Lindbergh entered the V-2 factory through the main tunnel, driving his jeep on the old railroad tracks. Moments later he reached a large production room, 100 feet or so below the earth's surface, "fully lighted," he wrote, "as though waiting only for a change in shift before starting operation again." Stepping out of his jeep into this ghostly place, Lindbergh found an assembly line for V-2 rocket engines—"dozens of them in the process of construction." A secondary tunnel led to a completed rocket, lying on it side. Another V-2 nearby had been cut open—dissected, really—by scientists for the Allies, who were trying to understand the rocket's anatomy.

Lindbergh wanted to follow the train tracks, but a boxcar blocked his way, so he reversed direction and walked back toward his jeep. Before he found it, he encountered the same rodentlike odor he'd smelled outside at Camp Dora. The smell told him that he was close to a small, enclosed space where inmates had been not very long ago. Soon he entered a room he hadn't seen before, where the smell was particularly pungent. It had been an office of some kind when the factory was in operation; now all the desks and file cabinets were overturned, and their contents were strewn on the floor. Thousands of identification cards—duplicates of the cards issued to the laborers—crunched beneath Lindbergh's shoes as he walked. A third of the men whose photographs appeared on those cards had died in these tunnels. The weakest among them never saw the sun again after they entered.

After spending the night in the town of Nordhausen, Lindbergh returned to the factory the next day with Lieutenant Uellendahl. Lindbergh's plan was to finish exploring the tunnels, but while they were driving through Camp Dora his attention was distracted. Once again it was by the smell, some of which seemed to be emanating from a building the two Americans hadn't noticed the day before. It was a "low, small, factory-like building," Lindbergh wrote, "with a brick smokestack of very large diameter for its height," atop a steep hill a few hundred yards from Camp Dora's barracks. There was no road leading to it, so Uellendahl shifted into four-wheel drive and drove up the hill, weaving around trees. As they got closer, Lindbergh could see a pile of equipment, stacked outside the building's entrance. When they arrived he saw that it was a pile of stretchers, "probably two dozen of them, dirty and stained with blood—one of them showing the dark red outline of a human body which had lain upon it," all of them having the same rodentlike stink.

The doors of the building were open, so Lindbergh and Uellendahl walked in. On the floor in front of them lay a corpse covered with a dirty sheet of canvas. The next room they entered contained two large furnaces built into a wall. The night before, when Lindbergh and Uellendahl were having a meal in Nordhausen with American soldiers, one of the Americans told Lindbergh that the only prisoners who left Dora before the Germans surrendered had "left in smoke." This, Lindbergh realized, was where that happened. Suddenly someone walked in, bringing more of the ubiquitous stench with him. He wore the striped uniform of a prisoner. "He is hardly old enough to call a man," Lindbergh wrote in his journal, using the present tense he favored. He is "a boy":

> The prison suit bags around him, oversize, pulled in at [the] waist and hanging loosely over [his] shoulders. He moves out of the brighter light so that I can see his face. He is like a walking skeleton; starved; hardly any flesh covering his bones; arms so thin that it seems only the skin is left to cover them.

Uellendahl spoke to the skeleton in German. Though the boy was Polish, he knew enough German to respond. He was seventeen; he'd survived both the forced labor and the deprivations of Camp Dora for three years. He pointed to the furnaces. "Twenty-five thousand [burned]," he said. He beckoned Lindbergh and Uellendahl to follow him into the other room. There he bent down and lifted the canvas off the corpse. "It covered an ex-prisoner just like himself, but thinner," Lindbergh wrote. "It is hard to realize that the one is dead, the other living, they look so much alike." The face of the living prisoner "screwed up in grief and anguish of his memories," Lindbergh wrote. "It was terrible," the inmate said. "Three years of it." He pointed a bony finger at the emaciated corpse on the floor. "He was my friend—and he [was] *fat!*"

Then the prisoner took the visitors outside, stopping a few feet from one of the building's walls. He gestured to a patch of ground colored differently from the soil around it. It wasn't soil, Lindbergh realized; it was ash. He was in front of "a large oblong pit, probably eight feet long and six feet wide, and, one might guess, six feet deep," he wrote. "It [was] filled to overflowing with . . . small chips of human bones." The prisoner walked over to a mound nearby that hadn't yet been dumped into the hole. He reached in and pulled something out that wasn't fully incinerated. Lindbergh had never taken an anatomy course, but he recognized it. It was a knee joint.

THE PRISONER CONTINUED TO speak while Lindbergh stared off into space. His mind was frozen, he later wrote, "still dwelling on those furnaces, on that body [on the floor], on the people and the system which let such things arise." It seemed impossible to him that the Nazis—with their crisp uniforms and clean streets, their love of Goethe and Beethoven, and, above all, their commitment to science—were capable of such uncivilized behavior. But there, right in front of him, was the proof. "Here was a place where men and life and death had reached the lowest form of degradation," he wrote. "How could any reward in national progress even faintly justify the establishment and operation of such a place?"

He thought of the man who had promised to deliver that "national progress" to Germany, to build a Reich that would last 1,000 years and serve as a bulwark against the threat of communism emanating from the Soviet Union. "A man who controlled such power, who might have turned it to human good, who used it to such resulting evil," Lindbergh wrote of the Führer. He remembered the one and only time he saw Hitler, at the 1936 Olympic Games in Berlin, where the Führer walked across the grass infield of the huge oval stadium to accept flowers from a young, blond German girl as 100,000 spectators cheered—nearly all of them, Lindbergh now realized, seduced and betrayed by Hitler's blend of "vision and blindness, hatred and patriotism, knowledge and ignorance."

Now, nine years later, Lindbergh saw the true legacy Hitler had left to Germany: "the best youth of his country dead; the cities destroyed; the population homeless and hungry; Germany overrun by the forces [Hitler] feared most, the forces of Bolshevism, the armies of Soviet Russia; much of his country a pile of rubble and flame-backed ruins." At Dora, Lindbergh saw the catastrophe Hitler had unleashed on the rest of the world, especially on those the Führer deemed "inferiors."

The Nazis hadn't invented evil; Lindbergh had read enough history to be certain of that. A litany of past cruelties came to mind as he stood by the ash pit: "As far back as one can go . . . , these atrocities have been going on," he wrote in his journal. "Not only in Germany with its Dachaus and its Buchenwalds and its Camp Doras, but in . . . the lynchings at home, . . . in [Russian] pogroms, . . . tearing people apart on the English racks, burnings at the stake for the benefit of Christ and God." Lindbergh thought of his time in combat in the South Pacific, where he'd seen a "load of garbage dumped on dead [Japanese] soldiers" by Americans, the shinbones of dead Japanese fighters used by GIs as letter-openers, and "Jap heads" buried in ant-hills by GIs "to get them clean for souvenirs."

Once again Lindbergh stared at the ash pit. Twenty-five thousand humans had been worked or beaten to death in the Nazis' V-2 factory; their skeletal corpses had been burned into pebbles, then shoveled into the hole in front of him. This was evil on a scale so huge that Lindbergh had trouble grasping it except as collective madness on the part of all humanity. That pit, and the scorched bone bits inside it, he wrote, were "not [a horror] confined to any nation or to any people":

> What the German has done to the Jew in Europe, we [have done] to the Jap in the Pacific. As Germans have defiled themselves by dumping the ashes of human beings into this pit, we have defiled ourselves bulldozing bodies into shallow, unmarked tropical graves. What is barbaric on one side of the earth is barbaric on the other. . . . It is not the Germans alone, or the Japs, but the men of all nations to whom this war has brought shame and degradation.

When Lindbergh's journal was published, these words were harshly criticized. To equate the organized, state-run mass murder of the Nazi concentration camps with the spontaneous brutality of the battlefield, as Lindbergh did, struck many as disingenuous. Others were even less charitable.

They saw it as yet another example of the moral obtuseness that Lindbergh had exhibited ever since he first praised Nazi Germany in 1936—and that had been at its worst five years later when he spoke against America's joining the war to stop the Nazi takeover of Europe (and the genocide that would surely accompany it), lest America make the mistake of choosing sides in a war where, according to Lindbergh, neither side had the moral high ground.

But a close reading of Lindbergh's journal entries for the period he spent in Germany after the Nazi surrender offers another interpretation of his moral awareness in the spring of 1945. This reading finds signs of a nascent consciousness in Lindbergh's mind that he had played a part as an individual in encouraging the moral deficit that made places like Camp Dora possible—and had done so long before he accepted the fateful invitation to visit Berlin in 1936. The evidence for this growing self-awareness lies in an unusual place: the repeated references made by Lindbergh in his journal to the rodentlike stench emanating from the prison camp inmates and other displaced persons whom he encountered during his travels through Germany for the technical mission.

At first, Lindbergh's comments on the smell of those unfortunates appear cruel and devoid of human sympathy. He seemed to be offended by their filthy, malodorous state and convinced that the prisoners' stench reflected an ugly internal truth about them: their degradation was a personal choice, Lindbergh's contempt implied, a sign that they'd willingly surrendered their human dignity, or, even worse, never had any dignity to begin with.

But after leaving Dora, Lindbergh came to understand the truth: the rodentlike stench he encountered there said nothing about the character of the camp's victims, and everything about the character of the victimizers. And the reason why he understood this was a sudden recognition about the stench that so disgusted him. Lindbergh realized he had smelled this odor before: in a sanctum of learning where he'd spent the most stimulating years of his life.

More specifically, he had smelled it in a small room inside the sanctum that was a kind of concentration camp in miniature—but not a place of punishment, a place of science. It was a place where, in the name of eugenics, the strong were allowed and even encouraged to dominate the weak. It was an environment where the "false" limitations of morality were irrelevant, as they are in nature. It was a place where natural selection enforced its will through intimidation, strength, violence, and even death. It was a place that, according to its creator, served as a living metaphor for how the "great races" were developed among humans: "The weak and the fools were not capable of surviving," that creator told Lindbergh, "but the strong and the persistent were: they became the great races of Europe." It was a place where, on many mornings, Lindbergh had stood watching the "pure" mechanism of natural selection at work—often seeing the "fit" eliminate the "unfit" by killing them—never wavering in his belief that his nonintervention in those encounters embodied the highest morality of all: the morality of science. It was a realm where the strong procreated and the weak died, without any coddling. And it was a place, Lindbergh noticed, where the smell of fear, cruelty, and death were ever-present.

Now, years later and thousands of miles away, Charles Lindbergh realized that the first place he had smelled that odor wasn't in Camp Dora, or any other place in Germany. It was in Alexis Carrel's mousery at the Rockefeller Institute.

CHAPTER 21

I FELT THE GODLIKE POWER

THE SIGHTS AND SMELLS OF CAMP DORA LED LINDBERGH TO examine his life with an urgency he hadn't felt since his epiphany seventeen years earlier in the Utah desert. Though the experiences were alike emotionally, they were totally dissimilar intellectually: in Utah Lindbergh committed himself to science and a quest for immortality; in Camp Dora he began to wonder if that vow hadn't made him complicit in a brutally efficient culture of death.

The second period of introspection began after Lindbergh returned from Germany in June 1945, and reached a crisis a few weeks later. On August 6, the atomic bomb, a weapon that, in the words of President Truman, "harness[ed] the basic power of the universe," was dropped by *Enola Gay*, an American B-29, on the Japanese city of Hiroshima, flattening five square miles and leaving approximately 100,000 people dead in a fiery instant. *Enola Gay*, Lindbergh saw with horror, was powered by four Wright Cyclone air-cooled engines—a larger version of the motor that had powered *Spirit of St. Louis* across the Atlantic.

The link between *Spirit* and *Enola Gay* gave Lindbergh pause, but it was the science behind the new weapon that he couldn't stop thinking about. History showed that man has "devised no weapon so terrible that he has not used it," Lindbergh wrote to a friend after Hiroshima. "With the key of science, man has turned loose forces he cannot re-imprison."

It's not surprising Lindbergh would restate the moral taught by the tale of the sorcerer's apprentice. The day after the first A-bomb was dropped, Lindbergh drove from his home in Westport to the Rockefeller Institute. There, for more than an hour, he opened boxes in a stuffy basement room where Carrel's papers and equipment had been stored after his death. While looking through his mentor's materials, Lindbergh began to wonder if he hadn't been a sorcerer's apprentice himself.

Several of those boxes contained Lindbergh's organ perfusion pumps. He removed one from its cardboard enclosure and used a handkerchief to wipe off a layer of dust. He held it in his hand, this machine that transformed Le Gallois's organ perfusion manifesto of 1812 from wishful thinking into fact. He then placed the device, which had once graced the cover of *Time*, on one of Carrel's black surgical tables—now stored in the same room—and walked around it, marveling again at the glassblowing expertise of his former coworker Otto Hopf. The efficient design of his perfusion machine still gave Lindbergh pride, but he had new doubts about the purpose of the work he'd undertaken with Carrel. The dropping of the atom bomb, Lindbergh wrote, put man "in the position of having challenged God"; now Lindbergh wondered if Carrel and he hadn't committed the same act of hubris.

Seeing the realities of Nazism firsthand in the summer of 1945 made Lindbergh uneasy about the elitist nature of the quest he'd begun with Carrel in late 1930. Though that quest started with the purest of motives— helping his sister-in-law—it soon evolved into something quite different. The future of the white race, the future of all western civilization, Carrel told him, was threatened by faster-breeding racial inferiors. (And after Lindbergh's trips to China and India in the 1930s the protégé was convinced his teacher was right.) Conventional politicians, Carrel said, were too weak, too foolish, and too shortsighted to deal with that crisis. It was left to scientists—scientists such as themselves—to ensure the continuing hegemony of the civilization-building white elite.

But how different was that ideology, Lindbergh began to wonder, from the one preached by the Nazis? Hadn't Hitler, in the name of saving western civilization, left it in ruins all over Europe, leaving behind death factories and slave-labor camps in its stead, places of unspeakable horror built in service to "science" and a racial utopia? Such questions surely intensified in Lindbergh's mind when he crossed the storage room and opened another series of boxes. Even before seeing the contents, he knew what was inside. It was Carrel's mousery, dismantled into its component parts, the rodents who once lived and fought inside it long dead, their usefulness to science now over. Though the mousery had been wiped clean before it was packed up, its aroma remained. For a sickening moment the odor transported Lindbergh back to Camp Dora.

The next morning he wrote to Frederic Coudert and Boris Bakhmeteff, urging that a "Carrel Association," cochaired by the three men and Madame Carrel, be established to find a proper repository for Carrel's papers and equipment. A few weeks later Lindbergh supervised the packing of Carrel's property, which included a 3,000-year-old Egyptian mummy the surgeon had tried to revivify in 1925. ("A small hole was made in the abdomen of the mummy about 3 cm. from the right iliac spine. The skin was hardened and very tough," Carrel wrote of his failed experiment.) All told, fifty-eight wooden crates were transported—temporarily, Lindbergh hoped—to a warehouse in Manhattan.*

THE BIRTH OF THE Lindbergh's second daughter—Reeve, on October 2, 1945 (Anne Spencer Lindbergh, called "Ansy," had been born five years earlier)—and the decision to give up the lease on the house the family was renting in Westport and purchase one in nearby Darien, caused Lindbergh to take a break from his self-examination. The new home, a three-story Tudor with seven bedrooms and, much to Lindbergh's delight, a screened-in sleeping porch just like the one in his childhood home in Little Falls, sat on a four-and-a-half-acre promontory jutting into a part of Long Island Sound known locally as Scott's Cove. From his new living room Lindbergh

---

* In August 1953, at the urging of Joseph T. Durkin, a Jesuit scholar at Georgetown University, Carrel's papers were moved to Georgetown's library, where they remain today. Father Durkin's monograph, *Hope for Our Time: Alexis Carrel on Man and Society*, was published in 1965.

could see a sprawling green lawn that led right to the water, where, less than 100 yards away, rose three tiny islands, connected to each other, and to the Lindberghs' backyard, at low tide by a sandbar. It reminded him of Illiec.

Though Lindbergh was at home in Darien for most of the fall, he instructed his secretary to inform callers that he was out of the country. He spent nearly all his time in his first-floor office, thinking about science and politics—most of all about the catastrophe that nearly always ensues when the two are mixed. On December 17 he spoke on this topic for the first time in public. The occasion was a dinner given by the Aero Club of Washington to mark the forty-second anniversary of the Wright brothers' flight. The pilots, politicians, and aviation executives in the audience were surely expecting a poetic tribute to Wilbur and Orville Wright, but Lindbergh had his own agenda. He told his listeners that man has "misused his inheritance from those pioneers." The Wrights' plane "held the magic of great construction in peace or great destruction in war. That we have used this inheritance primarily for war," Lindbergh said, "is a responsibility which rests squarely on our shoulders."

After his speech, Lindbergh returned to his book-filled office overlooking Scott's Cove to think about the tense new postwar world he found himself living in, a world that seemed to be hurtling toward another war, this one more "scientific" and "evil" than any that preceded it. His children were trained not to enter his office when the door was closed. Anne followed the same policy; she knew how much her husband valued his solitude. She also knew that his current period of self-analysis was causing him more pain than he cared to admit. Lindbergh finally emerged from his internal exile with a 12,000-word essay criticizing not only the excesses of science, but his own role in advancing that cause—including his support of National Socialism. Augustine had his *Confessions*; for Lindbergh it was *Of Flight and Life*, published by Scribner in August 1948.

This slender book, Lindbergh wrote in the preface, marked the third time he felt "an overwhelming desire to communicate [his] beliefs to others." On the first occasion, after his flight to Paris, his goal was to encourage America to "lead the world" in aviation. A decade later he felt compelled to argue that "it was best for America to keep out of Europe's internal wars." Now Lindbergh was urging Americans to break free from the "grip of scientific materialism," lest it lead them, shackled and helpless, to "the end of our civilization." The choice facing America, Lindbergh wrote, was as simple as it was stark: "If we do not control our science by a higher moral force, it will destroy us."

For proof he pointed to Germany. "Few nations contributed more to our civilization in the past—in art, music, religion, philosophy, science—in science above all in modern times. Millions of Germans devoted their lives to the discovery and development of scientific knowledge":

> With science, they felt, they could be supermen; they could rule the earth. . . . Instead of balancing science with other fields of wisdom, they let it dominate them, turned it loose in war and conquest. In their search for materialistic power, they set up science as their god, and science destroyed them.

What worried Lindbergh in 1948 was that he saw the same unbalanced thinking in the United States. "Scientific man," he wrote, has taken up residence in America, where he has "enthroned knowledge as his idol . . . and begun a ceremonial dance to which there is no end." Every advance in science, Lindbergh said, "demands more science to maintain it, more to improve it, more to keep it ahead of its use by our enemies. Scientific man is driven faster and faster by his system until he has no time left, no thought left, no appreciation left, for man himself."

When he was writing of "scientific man," Lindbergh was writing about someone he used to be. "I grew up a disciple of science," he wrote. "I know its fascination. I felt the godlike power man derives from [science,] the strength of a thousand horses at one's fingertips, the conquest of distance through mercurial speed, the immortal viewpoint of the higher air. . . . [S]cience was more important [to me] than either man or god. . . . I worshipped science."

Lindbergh wasn't merely confessing; he was repenting. "I have seen the science I worshipped, and the aircraft I loved, destroying the civilization I expected them to serve." He would worship it no more. "How can one work for the idol of science when it demands the sacrifice of cities full of children, makes robots out of men, and blinds their eyes to God?"

That false idol, Lindbergh knew, could even seduce scientists into the ultimate blasphemy: thinking of themselves as gods. Though he made no specific reference in *Of Flight and Life* to Carrel or to their research on perfusion, it seems certain that Lindbergh was remembering that work when he wrote, "By worshipping science, man becomes enmeshed in the complication of his own ideas and creations. Man sets up a system around his theories to compete with the divine plan." The theories that Carrel and he had advanced at the Rockefeller Institute, Lindbergh was reminding himself, had challenged the most inviolable rule of the divine plan: mortality.

The political ramifications Carrel and Lindbergh hoped for from their quest, Lindbergh now realized, were equally misguided. "Man has never been able to find his salvation in the exact terms of politics, economics, and logic," he wrote. "From Plato's *Republic* [on,] his planned Utopias have not proved the answer." This, he said, is because the answer lies at a much deeper—or higher—level. "Our only salvation lies in controlling science by a philosophy guided by the eternal truths of God." And of all those truths, the most eternal, and the most divinely ordained, is death.

Of Flight and Life sold out its 10,000-copy print run on the day it was published. Another 40,000 copies were hastily printed, earning Lindbergh a place on the *New York Times* best-seller list. All wasn't forgotten, but it was apparently forgiven. The man who was blackballed from officially serving in the U.S. military in January 1942 was invited in January 1950 to Arlington National Cemetery—the American military's holiest site—to be a pallbearer at the funeral of his onetime commanding officer in the army air corps reserve, General Henry H. "Hap" Arnold. Months later the school board of Little Falls, Minnesota, wrote to say that it had just named its newest educational facility Charles A. Lindbergh Elementary School.

The twenty-fifth anniversary of Lindbergh's flight to Paris, in May 1952, caused a further upsurge in Lindbergh's popularity, all the more so because Lindbergh politely declined to take part in any of the festivities, thus reminding his admirers of the modesty he had displayed in 1927. His heroic voyage across the Atlantic became even more of a national conversation topic in September 1953, when Scribner published his updated, fuller account of the flight: *The Spirit of St. Louis*, which sold several hundred thousand copies and was later turned into a Hollywood film starring James Stewart. Lindbergh wrote *Spirit* because he'd always been dissatisfied with his first account of the flight, *We*, which he wrote in less than three weeks in 1927 to fulfill a contract with the publisher George Putnam.

Lindbergh's return to glory was capped by two events in 1954. In February President Eisenhower nominated Lindbergh for appointment as a brigadier general in the U.S. Air Force Reserve. As the *New York Times* noted in its front-page story on the appointment, this was a significant turnaround for Lindbergh, who had resigned his military commission in anger fourteen years earlier because he believed that President Franklin Roosevelt had impugned his patriotism—a resignation that led much of the public to think Roosevelt was right. Then, on May 3, came more good news. The trustees of Columbia University announced that Lindbergh had been awarded the Pulitzer Prize for biography for *The Spirit of St. Louis.* "Boom days are here again," Anne wrote in her diary, after the news of the Pulitzer reached the Lindbergh home in Darien. "The Great Man—the Great Epic—the Great Author, etc, etc. [We] are living in the aura of 1929 again," 1929 having been the year of the Lindberghs' marriage.

But there was a difference. Despite all the praise, despite all the awards, and despite all the public displays of reverence—or maybe because of them—in 1954 Charles and Anne Morrow Lindbergh were unhappier, together and individually, than they'd ever been before.

THE WORLD HAD MADE peace with Lindbergh, but Lindbergh hadn't made peace with himself. The period of self-scrutiny that followed Lindbergh's encounter with the horrors of Camp Dora, as well as with his own scientific past at the Rockefeller Institute, was followed by a much longer period of rootlessness reminiscent of Lindbergh's childhood, when his unhappy mother moved him from Minnesota to California to Michigan to Washington, D.C., and then back to Minnesota—a cycle she repeated more than once. This time, however, it wasn't Lindbergh's unhappy mother who was the prime mover behind all the motion. It was Lindbergh.

Anne and the children didn't realize it right away, but it soon became obvious that Lindbergh had purchased their new home in Darien so that he would have a safe place to park them while he traveled the globe, virtually nonstop, as a consultant to Pan American Airways. Lindbergh's inspection tours for Pan Am led him to spend several months of 1953 island-hopping in the Caribbean and an even longer stretch of 1954 in South America. His biggest traveling year of the decade was 1955; he made eleven trips across the Atlantic and back, one round-trip over the Pacific, and several north-south trips within the Americas in just over ten months. It is a sign of Lindbergh's alienation during this period that all of those trips entailed participating in activities he loathed. When he was "flying" for Pan American, Lindbergh rarely entered the cockpit; instead, he sat in coach, a place where he felt less like a passenger than a package, and wrote detailed critiques of Pan Am's meal service and the merits of hot versus cold towels. Certainly that wasn't the future he had in mind when he pledged his life to aviation thirty years earlier.

Lindbergh's absences were always stressful for Anne, who was left behind in Darien to run the household with the help of servants. But those absences were especially painful for her in 1955, because that was the year of Anne's greatest professional success—yet her husband was nowhere to be found.

That success was *Gift from the Sea*, a collection of essays inspired by shells Anne found on the beach, many of them on Captiva Island, off the west coast of Florida, a place Anne had begun visiting in the late 1940s, often by herself. Those shells came to symbolize different stages of life for Anne—and, she suspected, for other women as well. A delicate bivalve shell called double-sunrise made her think of relationships. The two sides of the shell were marked by the same pattern, "translucent white, except for three rosy rays that fan out from the golden hinge binding the two together"—but each had its own function. In her own marriage, Anne realized, those separate but unequal functions had become calcified: her husband was active, she was passive; her husband was the critic, she was the criticized. Later, a bed of oysters prompted her to write about home and children. A meditation on an argonauta, a rare transparent mollusk shell, led Anne to conclude, "Woman must come of age by herself." Marriage and motherhood are milestones in a woman's life, she wrote in *Gift from the Sea*, but the true path to personal fulfillment is through solitude and quiet contemplation.

Gift from the Sea sold 600,000 copies in hardback, making it the number one best-selling nonfiction book of 1955; then it sold 2 million more copies in paperback. Clearly, Anne had given birth to one of the greatest publishing successes in the twentieth century, but her distress over her husband's frequent absences made her feel like a fraud. "There is a terrible irony in *Gift from the Sea* heading the best-sellers week after week, preaching 'Solitude!

Solitude!'" she told her diary. "Here I am, *having just what I say I want* & it does not seem to be the answer." Maybe, she worried, "the book [is] all 'hooey.'"

Making Anne's anxiety even worse was the fact that she was often un-happy when her husband *was* home. He would spend far more time than she liked in his office, behind a closed door, going over insurance policies, bank statements, and similar documents. But Lindbergh's greatest scrutiny was reserved for the detailed accounting books he required Anne to keep, a place where every household expenditure, from $300 for a winter coat to a dime for a box of rubber bands, had to legibly entered, and then, at the end of the month, transferred to larger balance sheets.

Her husband never left that room without a list of tasks for everyone in the house to perform, which he'd carry to meals and bark out to Anne and the children as if they were living in a military barracks. Only the three youngest children were living at home in the late 1950s (Jon was a married marine biologist in California and Land was studying at Stanford) so it was up to them—Ansy, Scott, and Reeve—to adjust to their father's comings, goings, and mood swings. It was certainly never dull having him around: he taught each of them to swim, sail, fish, and hike (somehow "hidden elves" nearly always remembered to leave candies along the trail for the youngest children to follow); rigged up a trapeze in the backyard; and urged them all to write stories and poems, which he took great pleasure in reading.

But there was a downside to all that activity, even in the best of times. Lindbergh insisted that his children join him in his diversions whether they were in the mood or not. As a result, some of the presumed "fun" was neither joyful nor spontaneous. ("This family is not a democracy. It's a benevolent dictatorship," he'd explain.) He censored his children's reading material—no comic books were allowed in the house—and he had strict rules about what they could eat: no candy, except for that left by the "elves." When he was happy with his children he gave them hugs, but no kisses. This was an inviolable rule: too much affection, Lindbergh believed, would make his children needy and less independent, a state of being he revered above all others.

Lindbergh was respected by his family—and loved as well—but there was no denying that he was a changed man in the 1950s. Whereas mealtime used to be an occasion for lively family chatter, now a typical dinner-table conversation topic was the possibility of a nuclear war with the Soviet Union. Because he was away so often, Lindbergh used mealtime to instruct his family on what he expected them to do if a war broke out while he was traveling. Scott's Cove was safe, he said, but New York must be avoided at all costs, because "it was a key target, and vulnerable both by air and sea." Lindbergh also ordered his family to cancel all trips to the dentist once hostilities began because the Soviets would surely tamper with the water supply. When Anne repeated some of these stories to her mother, Betty Morrow said that Charles was a "madman" in need of a steady job.

What Anne didn't tell her mother was that there was little or no respite for her in her private time with her husband. In their bedroom Charles berated his wife for seeing a psychiatrist, something Anne had begun doing in 1952. (The psychiatrist was Dr. John Rosen, who had achieved some success with his treatment of Anne's brother, Dwight Morrow Jr., for schizophrenia and hallucinations; he also treated Ansy for bouts of depression.) Lindbergh, Anne realized, had no model of a happy domestic union to draw on for guidance: his parents had lived apart for virtually their entire marriage. Charles's mother, Evangeline, certainly didn't encourage her son to "open up" emotionally: she died in September 1954, apparently without ever discussing her troubled marriage with her son. That Anne was now discussing her most intimate personal and marital issues with a complete stranger filled Lindbergh with anxiety. In fact, he moved out of the master bedroom in Darien for a while, and for one two-month period he stopped speaking to Anne altogether.

Maybe it's not surprising, then, that Anne looked elsewhere for comfort: in an affair with Dana Atchley, her physician. The two had met in 1946, when Dr. Atchley treated Anne for gallstones. He had a difficult marriage, too, and each found the other to be the soothing listener neither had at home. At first they exchanged phone calls, then long letters. By 1956, Atchley—whose other patients included Katharine Hepburn and Greta Garbo—told Anne he was in love with her. Anne wasn't ready to end her marriage, but she was ready to begin an affair. She rented a pied-à-terre on East Nineteenth Street in Manhattan, where she met Atchley for martinis, conversation, and an understanding she rarely encountered in Darien.

Whether her husband knew of this arrangement is a mystery to this day. Apparently he was similarly distracted. Almost fifty years later, in the summer of 2003, Astrid Bouteuil, then forty-two, announced that she was Lindbergh's daughter, the product of a seventeen-year-long affair between Lindbergh and Bouteuil's late mother, a milliner from Munich named Brigitte Hesshaimer, that produced two other children. All three children, said Bouteuil (who'd married a Frenchman), knew the lanky balding American as Careu Kent, a "family friend" who "visited often, could wiggle his ears, and cooked a mean omelet." (DNA tests performed at the University of Munich were said to confirm that the three children were Lindbergh's; it was also reported, in subsequent articles, that Lindbergh fathered two more children with Hesshaimer's older sister, Marietta.)

What we know for certain is that in late 1958, Anne decided to end her affair, but not her friendship, with Dr. Atchley and return to the marriage she often considered a sham. She told her sister Constance, one of her main confidants now that their mother had died (in January 1955), that Charles and she were "badly mated," but that she couldn't imagine herself with anyone else. What she hoped for most of all, as the 1950s came to an end, was that her husband would find something to commit himself to fully, even if it meant more traveling. It didn't have to be *her*. But it had to be something.

CHAPTER 22

ONLY BY DYING

LINDBERGH FOUND A NEW FOCUS WITH THE HELP OF A STRAN-
ger who'd never even heard of him. The stranger was Jilin ole "John" Konchellah, a Masai warrior every bit as tall and lean as Lindbergh, but with a
bushy black beard and extended earlobes, the latter pierced with shiny metal
disks. The two met in June 1961, near Vaud, Switzerland, where Lindbergh
and his wife had rented a chalet two years earlier. That real estate transaction followed a letter from Kurt and Helen Wolff, the publishers of *Gift
from the Sea*, who invited the Lindberghs to spend the summer of 1959 with
them at their Alpine getaway. When Charles read the letter he suggested to
Anne that they get their own place over there. It would be a perfect base for
his European travels for Pan Am, he said. That it wasn't far from Munich—
and Brigitte Hesshaimer—may also have had something to do with his suggestion.

Konchellah, a thirty-two-year-old father of five, came to Switzerland as
an African delegate to an international friendship conference sponsored by
Moral Re-Armament (MRA), a nondenominational group founded in 1938
by the American Protestant minister Frank Buchman to foster "a world free
of hate, fear, and greed." Lindbergh's friend James Newton, a businessman
who'd worked in the unsuccessful effort to find funding for Carrel's "Institute of Man," was a member of MRA. It was he who invited Lindbergh to

attend the MRA conference, held in June 1961, in the Swiss resort town of Caux.

Lindbergh was bored by the conference, but enthralled by Konchellah, who gave a short speech there on his tribe of about 300,000, who lived a seminomadic life with their cattle in the Great Rift Valley on the Kenya-Tanzania border, close to Mount Kilimanjaro. Through Newton, Lindbergh arranged a meeting with Konchellah the next day. Until Newton briefed Konchellah, the Masai had no idea who Lindbergh was.

Lindbergh was fascinated to hear Konchellah, who had learned English and mathematics at missionary schools, tell him that he received an equally valid education outside those classrooms. That other curriculum, taught by tribal elders, enabled him to learn the medicinal properties of plants, how to breed cattle, how to hunt wild game, how to survive in a drought, and how to perpetuate a society where everyone worked for the common good of the tribe. The defense of the Masai was left to its warrior class, for which Konchellah had qualified nearly twenty years earlier. The admission examination was as old as the tribe itself: to become a man a Masai boy had to kill a lion with a spear.

Such experiences, Konchellah said, showed that the Masai had an educational system in place long before the white man came to Africa. When Lindbergh asked if Konchellah thought the technological advances of modern civilization might make Masai life less arduous, and therefore better, the Masai shook his head. The traditional way of life is best, he said; of that he was sure. Then he invited "the great white flier" to visit him in Kenya, so that he could see that truth for himself.

Lindbergh arrived in Kenya in December 1962. For the first few days he traveled with a white game warden through the Kenyan portion of the Great Rift Valley, which extends from the Mediterranean Sea all the way to South Africa. Lindbergh was overwhelmed by the beauty of Kenya's vast, open grassland, dotted by large swamps and muddy ponds, an ecosystem sustaining untold thousands of antelopes, wildebeests, hippopotamuses, zebras, rhinoceroses, hyenas, leopards, giraffes, and, of course, lions, which could occasionally be seen shading themselves under acacia trees. One morning Lindbergh saw two male topi—large, hoofed, elklike creatures with long curved horns—butting heads in a ferocious battle for territorial supremacy. Not far away he watched jackals fighting for scraps of flesh on a gazelle carcass. On another day he saw a lioness eating a baby elephant that had been abandoned by its mother, as zebras drank from a nearby water hole. The vibrancy of nature in the wild, the stark rhythms of life and death, hunger and thirst, thrilled Lindbergh every bit as much as his first airplane flight four decades earlier.

Lindbergh was just as excited to arrive, alone in his rented Land Rover, at the Masai *boma* (village) that was Konchellah's home. For the next week Lindbergh lived there as a Masai male, joining other Masai, who were carrying spears and wore only red blankets slung over their shoulders, as they herded cattle. (Lindbergh wore the clothes he came with.) The American watched as the Masai played tag and listened as they sang to the trees, explaining that they did so because they knew God was inside all natural things. These charming practices hardly made the Masai pacifists: the herders gleefully told Lindbergh of a recent battle in which the Masai killed twenty-seven members of a rival tribe, removing the testicles of some of the defeated as war trophies—in several instances when their foes were still alive.

The freedom these Masai males exulted in without self-consciousness, the freedom of men living off the land, with no appointment books and no time clocks, made Lindbergh think of the complications of his own life. He always thought of himself as a free man, but was he really? Konchellah didn't think so. He felt sorry for his new friend—for most white men, in fact—because they'd lost contact with nature. "You speak of freedom in your country," Konchellah said, "but we know freedom far greater than yours." Lindbergh couldn't argue; the wild had much to teach civilized man, he now understood. Africa had changed him. He told Konchellah he would be back soon.

In fact, he bolted back. In February 1964 Lindbergh was attending a meeting at Pam Am's headquarters in midtown Manhattan. Along with Lindbergh and Pan Am's top executives, the conference room was filled with jet-propulsion engineers from Curtiss-Wright, Pratt and Whitney, and General Electric. The topic at hand was how Pan Am might compete with the British-French consortium then building the Concorde, a plane that would fly at twice the speed of sound (and thus at twice the speed of any plane operated by Pan Am). Would an American supersonic plane, if they built one, use turbojets or fan-jets? Would it have a delta wing or a variable sweep design? At what altitude should it be flying when it broke the sound barrier?

Not long ago those questions would have fascinated Lindbergh. But not anymore. As he listened to the engineers—men he knew and admired—batting their arguments back and forth, Lindbergh found himself feeling repulsed. He thought of his flight to Paris, a flight he had made all by himself, with an engine the men in the conference room would snicker at because it was only slightly larger than the engine in a compact car. But at least Lindbergh had flown that plane. Now planes flew themselves—and at speeds that seemed almost inconceivable. And to what end? To get from one overcrowded city to another while creating pollution unlike anything man had ever known? What kind of progress was that?

Those, not the issues being debated by the engineers sitting next to him, were the questions on Lindbergh's mind. So, at six p.m., when the meeting was adjourned, and a follow-up was scheduled for a few weeks later, Lindbergh jumped up from his padded leather chair, grabbed the briefcase he invariably carried whenever he left Darien (inside were one change of clothing, soap, a razor, a toothbrush, a New Testament, and a camera), took the elevator down to the street, hailed a cab to Kennedy airport, and got on the very next plane to Rome, where he transferred to another plane headed for Nairobi.

Less than forty-eight hours later he was driving a jeep in Kenya's Kajiado district when, near a clump of brush, he happened on a giraffe that was pulling leaves off a tree branch with its mouth. The spotted beast with the tasseled tail stopped for a moment, turned toward Lindbergh, looked at him silently with its doll-like eyes, then went on stripping branches. In that instant, under a bright Kenyan sun, Lindbergh felt at peace for the first time since he had left Africa more than a year earlier. It was an experience he needed to explain to the world he left behind in New York.

He did so in an article he wrote for *Reader's Digest*, "Is Civilization Progress?" published in July 1964. Watching that giraffe, Lindbergh said, made him understand the purpose of life with a clarity impossible to find in the modern world—surely not in a corporate conference room, and not even in a scientific laboratory.

"What progress that [giraffe] had made in its competition for survival—stretching its neck and forelegs, bit by bit, through countless generations until it could live on food above the reach of other creatures, condensing its gradual change in form into a genetic character that each individual protected and reblended in its time so the species might survive. There before me, tall and graceful," Lindbergh wrote, "were both a mortal being and the temporal emergence of genetic immortality."

Lindbergh's use of the word "immortality" was significant. His encounter with the giraffe had prompted an epiphany every bit as momentous as his experience in the Utah desert more than thirty years earlier. With a certainty that imbued every corner of his consciousness, Lindbergh now understood that the attempt he'd made with Alexis Carrel to tamper with natural selection to achieve unnatural immortality was as pointless as it was arrogant.

"When I watch wild animals on an African plain, my civilized [method] of measuring time gives way to a timeless vision in which life embraces the necessity of death," wrote the man who once questioned the very purpose of death. "I see individual animals as mortal manifestations of immortal life streams; and so I begin to see myself. I am not only one, I am also many, a man and his species. In death, then, is the eternal life which men have sought so blindly for centuries, not realizing they had it as a birthright." Though his readers didn't know it, Lindbergh was including himself among those blind seekers.

Now it was time for Lindbergh to state the life-changing truth he'd uncovered in the African bush: "Only by dying," he wrote, "can we continue living."

Lindbergh also used his article to reveal that he'd turned his back on racism, and especially on white supremacy. "How much have I really progressed beyond the black-skinned native who believes education should teach man to understand and cope with the things he finds around him, who says that God exists in every tree and mountain," and who hunts a lion with a spear rather than a rifle? "Is my framework of life superior to his," Lindbergh asked, "or have I allowed my thought to be conditioned by my civilization's teachings?"

There was no denying, Lindbergh knew, that the "African framework of life contains ideas [and] values which seem backward when measured by our Western scales. But who is to say that the record of future evolutionary ages will prove the black to be less progressive than the white?" History, he said, suggests the opposite.

"Had not the primitive survived the civilized over and over again?" he asked. "If civilization is progress in the basic sense of life, then why have past civilizations fallen—sixteen of them in the last few thousand years, according to [the historian] Arnold Toynbee?" Lindbergh remembered his walks through the ruins of the Roman forum and the Athenian Parthenon. "Maybe jungle natives sense destructive elements in civilization to which civilized men are blind. I believe there is wisdom in the primitive that the civilized cannot afford to ignore." Lindbergh certainly never heard Carrel praise the wisdom of primitives.

Is civilization progress? "Living in the jungle's primitive environment has made me doubt it," Lindbergh wrote. Of one thing he was sure: "The final answer," said the man who spent nearly five years in Carrel's laboratory at the Rockefeller Institute, "will be given not by the discoveries of our science, but by the effect our civilized activities as a whole have upon the quality of our planet's life."

Under an acacia tree in Africa, Lindbergh reinvented himself one more time. The man who explored outer space as a pilot, then inner space as a biologist, would now explore ways to preserve the planet's natural environment. The person who once tried to save the world by saving white civilization would now try to save the world *from* white civilization.

AND TRY HE DID, with the same mental commitment, physical strength, and imperviousness to fatigue he had exhibited as a young pilot and middle-aged scientist. After returning to the United States, Lindbergh joined the board of the World Wildlife Fund in Washington, while simultaneously offering his services to the International Union for the Conservation of Nature and Natural Resources in Switzerland. Before long Lindbergh was traveling the globe almost nonstop to advance environmental causes—in one year circumnavigating the earth five times.

He personally lobbied the president of Peru to prevent the overhunting of the vicuña in the Andes and the humpback whale in the Pacific, off Peru's coast. He met with President Ferdinand Marcos of the Philippines, urging him to enact measures to protect the monkey-eating eagle, the largest eagle in the world, of which barely two dozen were still alive; as well as the tamarau, a species of water buffalo whose number on the Philippine island of Mindoro, the only place they were known to live, had shrunk to about thirty. In Taiwan Lindbergh persuaded the government to create national parks to ensure that the island's cypress trees would be saved from eradication by over-logging. In Indonesia he urged a ban on the hunting of the one-horned Javanese rhinoceros, a beast facing extinction because of the belief that a powder made from its horn gave human males special sexual powers. (Fewer than forty rhinos were left when Lindbergh arrived.) He met with the king of Tonga, Taufa-ahau Tupou IV, to encourage him to create wildlife reserves there.

Back in the United States, Lindbergh led a movement to expand local wetlands for the accommodation of indigenous waterfowl in Darien. He inspected the site of an oil spill in Santa Barbara, California, then spoke out for more government regulation of the oil-tanker industry. He helped establish the Voyageurs National Park on Minnesota's Kabetogama peninsula. He made a speech before the state legislature of Alaska—his first before an elected body since he had testified in Congress against America's participation in World War II—to argue that Alaska's wilderness not be developed, either for tourism or for natural resources, and that it be protected as a habitat for the arctic wolf.

This traveling, like the flying he did for Pan Am, put stress on Lindbergh's marriage. In 1963 he sold the family's house in Darien, but kept part of the property it was located on and built a smaller house there, tucked amid the marsh grasses even closer to the Long Island Sound. Anne liked the more intimate quarters, but her solitude soon blurred into loneliness. Her five children were all married or away at school and her husband was off, well, she didn't even know where. Twice she had to turn down invitations from President and Mrs. Johnson to the White House—once for a reception for Britain's Princess Margaret, the other for a state dinner with the shah of Iran—because, as Anne instructed her secretary to reply, "of the absence of Mr. Lindbergh from the country and Mrs. Lindbergh's lack of knowledge of the date of his return or where to reach him." Even Reeve Lindbergh, who adored her father, was troubled by his seeming abandonment of her mother. "What does the monkey-eating eagle got that you don't got, I'd like to know," she asked Anne.

As the "benevolent dictator" of his family, Lindbergh wasn't about to let any comments from his wife or children interfere with his commitment to environmentalism. Even so, there were two occasions in the late 1960s when he was momentarily distracted from that commitment, in each instance by the siren song of a cause he thought he'd become deaf to: first experimental biology, then aeronautics.

His unexpected return to the biology lab followed a call for help from two American scientists—Vernon P. Perry and Theodore I. Malinin—then working at the Navy Medical Research Institute in Bethesda, Maryland. They were part of a team that had already demonstrated the feasibility of freeze-drying skin, bone marrow, and blood for use in emergency transplants performed at military hospitals in the field. Before moving on to the next step—freeze-drying whole organs—they read through the scientific literature, where they encountered the articles on organ perfusion cowritten by Lindbergh and Carrel some thirty years earlier. After reading those articles, Dr. Perry and Dr. Malinin obtained one of Lindbergh's perfusion pumps, which they immediately tested in their laboratory.

"Three weeks ago," Perry wrote to Lindbergh in September 1966, "we [took] a heart from a monkey . . . and placed it in [your] pump. We were interested to see if a pH change could be observed in the media after prolonged perfusion":

> [A]fter one hour of perfusion at room temperature the heart began to beat independently of the pulsation of your pump. I don't mean that the heart merely fibrillated (i.e. twitched); there were strong synchronous auricular contractions. The heart continued to beat for six hours.

To the amazement of Perry and Malinin, two experienced bench scientists who were not easy to amaze, Lindbergh's glass machine, with its simple "oil flask" mechanism and cotton-ball air purifiers, had brought the monkey heart back to full operation, functioning just as it would inside a monkey.

Lindbergh wasn't surprised. He'd always believed his machine was capable of maintaining the viability of primate hearts. Unfortunately, politics and war separated Carrel and Lindbergh before they reached that stage of their work. What did surprise Lindbergh was that Perry and Malinin wanted him to join them at the navy research center to redesign his pump so that it could house the larger human organs they wanted to work with and withstand the lower temperatures required for their freeze-drying experiments. Lindbergh begged off, writing to Perry that "it has been so many years since I have done any lab work in connection with tissue or organ culture that I am sure I would have very little to contribute."

But Perry thought otherwise and continued his letter-writing campaign; he was as persuasive as he was persistent. In early 1967 Lindbergh told Perry he would be in Washington in March to meet with Father Joseph Durkin at Georgetown University, where Durkin was a professor of history researching Carrel's life. Perry volunteered to fetch Lindbergh after his meeting, then drive him to his laboratory in nearby Bethesda, where he would show him the work Malinin and he were doing.

Lindbergh's expectation was that he'd take a quick look, make some recommendations, and leave, never to return. Once there, however, the problem solver in him took over. The issues facing Perry and Malinin were mainly created by the low temperatures they were working with. Lindbergh looked at the device he designed three decades earlier and guessed that a few simple improvements—replacing some of the glass with plastic—might enable Perry and Malinin to achieve their goals. So instead of leaving, Lindbergh accepted a nonsalaried appointment as a guest scientist at the Navy Medical Research Institute, where he went to work to see if his hunch was right.

It was, but what really impressed Perry and Malinin was Lindbergh's tireless commitment to the task of finding out. "His industry in the laboratory was overwhelming," Perry later wrote:

When we were forced to retreat from exhaustion, he would carry on, monitoring the apparatus, answering our phone calls, and writing in his journal. When we, after a good sleep, would next see him, he was as fresh as we and would comment that he had caught "a few winks here and there." We really understood how this unusual man had managed to stay awake to make the first solo flight to Paris in May 1927.

Lindbergh's stay in Bethesda lasted several weeks and resulted in two articles: "An Apparatus for the Pulsating Perfusion of Whole Organs," published in *Cryobiology*; and "Observations on Contracting Monkey Hearts Maintained in Vitro at Hypothermic Temperature," published in the *Johns Hopkins Medical Journal*.

LINDBERGH ENJOYED HIS STAY at the navy laboratory, but that didn't mean he planned to return to a life of science. He made this clear in an essay he wrote for *Life* magazine after he left Bethesda, titled "The Wisdom of Wildness." The article was published in December 1967 with a photo of Lindbergh, then sixty-five, standing on the slopes of the volcano Krakatau, in Indonesia, his still slim frame tall and erect, his pate partially covered by strands of gray hair blowing in the breeze. "If I were entering adulthood now," Lindbergh told *Life*'s readers, "I would choose a career that kept me in contact with nature more than science. . . . In wildness I see the miracle of life, and beside it our scientific accomplishments fade to trivia."

Lindbergh's belief in the triviality of scientific accomplishment was put to the test when he was invited by the astronauts Frank Borman, James Lovell, and William Anders to lunch with them at Cape Kennedy in Florida on December 20, 1968. That lunch was the astronauts' last meal on earth before taking off the next morning in Apollo 8, the first lunar orbit mission. Lindbergh and his wife witnessed the takeoff from a special VIP section. Lindbergh accepted the invitation out of respect for his late friend Dr. Robert Goddard, the American physicist who had predicted to him nearly forty years earlier that one day man would fly to the moon in a multistage rocket.

"I have never experienced such a sense of power," Lindbergh wrote to a NASA official after watching the launch, noting that Apollo 8 burned more fuel in its brief takeoff than he had used in his entire flight from New York to Paris. After the astronauts splashed down into the Pacific on December 27, Lindbergh wired the following message: "YOU HAVE TURNED INTO REALITY THE DREAM OF ROBERT GODDARD."

When the editors of *Life* learned of that telegram, they asked Lindbergh to be their special correspondent when NASA's first attempt at a lunar landing—Apollo 11, manned by Neil A. Armstrong, Michael Collins, and Edwin E. Aldrin Jr.—took off from Cape Kennedy the following July. "Forty-two years ago you completed one of the great adventures of modern time, the first solo flight from New York to Paris," an editor at *Life* wrote to Lindbergh. "Would you consider writing an article for a special issue of *Life* about the motivation behind man's great adventures in history, [including] the forthcoming landing on the moon?"

Lindbergh declined, and explained his reasons, in a letter stretching to nearly 5,000 words that *Life* published in its entirety, with his permission, on July 4, 1969, as "Letter from Lindbergh." It was a remarkable document for many reasons, not the least of which is its literary merit as an autobiographical meditation.

"What motivates man to great adventures? I wonder how accurately these motives can be analyzed, even by the participants themselves," Lindbergh wrote. "When I think of my own flights in the early years of aviation, I realize that my motives were as obvious, as subtle and as intermixed as the ocean waves I flew over. But I can say quite definitely that they sprang more from intuition than from rationality, and that the love of flying outweighed practical purposes—important as the latter often were."

Among the "practical purposes" Lindbergh envisioned in the 1920s were his conviction that "a nonstop flight between New York and Paris would advance aviation's progress and add to my prestige as a pilot—with ensuing material rewards" for both the industry and himself. He was right on both scores. Even so, Lindbergh found his passion for airplanes decreasing over time "as the art of flying transposed into a science. Rationally," he wrote, "I welcomed the advances that came with self-starters, closed cockpits, radio and automatic pilots. But intuitively I felt revolted by them, for they upset the balance between intellect and senses that had made my profession such a joy."

Thoughts like that surely surprised Lindbergh's readers, but the biggest revelation was yet to come—specifically, in the section of the letter in which the author chronicled his own past allegiance to science. For the first time, Lindbergh publicly confessed that the goal of his experiments with Carrel at the Rockefeller Institute was immortality.

"How mechanical is man?" "Is death an unavoidable portion of life's cycle or might physical immortality be achieved through scientific methods?" Those, Lindbergh wrote, were the questions motivating his research with Carrel some thirty years earlier.

External events aborted their quest, and this, for a while, caused Lindbergh considerable regret. But not anymore. Now, after spending time in East Africa and the South Pacific, Lindbergh understood that the "mechanical" approach to life, which had led him to see the human body as a living machine made of constantly reparable and replaceable parts—and therefore potentially immortal—was wrong. Wildness, Lindbergh wrote in *Life*, had shown him the truth: "The cycle of life and death is essential to progress. Physical immortality would be undesirable even if it could be achieved."

—

CHAPTER 23

NO, SCIENCE HAS ABANDONED *ME*

THOUGH LINDBERGH DIDN'T WRITE ABOUT APOLLO II, HE DID attend the launch on July 16, 1969. He arrived several days early, which enabled him to meet the three men hoping to make aeronautical history. The NBC News anchor David Brinkley was stunned to see that Neil Armstrong, Edwin "Buzz" Aldrin, and Michael Collins—men who were about to fly 250,000 miles to the moon—"stood in the utmost respect, and even awe, of a man who had flown to Paris."

It wasn't just astronauts and anchormen who noticed Lindbergh's return to the status of hero. In the period just before and after his trip to Cape Kennedy, Lindbergh accepted the Bernard M. Baruch Conservation Prize, a National Institute of Social Sciences Gold Medal, a National Veterans Award, and an honorary doctorate from Georgetown University. The Samoan high chief Tufele-Faiaoga was so moved by Lindbergh's conservation efforts on behalf of his nation that he bestowed on the American the ancient title of Tuiaana-Tama-a-le-Lagi ("son of heaven"). President Marcos presented Lindbergh with a Presidential Plaque of Appreciation for "inspiring and spurring action to save the Philippines' prized fauna from extinction and his pace-setting advancement of the cause of wildlife conservation throughout the world."

This shower of praise and plaques slowed down a bit in September 1970, when Harcourt Brace Jovanovich published *The Wartime Journals of Charles A. Lindbergh*. As Lindbergh made clear in his introduction, he would not be apologizing for his stand against American participation in World War II. To the contrary, Lindbergh asserted that history had proved him right.

"We won the war in a military sense, but in a broader sense it seems to me that we lost it, for our Western civilization is less respected and secure than it was before." He wrote:

> To defeat Germany and Japan we supported the still greater menaces of Russia and China—which now confront us in a nuclear-weapon era. Poland was not saved. The British Empire has broken down with great suffering. . . . France had to give up her major colonies and turn to a minor dictatorship herself. Much of our Western culture was destroyed. We lost the genetic heredity formed through aeons in many million lives. Meanwhile, the Soviets have dropped their iron curtain to screen off Eastern Europe, and an antagonistic Chinese government threatens us in Asia.

Wartime Journals also contained passages that seemed indifferent to the regimented brutality of Hitler's Nazi Germany, and some passages that seemed clearly anti-Semitic. What the public didn't know was that the most egregiously bigoted remarks in Lindbergh's book—including those suggesting that America already had "too many" Jews—had been excised by Lindbergh and his editor and publisher, William Jovanovich. This was one of many heretofore unknown facts uncovered by A. Scott Berg and reported in his Pulitzer Prize–winning biography, *Lindbergh*, published in 1998.

A front-page story in the *New York Times* on the release of *Wartime Journals* was illustrated by the famous photograph of Lindbergh and his wife standing next to the Nazi air marshal Hermann Göring in his Berlin office in 1938, not far from a portrait of Adolf Hitler on Göring's desk. The Pulitzer Prize–winning historian James MacGregor Burns told the *Times*, "Those who think it would be better to be living today in a world dominated by Hitlerism than in the existing world simply do not remember Hitlerism—or never understood it."

An essay in the *New York Review of Books* by the novelist Jean Stafford praised Lindbergh's earlier book, *The Spirit of St. Louis*, as a "work in the first rank of suspense and adventure stories." But Stafford mocked *Wartime Journals* for being filled with "comic strip American clichés" and "statements of fact that cause the eye to leave the page in embarrassment." As an example of the latter Stafford cited this passage: "At lunch today we were introduced to a new drink—carrot juice. It was very good, I thought. [Henry] Ford said it was made by crushing carrots and collecting the juice." Stafford's most perceptive criticism of *Wartime Journals*, however, was that it gave short shrift to Lindbergh's time with Alexis Carrel.

"Although Lindbergh makes frequent references to his work with Carrel at the Rockefeller Institute," she wrote, "he is disappointingly vague":

> If [only] Lindbergh had slashed out sufficient undergrowth to make way for more on Carrel's memorable research and Lindbergh's contribution to it. (He constructed a pump that supplied pure oxygen to a chicken heart which was artificially animated for an impressive length of time.) It would be infinitely more interesting to read a description of the pump than it is to read an impassioned defense of Carrel's activities as a Pétainist, after he had returned to France, where he was eventually jailed as a collaborationist.*

Lindbergh ignored the bad reviews and kept traveling. In November 1970, he stopped in Pasadena, California, to call on a scientist who had worked with him on the experiments Stafford was so curious to read more about. This was Dr. Richard Bing, the German Jew who was assigned to assist Carrel and Lindbergh in demonstrating their organ perfusion device at the International Congress of Experimental Cytology, held in Copenhagen in 1936, when Bing was a recent medical school graduate.

It was a wonderfully nostalgic afternoon, Bing said in a recent interview. Each man laughed as he recalled his first meetings with Carrel. (Bing too had endured a "face test" from the Frenchman.) "I was reminded of what a likable, unpretentious, and even folksy man Lindbergh could be, despite the coldness he exhibited in public," Bing said. Lindbergh and Bing drove from Bing's home to his cardiology laboratory, where Bing was

---

* Stafford was wrong on two points: Lindbergh's pump supplied the perfused organ with more than oxygen (i.e., other nutrients); and Carrel was never jailed.

using Lindbergh's pump to perfuse atherosclerotic blood vessels (vessels showing progressive hardening and narrowing) in experiments aimed at devising new treatments for that disease. "The noise made by Lindbergh's pump—a kind of 'sheesh, sheesh' sound made by the valves sending the perfusion fluid into the organ—brought him right back to Carrel's lab," Bing said.

Lindbergh said Carrel saved his life—or, at least, his sanity—by providing him with a haven at the Rockefeller Institute from the trauma caused by the murder of his first child. Bing said he felt similarly: without Carrel's help, would he ever have been able to leave Nazi Germany? Would he have had a medical career at all? It pained both men to realize that the Carrel they knew—a man capable of great warmth and kindness—was now, twenty-five years after his death, regarded by the public as having had little or no humanity because of his association with fascism, if he was remembered at all.

There were just too many contradictions in Carrel for the public, or even many scientists, to understand, Bing said. Carrel was a brilliant physiologist who believed in prayer cures and parapsychology. He was a Catholic who supported birth control and euthanasia. He revered his wife, yet believed higher education was wasted on women. He loved children, yet never had any. He was a laboratory scientist who thought himself a philosopher. He was a Nobel Prize–winning surgeon who wrote articles on breast-feeding in *Reader's Digest*. "If [Carrel] were still alive," Bing told his old friend, "he'd probably be the host of a TV talk show." Lindbergh, who rarely watched television, smiled at the idea.

Lindbergh said he considered his time with Carrel to be the most thrilling intellectual adventure of his life. Even so, he told Bing that he had moved on—or maybe "moved back" was a better term. This didn't mean that Lindbergh had lost his respect for scientists. Far from it; he had only admiration for their hard work and diligence. Hearing this, Bing volunteered his opinion that Carrel was a genius whose breakthroughs in surgery and tissue culture were mutations in the evolution of those disciplines every bit as important as the physical changes occurring in the natural world through natural selection.

Lindbergh agreed, but said his argument was less with scientists than with science itself. Science, he decided, had led mankind astray—toward complication and away from simplicity, toward urban sprawl and away from nature. Lindbergh told Bing he was trying to reverse that trend by committing himself to conservation.

The subject of immortality never came up when the two old friends spoke, but Lindbergh did write these words in a letter he sent to Bing a few days later:

> How time collapses under the circumstances of our visit—the thirty years between Carrel's laboratory and your own. There are moments when it seems to me that the time-gap disappears, and that you, Carrel and I are still together—without the separation that we think so obvious in death. Maybe if man had deeper awareness, life and death would make less difference. I am inclined to think so.

The subject of Nazism didn't come up either when Bing and Lindbergh met, even though the publication of *Wartime Journals* had certainly reminded Americans of the aviator's infatuation with Hitlerism. Lindbergh ignored the criticism of his writings on Nazi leaders, but his wife found it difficult to do so. This was because Anne was rereading her own journals with the idea of publishing them, and she was pained by the enthusiastic comments she herself had made in that private space about Hitler, Göring, and the political system they'd created.

As the 1970s began, however, the "system" that still troubled Anne the most was the one defining her marriage: her husband felt free to travel the world nonstop to save the monkey-eating eagle, the tamarau, and other endangered species she'd never heard of, while Anne felt trapped in a marriage without a husband. Finally, Lindbergh agreed to spend more time with her, but only after he found a new place for them to live.

That place was Maui, where Lindbergh had spent considerable time in the past few years while crisscrossing the Pacific for environmental groups. An old business friend of his from Pan American, Sam Pryor, had retired there with his wife, near the small village of Kipahulu, where he built a large A-frame structure on a hundred acres of rolling grassland, an area more than large enough for Pryor's pet gibbons, some of whom he dressed up as children, to roam undisturbed among the orchids and bougainvillea, as well as the mango and guava trees. A figure of some influence on the island,

Pryor lobbied successfully to ensure that Kipahulu would remain largely undeveloped: there was no electricity, and there were very few paved roads.

Lindbergh loved the astonishing plant life, the warm breezes, and the rustic conditions so much that he asked Pryor to let him know if a plot of land nearby came on the market. Pryor offered Lindbergh five of his own acres, which Lindbergh bought for $25,000. In January 1971, Anne and he moved into a simple two-story, boulder-and-cement A-frame without air conditioning, a telephone, or electrical wiring; propane motors powered the appliances and kerosene lamps provided the lighting. It was half an hour's walk on a mud path to the nearest store. Finally, Anne thought, she and her husband would retire in peace, in one of the most beautiful settings on earth.

Things didn't quite happen that way. It rained steadily during their first week in their Kipahulu home, which they named Argonauta after one of the shells Anne wrote about in *Gift from the Sea*. Argonauta's roof leaked; even worse was the fact that the builder of the house hadn't created proper drainage, which meant that muddy streams of rainwater sluiced through the structure itself. Charles and Anne spent one of their first nights in Kipahulu outside their home, in the pouring rain, as he dug drainage channels with a bucket and she built a mud dam. Before the house was dry it was invaded by ants, spiders, cockroaches, lizards, rats, and an adult mongoose. Just as the Lindberghs were about to address that infestation, Charles left Hawaii for New York to attend a Pan Am board meeting.

In fact, Lindbergh spent barely two months of 1970 with his wife at Argonauta, a pattern that he repeated in 1971 and 1972. The rest of the time he spent traveling on behalf of conservation groups all over the world. Anne eventually made her peace with Kipahulu—she put down rat poison, covered the walls with insecticide, and cut down the wild bamboo that was crowding out her fruit trees—and she even came to value her solitude, which enabled her to work on editing her diaries for publication. Still, she never gave up hoping that one day her husband would finally decide to stop traveling and stay at home. That day finally did come, but it wasn't the result of a decision. It was the result of a diagnosis.

IN OCTOBER 1972, WHILE Lindbergh was having his annual medical checkup, his physician, George Hyman, found nodes under Lindbergh's armpits that suggested lymphoma. Dr. Hyman had Lindbergh check into Columbia-Presbyterian Medical Center in upper Manhattan (as "Mr. August") to have biopsies done on the suspicious growths. The report back from the lab was unequivocal: Lindbergh had cancer.

On January 31, 1973, he began three days of radiation therapy, which left him weaker than he'd ever felt before. He dropped thirty pounds from his already lean frame and, for the first time in his life, looked older than his chronological age which, as of February 4, was seventy-one. When his children and friends inquired about those physical changes, Lindbergh said he'd caught a strange virus on one of his environmental trips. To regain what was left of his health he returned to Maui. It was a good decision: in Kipahulu his appetite returned and he started sleeping through the night without attacks of nausea.

But he had no illusions. One day he walked down a dirt path from Argonauta to his friend Sam Pryor's house and asked, as offhandedly as he could, so as not to arouse any suspicion, "Sam, where are you going to be buried?"

Apparently Lindbergh's nonchalance was convincing. "Right behind the little church I've restored about twenty minutes from here," Pryor said.

Lindbergh asked to see it, so the men walked down an unpaved road to the Ho'omau Congregational Church, a small white building made of lava rock and stucco, with ten pews, nestled among coconut palms and banyan trees just in from a cliff. The simple structure, with a large wooden cross at one end, had been built by missionaries from Connecticut more than a century earlier. Lindbergh walked behind the church, past the Pryors' plot, close to a cliff overlooking Kipahulu Bay. Later, after consulting with church and state officials, he secured a thirty-foot plot atop that cliff, then marked it off with stakes and twine.

The Lindberghs returned to Darien in the spring of 1973. Shortly after arriving, Lindbergh came down with a case of shingles, then other maladies: a serious bout of flu, coupled with anemia. These led him to cancel many of the appointments he'd planned. But there was one commitment he was determined to honor, no matter how ill he felt.

"The Alexis Carrel Centennial Conference," organized by Father Joseph T. Durkin to honor the Nobel laureate, took place on June 28, 1973, in an auditorium at the Georgetown University School of Medicine. Eight papers were read by the participants, who sat at a conference table. Most, but not all, were surgeons. Dr. Theodore I. Malinin, who (with Dr. Vernon P. Perry) had been so impressed at the Navy Medical Research Institute in 1967 by the efficiency of Lindbergh's pump thirty years after Lindbergh designed it, spoke on "Organ Culture and Perfusion by the Carrel Method." Dr. Charles A. Hufnagel, who in 1952 became the first surgeon to implant an artificial valve in a human heart—the very goal that led Lindbergh to seek out Alexis Carrel in 1930—spoke on "Alexis Carrel's Contributions to Surgery." Dr. André Gros, the former vice-regent of the French Foundation for the Study of Human Problems, came from Paris to remind his listeners, in "Reflections on Alexis Carrel," of the extraordinary goals of that wartime organization, established and run by Carrel in occupied France.

But there was no doubt which speaker the audience of surgeons and students wanted to hear most. Looking as tall and lean as he did when he became, overnight, the most famous man in the world, Lindbergh read his own presentation, the last thing he ever wrote for publication: "I first met Doctor Carrel in New York, in 1930, at the Rockefeller Institute for Medical Research," he said in the hushed room. "I came to ask his advice about problems I had encountered in designing an artificial heart for use during operations. Our meeting resulted in a collaboration and friendship that ended only with his death, in 1944.

"I soon found Carrel himself even more fascinating than the laboratory projects I pursued in his department of experimental surgery. There seemed to be no limit to the breadth and penetration of his thought. One day he might discuss the future of organ perfusion, for which I was building an apparatus. On another, he would be talking to a professional animal trainer about the relative intelligence of dogs and monkeys, and the difficulty of teaching a camel to walk backward. I once looked up from my work to see him step into the room with Albert Einstein, discussing extrasensory perception. . . .

"According to his mood, Carrel could work with a precision that caused the admiration of the scientific world, or he could speak with an abandon that brought criticism heaping on his shoulders. He might straighten his back and assert that 'all surgeons are butchers,' and that 'all people are fools,' or sit at his desk and write that 'on the scale of magnitudes man is placed midway between the atom and the star.' He had a character that attracted the love of those who knew him well, and a blunt tactlessness that created many enemies. . . .

"In his latter years, [Carrel] became more and more interested in creating an institute where experiments with life could be planned to extend through centuries, where a contemplative environment would be made available to scientific workers as well as the facilities of modern civilization. It was to be called 'The Institute of Man.' World War II intervened and his death, near the end of the war, came before such an institute had been more than partially established.

"In eulogizing Carrel, one might emphasize his skill as a surgeon, his pioneering work in the fields of tissue and organ culture, his treatment of the wounded in World War I, his suturing of blood vessels which brought him the Nobel Prize, his perception and his depth of vision. Personally, I can say that he had the most stimulating mind I have ever met."

LINDBERGH RETURNED TO DARIEN after the conference, happy to have helped in keeping Carrel's name alive. Biopsies and blood tests performed at Columbia-Presbyterian indicated that Lindbergh's cancer wasn't spreading, but his vitality never fully returned. For the first time in years Charles and Anne actually lived together under the same roof, quietly discouraging visitors, though they did entertain Imelda Marcos for lunch one afternoon. Lindbergh was able to fly to Maui on several occasions to rest in the tropical climate, on one trip taking steps to establish Kipahulu as his legal residence.

Several months later, while back in Darien, Lindbergh came down with a fever so high and persistent that he returned to Columbia-Presbyterian, where his doctors insisted he stay until his temperature returned to normal. That took two weeks, and even then the doctors were alarmed at their inability to improve his blood count, so Lindbergh remained in the hospital, ordering Anne to tell no one of his worsening health. Eventually Lindbergh began to respond positively to blood transfusions, so, on July 7, he was allowed to return to Darien. He'd also begun a new round of chemotherapy, but the results wouldn't be clear for months. Once at home in Connecticut, Lindbergh informed his children of the true nature of his illness. His plan was to rest there, then travel in August to the family's summer home in

Switzerland. But that never happened. After a new round of tests doctors told Charles and Anne they could offer no hope for Lindbergh's recovery.

Lindbergh's children flew into New York and gathered around his bed in Room 1148 at Columbia-Presbyterian's Harkness Pavilion. Somehow, word leaked that there was very famous American in the intensive care unit; before long, messages wishing him a speedy recovery arrived from President and Mrs. Nixon, Mr. and Mrs. DeWitt Wallace (of *Reader's Digest*), and dozens of other luminaries.

On August 14 Lindbergh called the editor and publisher of *Wartime Journals*, William Jovanovich, and said he had another project to discuss. When Jovanovich arrived in Lindbergh's room, the patient pointed to a brown leather bag on the floor. Inside were about 400 pages of the autobiography Lindbergh had been writing off and on for the past several decades. Another 1,000 pages, Lindbergh said, were in two other locations: at his office in Darien; and in locked files at Yale University, where Lindbergh had decided to leave his personal papers. Lindbergh asked Jovanovich to read the pages in the bag and tell him "whether it is any good and if it should be published."

Jovanovich took the pages home, read them that night, and responded in the affirmative. The two men drew up an agreement, calling for Jovanovich to edit and publish the pages as a book, "establishing a sequence" out of Lindbergh's sometimes disjointed recollections, and informing the book's readers, once it was published, that the writing was all Lindbergh's, even though "parts of the text were subject to editing consistent with his purpose."*

On August 16, 1974, Lindbergh reviewed his will. When he was finished, he told Anne, "I want to go home to Maui." This was not a request Lindbergh's doctors could approve; returning home to Darien was a possibility, they said, but even that short journey would gravely endanger his health. But Lindbergh was adamant. Ironically, the only doctor at Columbia-Presbyterian who would sign him out of the hospital was his wife's former lover, Dana Atchley.

Lindbergh called his physician in Hawaii, a Minnesotan by birth named Milton Howell. (Dr. Howell's father had been mayor of Glencoe, a small

---

* *Autobiography of Values* was published by Harcourt Brace Jovanovich in February 1978.

Minnesota town where Lindbergh and his father had dropped campaign literature from Lindbergh's plane when C.A. was running in a primary election for the U.S. Senate in 1923.) Howell tried to talk Lindbergh out of his plan, telling him that the hospital physicians knew best, but to no avail.

After Howell backed down, Lindbergh spoke to him about where he would spend his final days in Maui. The Lindberghs' A-frame in Kipahulu was out of the question because it lacked electricity; perhaps Howell could find another house, with wiring, in or near Hana, the largest nearby town. After a few telephone calls, Howell found one: a guest cottage belonging to two friends of his, Edwin and Jeannie Pechin, who'd just left Maui on a vacation cruise to Alaska.

On August 17, doctors at Columbia-Presbyterian administered two blood transfusions, which seemed to revitalize Lindbergh. He spent several pleasant hours reminiscing with his son Jon about hunting trips Charles had taken with his father in Little Falls, until they were interrupted. One of Lindbergh's physicians entered the room and announced that he and the rest of Lindbergh's medical team at the hospital were unanimous in their conviction that Lindbergh's trip was "medically unsound" and "incompletely thought out." If Lindbergh insisted on going to Maui, the doctors wanted thirty-six hours to put together a small medical team to fly with him. Without such assistance, the doctor said, Lindbergh was risking infection, hemorrhage, breathing problems—the list went on and on, culminating in death. Lindbergh listened, thanked the doctor for his concern, but said his mind was made up. He was flying to Maui with his wife and three sons the next morning.

"But you're abandoning science!" the doctor said.

"No," said Lindbergh, "science has abandoned *me*."

The next morning Lindbergh was carried on a stretcher by two airline employees into the first-class section of a United flight traveling from New York to Honolulu, via Los Angeles. A makeshift bed was fashioned out of two seats, and then enclosed by a curtain. In Honolulu, Lindbergh would be transferred to an ambulance plane, which would take him to Hana, on Maui. Milton Howell met that plane at the Hana airstrip.

He was surprised that Lindbergh had made it alive, but, as Anne reminded him, no one had expected the "Lone Eagle" to survive the flight from New York to Paris, either. Once out of the plane, Lindbergh told his physician he didn't want any heroics. He'd come to Maui to die and he hoped his friend would help him make that end "a constructive act." Howell pledged he would, then spoke to Anne and the Lindberghs' sons about what they should expect.

The malignancy in Lindbergh's body had already blocked one lung, Howell said; it was likely to spread to the other. If the cancer attacked the lungs' covering—the pleura—as was also likely, Lindbergh would experience severe pain. There would be sedatives and opiates to deal with that, Howell said, and a large oxygen pump, with facial mask and pressure regulators, to address all respiration issues.

Ever the engineer, Lindbergh examined the oxygen equipment Howell installed in the room he'd be staying in at the Pechins' guest cottage, in Hana. "Now, doctor," Lindbergh said, "is the caliber of the oxygen tube really large enough to supply me with the oxygen I need?" Howell assured him that it was.

For the next week Lindbergh planned every detail of his death. The grave would be dug in the traditional Hawaiian style, blessed by a local kahuna (holy man), and lined with lava rock. The coffin would be made by hand from indigenous trees, rough-sawn, flat-sided, flat-topped, and with no curves. The coffin lining would be fashioned from biodegradable materials, starting with a canvas sheet on the bottom. His corpse would be dressed in old gray cotton pants and a tan shirt, without shoes or a belt, then covered in a Hudson Bay blanket that Charles had once given his mother. There would be no embalming or mortuary services of any kind. The headstone would come from the Rock of Ages quarry in Barre, Vermont. Below his name would be the dates and places of his birth and death, above a short passage from Psalm 139: "If I take the wings of the morning, and dwell in the uttermost parts of the sea." There would be a short graveside service, but no eulogy.

When he wasn't going over his checklist, napping, or taking oxygen from his respirator, Lindbergh reminisced with his wife and sons about his life. He told stories about Little Falls, driving his dad's car on campaign trips through central Minnesota, the early barnstorming years, his training in the army air section, and, most of all, his time with Carrel at the Rockefeller Institute and on Saint-Gildas. He had no regrets, Lindbergh said, except for the complications his unexpected fame had created.

One night Anne asked him to describe what he was feeling, now that he was on his deathbed. "You're going through an experience all of us will have to go through," she said. "What is it like?" Lindbergh said the most surprising part was his awareness that "death is so close all the time—it's right there next to you," even when you're young and healthy. He wasn't happy that his body was weakening, but it didn't frighten him, he said. "It's harder on you watching than it is on me."

On August 25, Lindbergh's grave and coffin were completed. That morning Lindbergh felt chest pains with an intensity he hadn't felt before, and his breathing became extremely labored. He had trouble standing up and found himself relying more and more on his respirator. After drifting off to sleep in the afternoon, he awoke in the evening, feeling short of breath. He saw that Anne was in the room with him, smiled at her, and put on his oxygen mask.

But this time the machine seemed to be malfunctioning. Lindbergh reached over to the respirator to adjust the pressure, hoping to get more oxygen. The man who had spent five years with Alexis Carrel trying to invent a machine to defeat death turned a knob and was puzzled by the lack of response. He tried other valves—still no change. Then he looked over at Anne. They locked eyes for a moment, each realizing in this instant that the machine was functioning perfectly. It wasn't failing. He was. Lindbergh moved to pull off his mask, but never got that far. His arm, pulled by gravity, dropped toward the floor. The immortalist had finally become immortal.

ALL OF US ARE FOLLOWING

WHEN LINDBERGH SAID "SCIENCE HAS ABANDONED *ME*" HE was stating a personal truth, not a historical one.

In a 1937 article in *Archives of Surgery* that cited Lindbergh's pioneering work in perfusing whole organs, Dr. John H. Gibbon reported achieving whole-body perfusion with a heart-lung machine that enabled him to perform open-heart surgery on a cat. Sixteen years later Gibbon used a larger version of his device to perform the world's first open-heart surgery on a human.

The first implantation of an artificial heart—a remedy Lindbergh conceived of for his sister-in-law in 1930—was achieved in 1969 by Dr. Denton Cooley in a heart-transplant patient waiting for a donor. (It was removed sixty-four hours later when an organ became available.) The first artificial heart intended as a permanent replacement, the Jarvik-7 designed by Dr. Robert Jarvik, was implanted in 1982 by Dr. William C. DeVries. The patient survived for 112 days. An updated version of Jarvik's device, called the CardioWest, is the only FDA-approved artificial heart—but only for use as a bridge to transplantation.

Lindbergh and Carrel's concept of the body as a living machine made of replaceable parts took a giant step forward in 2006 when Dr. Anthony Atala announced that he'd cultured seven complete urinary bladders in his laboratory at Wake Forest University and implanted them in seven patients. These "neobladders," each grown from a dollop of the patient's own cells, are shaped by an artificial scaffold made of biodegradable materials. Once the patient's blood nourishes those cells and the tissue grows, the structure dissolves, leaving behind a totally functional organ immune from attack by the patient's immune system—because it's made from the patient's own body.

Tissue-engineering projects similar to Atala's are now under way in his and other medical centers where research is being done on culturing kidneys, livers, tendons, ligaments, and even the human heart. None of this will be easy, and whatever progress is made will be years in the making. Still, there's no doubt that Atala's work marks a huge breakthrough in bioengineering—one conceived by Carrel and Lindbergh at the Rockefeller Institute in the 1930s.

That truth, Dr. Atala says, is often forgotten. "I give sixty lectures a year on my work. In one of the first slides in my presentation I show the cover of *The Culture of Organs*." The reaction, Atala said, is always the same: an audible gasp. "These are experienced scientists, many of them working in regenerative medicine, but they have no idea that some of the truly groundbreaking work in our field was done by Lindbergh and Carrel seventy years ago." The experiments done by those two men were not sophisticated by current standards, Atala concedes. Even so, "their work is the beginning of it all. It's where regenerative medicine starts. All of us are following in their footsteps."

ACKNOWLEDGMENTS

—

Much is owed to many. I'll start with David "Don Hirsheone" Hirshey, the senior vice president of HarperCollins, who guided me to Ecco and the legendary Dan Halpern. After my peerless agent, David Black, sealed the deal, my manuscript was edited by "the Don," ably assisted by Nick Trautwein and Kate Hamill.

Several friends were kind enough to read my manuscript in varying stages of completion. Ben Yagoda—the smartest person I know—provided invaluable help. Every author should have a friend so generous and astute. I also benefited from the comments of John Capouya, Paul Gardner, Stanley Mieses, Clive Priddle, and Joel Rose. Several scientists were equally generous. Chief among them were Dr. Richard Bing, who worked with Lindbergh and Carrel; Dr. Anthony Atala, who continues in their footsteps today at Wake Forest University; Dr. Leo Furcht of the University of Minnesota; and Dr. Stuart Lewis of New York Universtity. Dr. Hannah Landecker, a historian of science at Rice University, was also a great friend of this project, as were William Hoffman, a science writer at the University of Minnesota, and Brian Horrigan, exhibits curator at the Minnesota Historical Society in St. Paul, and at the Charles A. Lindbergh Historic Site in Little Falls, Minnesota.

Librarians in several institutions were invariably helpful and patient. I want to especially salute those who guided me through the Lindbergh collection at Yale University—and to Reeve Lindbergh and Land Lindbergh, for granting me access to their father's papers. Identical thanks go to the staff overseeing the Carrel papers at Georgetown University, in Washington, D.C. I am also grateful for the assistance I received at the American Philosophical Society Library in Philadelphia, the Bobst Library of New York University, the Humanities and Social Science Library of the New York Public Library, the Minnesota Historical Society Library in St. Paul, the New York Academy of Medicine, and the Rockefeller Archive Center in Sleepy Hollow, New York.

Two service providers were extremely helpful: Devorah Cohen, photo researcher extraordinaire; and Antony Shugaar, whose graceful translations, from the French, of the many scholarly journal articles and biographies written about Carrel in that language made my job a lot easier.

I also honor the memory of Miss Appelgate, my American history teacher in tenth grade at Mamaroneck High School, who taught me the two most important lessons any historian can learn: "Have thoughts. Organize them."

But thanks most of all to Marion Ettlinger, who has made me a better writer and a better man.

NOTES

—

WWW   AML, *War Within and Without: Diaries and Letters, 1939–1944* (New York: Harcourt Brace Jovanovich, 1980)

CHAPTER I **I Will Show You What I'm Doing Here**

1   *That CAL was the world's most famous man:* Daniel J. Boorstin, *Hidden History* (New York: Harper and Row, 1987), pp. 291–293; Leo Braudy, *The Frenzy of Renown* (New York: Oxford University Press, 1986), p. 25.

1   *"The New York Times":* "Lindbergh Does It! To Paris in 33 1/2 Hours; Flies 1,000 Miles through Snow and Sleet; Cheering French Carry Him Off Field," *NYT,* May 22, 1927.

1   Many books and journals have documented the frenzy that greeted Lindbergh in Paris and in the weeks and months that followed. Lindbergh's own rather brief account can be found in *AOV,* pp. 12–14, and then again in *SSL,* pp. 493–498. A. Scott Berg's magisterial biography, *Lindbergh* (New York: Putnam, 1998), does an excellent job, pp. 128–170. So do the following: Kenneth S. Davis, *The Hero: Charles A. Lindbergh and the American Dream* (Garden City, N.Y.: Doubleday, 1959), pp. 205–244; Joyce Milton, *Loss of Eden: A Biography of Charles and Anne Morrow Lindbergh* (New York: HarperCollins, 1993), pp. 117–131; Leonard Mosley, *Lindbergh: A Biography* (Garden City, N.Y.: Doubleday, 1976), pp. 113–123; Walter S. Ross, *The Last Hero* (New York: Harper and Row, 1976), pp. 120–169; and Robert Wohl, *The Spectacle of Flight* (New Haven, Conn.: Yale University Press, 2005), pp. 19–46. Useful magazine accounts include Raymond H. Fredette, "The Making of a Hero: What Really Happened after Lindbergh Landed at Le Bourget Seventy-Five Years Ago," *Air Power History* (Summer 2002): 6–19; Curtis Wheeler, "Lindbergh in New York," *Outlook* (June 22, 1927): 60–63; "Colonel Lindbergh's Homecoming," *Aero Digest* (July 1927): 31–32; and "Why the World Makes Lindbergh Its Hero," *Literary Digest* 93 (June 25, 1927): 5–11.

2   *"He heard it":* *AOV,* p. 132; CAL to Joseph T. Durkin, S.J., May 20, 1966, CALPY; Robert Soupault, *Alexis Carrel* (Paris: Les Sept Couleurs, 1972), p. 169, trans. for this project by Antony Shugaar.

2   *"Anne, then":* Milton, p. 153.

3   *"Lindbergh 'moved so'":* Berg, p. 187; Milton, p. 153.

3   *"Most of the":* Milton, p. 178.

3   *"Lindbergh spoke to":* Dr. Paluel Flagg to CAL, Dec. 1, 1930; and June 22, 1935, CALPY.

3   *"'Would you show'":* Flagg, "Lindbergh's Introduction to Medicine," in Flagg's unpublished memoir.

4   *"When Dr. Carrel won":* http://nobelprize.org

4   *"This new ability":* Numerous articles in medical journals acknowledge Carrel's crucial contribution to vascular surgery. Among the most informa-

tive are Lyman A. Brewer III, "Alexis Carrel: A Cardiovascular Prophet Crying in the Wilderness of Early Twentieth Century Surgery," *Journal of Thoracic Surgery* 94 (1987): 724–726; Julius H. Comroe Jr., "Who Was Alexis Who?" *American Review of Respiratory Disease* 118 (1978): 391–402; John Marquis Converse, "Alexis Carrel: The Man, the Unknown," *Plastic and Reconstructive Surgery* 68 (1981): 629–639; Denton A. Cooley, "The History of Surgery of the Thoracic Aorta," *Cardiology Clinics of North America* 17 (1999): 609–613; Robert Cousimano, Michael Cousimano, and Steven Cousimano, "The Genius of Alexis Carrel," *Canadian Medical Association Journal* 131 (1984): 1142–1150; Ralph A. Deterling Jr., "Alexis Carrel: The Man and His Contributions to Vascular Surgery," *Journal of Cardiovascular Surgery* 2 (1961): 81–88; W. Sterling Edwards, "Alexis Carrel's Contributions to Thoracic Surgery," *Annals of Thoracic Surgery* 35 (1983): 111–114; Steven G. Friedman, "Alexis Carrel: Jules Verne of Cardiovascular Surgery," *American Journal of Surgery* 155 (1988): 420–424; G. M. Lawrie, "The Scientific Contributions of Alexis Carrel," *Clinical Cardiology* 19 (1987): 428–430; Theodore I. Malinin, "Remembering Alexis Carrel and Charles A. Lindbergh," *Texas Heart Institute Journal* 23 (1996): 28–35; Jack Moseley, "Alexis Carrel, the Man Unknown," *JAMA* 244 (1980): 1119–1121; F. T. Rapaport, "Alexis Carrel, Triumph and Tragedy, *Transplantation Proceedings* 19 (1987): 2–8. Also extremely helpful are Robert W. Chambers and Joseph T. Durkin, eds., *Papers of the Alexis Carrel Centennial Conference* (Washington, D.C.: Georgetown University Press, 1974); and Shelley McKellar, "Innovation in Modern Surgery: Alexis Carrel and Blood Vessel Repair," in Darwin Stapleton, ed., *Creating a Tradition of Biomedical Research* (New York: Rockefeller University Press, 2004), pp. 135–150.

4   *"Carrel's office":* Interview with Dr. Richard Bing, February 24, 2002.

5   *"Carrel was so convinced":* Joseph T. Durkin, S.J., *Hope for Our Time: Alexis Carrel on Man and Society* (New York: Harper and Row, 1965), pp. 10–11; David Le Vay, *Alexis Carrel: The Perfectibility of Man* (Rockville, Md.: Kabel, 1996), p. 163; Ross, p. 230; Frederic Sondern, Jr., "The Amazing Dr. Carrel," *This Week* Magazine, *New York Herald Tribune* (June 18, 1939): 5, 22.

5   *"Carrel's probing eyes":* Durkin, p. xix; Le Vay, p. 16; Ross, p. 230.

5   *"Lindbergh was not":* Berg, p. 154.

6   *"He'd performed the world's":* AC, "On the Experimental Surgery of the Thoracic Aorta and the Heart," *Annals of Surgery* 52 (1910): 83–95; Comroe, "Who Was Alexis Who?"

7   *"To be treated":* CAL, preface to AC, *The Voyage to Lourdes* (New York: Harper, 1950), p. vi.

7   *"The pilot was a hero":* Berg, p. 223; *MTU*, p. 296.

7   *"If you like'":* Davis, p. 344.

8   *"Several of the animals":* Hannah Landecker, "Building a New Type of Body in Which to Grow a Cell," in Stapleton, pp. 151–174.

8 *"This study built":* W. Sterling Edwards and Peter D. Edwards, *Alexis Carrel: Visionary Surgeon* (Springfield, Ill.: Charles C. Thomas, 1974), p. 69; Le Vay, pp. 116–117; Theodore I. Malinin, *Surgery and Life: The Extraordinary Career of Alexis Carrel* (New York: Harcourt Brace Jovanovich, 1979), p. 75.

8 *"After five minutes'":* Le Vay, pp. 75–76.

9 *"In 1907 Carrel":* AC, "Heterotransplantation of Blood Vessels Preserved in Cold Storage," *JEM* 9 (1907): 226–228; AC, "Latent Life of Arteries," *JEM* 12 (1910): 460–486.

9 *" 'Most of the grafts' ":* AC, "The Preservation of Tissues and Its Application in Surgery," *JAMA* 59 (1912): 523–527.

9 *On Carrel's "mousery":* George W. Corner, *A History of the Rockefeller Institute, 1901–1953* (New York: Rockefeller Institute Press, 1964), pp. 228–229; Le Vay, pp. 164–165; Alleyne Macnab, "The Mousery of the Rockefeller Institute for Medical Research, Under the Direction of Alexis Carrel: 1930 Mousery Report," ACPG; Malinin, *Surgery and Life*, p. 102; Soupault, p. 168. Soupault writes that the mouse population in Carrel's experiment at times surpassed 50,000.

9 *On Claude Bernard and physiology:* Claude Bernard, *An Introduction to the Study of Experimental Medicine*, trans. H. C. Greene (New York: Dover, 1957), p. 99. For an excellent overview of Bernard's views on vivisection see Joseph Schiller, "Claude Bernard and Vivisection," *Journal of the History of Medicine and Allied Sciences* 22 (1967): 246–270; and Alan G. Wasserstein, "Death and the Internal Milieu: Claude Bernard and the Origins of Experimental Medicine," *Perspectives in Biology and Medicine* 39 (1996): 313–326.

10 *"This happened on March 8":* T. Wood Clarke, "The Birth of Transfusion," *Journal of the History of Medicine and Allied Sciences* 4 (1949): 337–338; Herman O. Mosenthal, "Transfusion as a Cure for Melena Neonatorum, with Report of a Case," *JAMA* 54 (1910): 1613; L. G. Walker Jr., "Carrel's Direct Transfusion of a Five Day Old Infant," *Surgery, Gynecology, and Obstetrics* 137 (1973): 494–496.

11 *"They were microscopic":* AC, "Cultivation of Adult Tissues Outside of the Body," *JAMA* 55 (1910): 1554.

11 *"He did that by":* R. G. Harrison, "Observations on the Living Developing Nerve Fiber," *Proceedings of the Society for Experimental Biology and Medicine* 4 (1907): 140–143; Hannah Landecker, "New Times for Biology: Nerve Cultures and the Advent of Cellular Life in Vitro," *Studies in History and Philosophy of Biological and Biomedical Sciences* 33 (2002): 667–694.

12 *"That last innovation":* J. Michael Bishop, "What Causes Cancer: Genetic Sloppiness, the Cellular Social Contract, and Malignancy," *Harvard Magazine* 105 (March–April, 2003), Internet.

12 *"What the pilot saw"* (regarding granulation): Burton J. Hendrick, "On the Trail of Immortality," *McClure's* 40 (January 1913): 304–317.

12 *"This experiment began":* Albert H. Ebeling, "Dr. Carrel's Immortal Chicken Heart," *Scientific American* (January 1942): 22–24; J. A. Witkowski, "Dr. Carrel's Immortal Cells," *Medical History* 24 (1980): 129–142.

13 *"My results'":* AC, "Rejuvenation of Cultures and Tissues," *JAMA* 57 (1911): 1611.

13 *"It's no accident":* Landecker, "Building a New Type of Body."

13 *"And this he'd done":* AC, "On the Permanent Life of Tissues Outside the Organism," *JEM* 15 (1912): 516–528.

13 *"CARREL'S NEW MIRACLE":* *NYT,* September 14, 1913.

13 *"FLESH THAT IS IMMORTAL":* *World's Work* 28 (October 1914): 590–593.

13 *"The* New York World*":* *New York World,* June 12, 1921.

14 *On Carrel and light: MTU,* 214; Alain Drouard, *Alexis Carrel (1873–1944): De la memoire à l'histoire* (Paris: L'Harmattan, 1995), trans. for this study by Antony Shugaar, pp. 166–167; Christopher Hallowell, "Charles Lindbergh's Artificial Heart," *American Heritage of Invention and Technology* 1(1985): 58–62.

14 *"To demonstrate these":* Malinin, *Surgery and Life,* pp. 48–49.

15 *"In trying this":* AC and CAL, "The Culture of Organs, *Science* 81 (1935): 621–623; *TCOO,* p. 6; A. E. Belt, H. P. Smith, and G. H. Whipple, "Factors Concerned in the Perfusion of Living Organs and Tissues," *American Journal of Physiology* 52 (1920): 101–120; Wolfgang Boettcher, Frank Merkle, and Heinz-Hermann Weitkemper, "History of Extracorporeal Circulation: The Conceptual and Developmental Period," *Journal of the American Society of Extra-Corporeal Technology* 35 (2003): 172–183; W. B. Fye, "Julien-Jean-César Le Gallois," *Clinical Cardiology* 18 (1995): 599–560; J. J. C. Le Gallois, *Experiments on the Principle of Life* (Philadelphia: M. Thomas, 1813), pp. 130–131 (originally published in French in 1812, as *Expériences sur le principe de la vie*).

15 *"A particle in":* Corner, pp. 231–232.

CHAPTER 2 **If Man Could Learn to Fly**

17 *"Two years earlier":* *AOV,* 319; James Newton, *Uncommon Friends: Life with Thomas Edison, Henry Ford, Harvey Firestone, Alexis Carrel, and Charles Lindbergh* (New York: Harcourt Brace, 1987), pp. 242–244.

17 *"Lindbergh always referred":* Charles L. Ponce de Leon, "The Man Nobody Knows: Charles A. Lindbergh and the Culture of Celebrity," *Prospects* 21 (1996): 347–372.

18 *"I wonder if'":* "Lindbergh Wonders 'If I Deserve All This,'" *NYT,* June 11, 1927; "Why the World Makes Lindbergh Its Hero," *Literary Digest* 93 (June 25, 1927): 5–8; Edward T. Folliard, "Ships and Aircraft Welcome Memphis Bringing Flier Here," *WP,* June 11, 1927.

18 *"A young Minnesotan'":* F. Scott Fitzgerald, "Echoes of the Jazz Age," in *The Crack-Up* (New York: Scribner, 1931), p. 20.

18   *"Women see'":* "Why the World Makes."

18   *"Lindbergh's flight":* Ibid.

18   *"Six men died":* A. Scott Berg, *Lindbergh* (New York: Putnam, 1998), pp. 91, 104–105.

19   *"Dewey's world'":* "Why the World Makes."

19   *"As the* New Yorker's*":* John Lardner, "The Lindbergh Legends," in Isabel Leighton, ed., *The Aspirin Age* (New York: Simon & Schuster, 1949), pp. 190–213.

19   *"Sure, his celebrity":* Berg, pp. 189–192.

20   *"When Lindbergh arrived":* Joyce Milton, *Loss of Eden* (New York: Harper-Collins, 1993), p. 110; Ethan Mordden, *That Jazz!* (New York: Putnam, 1978), p. 245.

20   *"As a child":* AOV, p. 129; CAL to Durkin, May 20, 1966, CALPY.

20   *On Lindbergh's flunking out of college:* Berg, p. 60.

20   *"There was the earth":* AOV, p. 9.

21   *"I lived on a higher'":* SSL, pp. 261–262.

21   *"The great and rapid'":* AOV, pp. 129–130.

21   *"While making his":* Ibid., pp. 11, 394–395; SSL, pp. 389–390.

21   *"If man could learn'":* AOV, p. 130.

21   *On CAL's home in Little Falls:* Donald H. Westfall, *Charles A. Lindbergh House*, Minnesota Historic Sites Pamphlet Series, No. 22 (St. Paul: Minnesota Historical Society Press, 1994).

22   *"It was the swollen":* AOV, p. 129.

22   *"The difference between'":* Ibid.

22   *"I [saw] chickens'":* Ibid., p. 5.

22   *"I crept up'":* BUM, p. 15.

23   *"But I wondered'":* AOV, pp. 5–6, 384.

23   *"What would be":* Ibid., p. 16.

23   *"Suppose an old head":* CAL, "Alexis Carrel," Foreword to W. Sterling Edwards and Peter D. Edwards, *Alexis Carrel* (Springfield, Ill.: Charles C. Thomas, 1974), p. vi.

23   *"With science at'":* AOV, p. 130.

24   *"These organs, after":* Kenneth S. Davis, *The Hero* (Garden City, N. Y.: Doubleday, 1959), pp. 344–345; Theodore I. Malinin, *Surgery and Life* (New York: Harcourt Brace Jovanovich, 1979), pp. 103–104.

24   *"'Infection,' Carrel said":* Berg, p. 224.

24   *"As impressed as'":* AOV, p. 133.

24   *"If I could design'":* Ibid., p. 135.

24   *"Is death an'":* CAL, "A Letter from Lindbergh," *Life* (July 4, 1969): 60A–60C.

24   *"'Suppose we could'":* AOV, p. 138.

CHAPTER 3 A Student Who May Amount to Something

25   *"More often than not"*: Arthur Train Jr., "More Will Live," *Saturday Evening Post* (July 23, 1938): 5–7, 67–70

26   *"Probably the first"*: Kenneth S. Davis, *The Hero* (Garden City, N.Y.: Doubleday, 1959), pp. 62–63.

26   *"Engstrom was often"*: Joyce Milton, *Loss of Eden* (New York: HarperCollins, 1993), p. 27.

26   *"Using tongs"*: *BUM*, pp. 19–20.

26   *"This happened when"*: *AOV*, pp. 6, 59–60; *BUM*, pp. 18, 25–31; *SSL*, p. 246.

26   *" 'I took care' "*: *AOV*, p. 60.

26   *"Lindbergh fixed"*: Davis, p. 63.

27   *"That there were"*: Davis, pp. 55–57; Milton, pp. 15–18.

27   *"This was Land's laboratory"*: *AOV*, p. 309; A. Scott Berg, *Lindbergh* (New York: Putnam, 1998), pp. 39–41; Davis, p. 57; *SSL*, pp. 317–318.

27   *"This gave Lindbergh"*: *SSL*, p. 319.

27   *"The 'clear-cut language' "*: *SSL*, p. 318.

27   *"That happened in 1899"*: Jerry J. Herschfeld, "Charles H. Land and the Science of Porcelain in Dentistry," *Bulletin of the History of Dentistry* 34 (1986): 48–54; Melvin E. Ring, "Oddments in Dental History: Charles Land and His Ingenious Maxillofacial Prosthetic Appliance," *Bulletin of the History of Dentistry* 34 (1986): 46–47; L. Laszlo Schwartz, "The Life of Charles Henry Land (1847–1922)," *Journal of the American College of Dentists* 24 (1957): 33–51.

28   *"But before Lindbergh could"*: CAL to Durkin, May 20, 1966, CALPY.

28   *"Actually, Carrel had proved"*: L. Hayflick and P. S. Moorhead, "The Serial Cultivation of Human Diploid Cell Strains," *Experimental Cell Research* 25 (1961): 585–621.

28   *"Robert Hay and"*: R. J. Hay and B. L. Strehler, "The Limited Growth Span of Cell Strains Isolated from the Chick Embryo," *Experimental Gerontology* 2 (1967): 123–135.

29   *"The biologist"*: J. A. Witkowski, "Dr. Carrel's Immortal Cells," *Medical History* 24 (1980): 129–142.

29   *"One day in"*: *AOV*, pp. 392–393.

29   *"Weather permitting"*: George Kent, "Dr. Alexis Carrel Believes We Can Read Each Other's Thoughts," *American Magazine* 121 (March 1936): 20–21, 142–144; David Le Vay, *Alexis Carrel* (Rockville, Md.: Kabel, 1996), p. 157; Robert Soupault, *Alexis Carrel* (Paris: Les Sept Couleurs, 1972), p. 148.

29   *"Carrel was married"*: Le Vay, pp. 157–158; Soupault, pp. 89–90, 103.

30   *"These gatherings"*: Soupault, pp. 91–93.

30   *"Other 'philosophers' "*: Le Vay, p. 173.

30   *"Americans have"*: Soupault, p. 152.

30   *"The white race"*: *MTU*, pp. 110, 277, 291, 299; Le Vay, p. 329.

30   *"Democracy is"*: Soupault, p. 153.

30   *"Socialist governments"*: *MTU*, p. 20.

31   *"Most surgeons"*: CAL, "A Tribute to Alexis Carrel," in Robert W. Chambers and Joseph T. Durkin, eds., *Papers of the Alexis Carrel Centennial Conference* (Washington, D.C.: Georgetown University Press, 1974).

31   *"In May 1902"*: Le Vay, p. 24. Soupault gives the date as May 1903. Newspaper accounts cited by Le Vay, however, show that the controversy over Carrel's visit to Lourdes was already well under way by June 1902. AC wrote his own version of his experience at Lourdes, which was published posthumously in 1950 as *The Voyage to Lourdes* (and a year earlier, in French.)

31   *"While still on"*: Le Vay, pp. 25–29; Soupault, pp. 31–35.

33   *"The color black"*: *AOV*, pp. 134–135.

33   *"In front of him"*: Ibid.

34   *"Carrel removed"*: W. Sterling Edwards and Peter D. Edwards, *Alexis Carrel* (Springfield, Ill.: Charles C. Thomas, 1974), p. 91; Theodore I. Malinin, *Surgery and Life* (New York: Harcourt Brace Jovanovich, 1979), p. 129.

35   *" 'We were, for the' "*: CAL to Durkin, May 20, 1966, CALPY.

35   *"Even so, Carrel thought"*: "Apparatus to Circulate Liquid under Constant Pressure in a Closed System," *Science* 73 (1931): 566.

36   *" 'I have received' "*: AC to W. M. Weishaar, September 15, 1931, CALPY.

36   *"In July 1930"*: Marlen Pew, "Shop Talk at Thirty," *Editor and Publisher* 63 (July 26, 1930): 60–64. The following are worth reading on Lindbergh's problems with the press: Silas Bent, "Lindbergh and the Press," *Outlook* 160 (1932): 212–214, 240; John S. Gregory, "What's Wrong with Lindbergh," *Outlook* 156 (1930): 532–534; Donald E. Keyhoe, "Has Fame Made Lindy 'High Hat'?" *Popular Science Monthly* (July 1929): 32, 142–144; John Lardner, "The Lindbergh Legends," in Isabel Leighton, ed., *The Aspirin Age* (New York: Simon & Schuster, 1949), pp. 190–213; Charles L. Ponce de Leon, "The Man Nobody Knows: Charles A. Lindbergh and the Culture of Celebrity," *Prospects* 21 (1996): 347–372; Constance Lindsay Skinner, "Feet of Clay—or Eyes of Envy?" *North American Review* 228 (1929): 41–46; Dixon Wecter, *The Hero in America* (New York: Scribner, 1941).

36   *"Three years after"*: Susan Hertog, *Anne Morrow Lindbergh: Her Life* (New York: Nan A. Talese/Doubleday, 1999), p. 239.

36   *"The real Lindbergh"*: Morris Markey, "Young Man of Affairs—II," *New Yorker* (September 27, 1930): 30–33.

37   *"While dining"*: Soupault, p. 171.

37   *"Two cats"*: AC, "Visceral Organisms," *JAMA* 59 (1912): 2105–2106; AC, "Concerning Visceral Organisms," *JEM* 18 (1913): 155–161. For a good analysis of AC's visceral organism experiments, see Hannah Landecker, "Building a New Type of Body in Which to Grow a Cell," in Darwin Stapleton, ed., *Creating a Tradition of Biomedical Research* (New York: Rockefeller University Press, 2004), pp. 151–174.

38   *"The heart still'"*: AC, "Visceral Organisms."

39   *"An account of"*: Arthur Train Jr., unpublished biography of Alexis Carrel, CALPY.

40   *"In the spring"*: Donald E. Keyhoe, "Lindbergh Four Years After," *Saturday Evening Post* 203 (May 30, 1931): 21, 46, 48, 53.

40   *"But something else"*: Milton, p. 204.

CHAPTER 4 Isn't He in the Crib?

41   *"The first Oriental capital"*: Hugh Byas, "Lindberghs in Tokyo Cheered by 100,000 In Triumphant Ride," *NYT*, August 27, 1931.

42   *"There were tea ceremonies"*: "Lindberghs Feted at Dinner in Tokyo," *NYT*, August 29, 1931; "Lindberghs Pick Up Knack of Chopstick," *NYT*, August 30, 1931.

42   *"Lindbergh had pontoons"*: *AOV*, p. 109.

42   *"While drinking tea"*: A. Scott Berg, *Lindbergh* (New York: Putnam, 1998), p. 230.

43   *"Whole families'"*: *AOV*, p. 254.

43   *"On September 21"*: Berg, p. 230.

43   *"After Lindbergh touched"*: *AOV*, p. 252–254; CAL to Grace Nute, June 27, 1923, MHS.

44   *"Lindbergh thought of"*: *AOV*, p. 253.

44   *"A sense of being"*: Ibid., p. 255.

45   *"There was not'"*: Ibid.

45   *"After all, Lindbergh"*: Ibid.

45   *"Just before he"*: AC to CAL, April 30, 1931, CALPY.

45   *"As a precocious"*: David Le Vay, *Alexis Carrel* (Rockville, Md.: Kabel, 1996), p. 9.

45   *"When this way"*: Francis Galton, *Inquiries into the Human Faculty* (London: Macmillan, 1883), pp. 24–25.

46   *"There is no'"*: "Everybody Has Telepathic Power, Dr. Carrel Says after Research," *NYT*, September 18, 1935.

46   *"In 1928 Carrel"*: Le Vay, p. 215.

46   *"Carrel presented"*: AC, "The Immortality of Animal Tissues and Its Significance," in *Proceedings of the Third Race Betterment Conference* (Battle Creek, Mich.: Race Betterment Foundation, 1928), p. 314; Eugene Lyman Fisk, "Possible Extension of the Human Life Cycle," *Annals of the American Academy of Political and Social Science* 145 (1929): 153–201.

46   *"Most of the girls'"*: *AOV*, p. 119; Brian Horrigan, " 'My Own Mind and Pen': Charles Lindbergh, Autobiography, and Memory," *Minnesota History* 58 (2002): 2–15.

47   *"Among those attending"*: Berg, p. 231.

48   *"Charles and Anne had been gone"*: Ibid., p. 232.

49   *"The inner tube"*: Raymond C. Parker, *Methods of Tissue Culture* (New York: Paul B. Hoeber, 1938), pp. 64–65.

49   *"This simple technique"*: Ibid., p. 63.

49   *" 'The piece of chicken heart' "*: "Topics of the Times," *NYT*, January 19, 1932.

50   *" 'The world is left' "*: Berg, p. 234.

50   *"She was writing"*: Ibid., p. 237.

51   *"Sitting on 400 acres"*: Ibid., p. 233; Jim Fisher, *The Lindbergh Case* (New Brunswick, N.J.: Rutgers University Press, 1998), p. 9; Joyce Milton, *Loss of Eden* (New York: HarperCollins, 1993), p. 210.

51   *"A photograph of "*: Milton, p. 210.

52   *"He'd be so busy"*: Berg, p. 237.

52   *"A half hour later"*: Ibid., pp. 237–240; Fisher, pp. 10–12; Milton, pp. 211–214.

CHAPTER 5   The Chamber of Life

54   *"LINDBERGH BABY"*: *NYT*, March 2, 1932.

54   *"The Hearst Corporation's"*: Silas Bent, "Lindbergh and the Press," *Outlook* 160 (1932): 212–214, 240.

54   *"One wire syndicate"*: Ibid.

54   Regarding *"the average American's yearly income"*: Joyce Milton, *Loss of Eden* (New York: HarperCollins, 1993), p. 233.

54   *"The abduction of "*: "100,000 in Manhunt," *NYT*, March 3, 1932; A. Scott Berg, *Lindbergh* (New York: Putnam, 1998), p. 246; Susan Hertog, *Anne Morrow Lindbergh* (New York: Nan A. Talese/Doubleday, 1999), p. 172.

54   *"The Federal Bureau"*: Berg, p. 246.

55   *"Once inside"*: Berg, pp. 252–253; Jim Fisher, *The Lindbergh Case* (New Brunswick, N.J.: Rutgers University Press, 1998), pp. 27–28.

55   *"The tension increased"*: Fisher, pp. 79–83.

55   *" 'The boy is' "*: Berg, p. 267.

56   *"Found facedown"*: Fisher, p. 108, 116.

56   *"Soon human scavengers"*: Ibid., p. 119.

56   *"The coroner ruled"*: Ibid., p. 118.

56   *"The Lindberghs' grief "*: Milton, p. 251.

56   *" 'I'll never believe' "*: Hertog, p. 206.

56   *"Even worse"*: Ibid., p. 204.

56   *"In one theory"*: Jim Fisher, *Ghosts of Hopewell* (Carbondale Ill.: Southern Illinois University Press, 1999), pp. 92–96.

56   *"In another theory"*: Ibid., pp. 57–58.

57   *" 'There are no words' "*: David Le Vay, *Alexis Carrel* (Rockville, Md.: Kabel, 1996), p. 160.

57   *"Rather than working"*: Theodore I. Malinin, "Remembering Alexis Carrel and Charles A. Lindbergh," *Texas Heart Institute Journal* 23 (1996): 28–35.

57   *"Lindbergh's final design":* Ibid.
58   *"So in early 1933":* CAL to Durkin, May 20, 1966, CALPY; W. Sterling Edwards and Peter D. Edwards, *Alexis Carrel* (Springfield, Ill.: Charles C. Thomas, 1974), p. 91; Malinin, "Remembering Alexis Carrel."
59   *" 'As soon as we' ":* Christopher Hallowell, "Charles Lindbergh's Artificial Heart," *American Heritage of Technology and Invention* 1 (1985): 58–62.
59   *"Many of Lindbergh's chits":* personal communication from Carol Moberg of Rockefeller University.
59   *"One day at":* CAL, "A Tribute to Alexis Carrel," in Robert W. Chambers and Joseph T. Durkin, eds., *Papers of the Alexis Carrel Centennial Conference* (Washington, D.C.: Georgetown University Press, 1974), pp. 29–33.
59   *"Another time he heard":* Ibid.
59   *" 'The American is a' ":* "Americans Stake Cash against Death," *NYT,* January 12, 1913; Joseph T. Durkin, *Hope for Our Time,* (New York: Harper and Row, 1965), p. xviii.
60   *"He began his reading":* AOV, p. 137.
60   *" 'If one could' ":* J. J. C. Le Gallois, *Experiments on the Principle of Life* (Philadelphia: M. Thomas, 1813), pp. 130–131 (originally published in French in 1812, as *Expériences sur le principe de la vie*).
60   *"Brown-Séquard began":* J. M. D. Olmsted, *Charles-Éduoard Brown-Séquard: A Nineteenth Century Neurologist and Endocrinologist* (Baltimore, Md.: Johns Hopkins University Press, 1946), pp. 41–43.
61   *"Working at Carl Ludwig's":* Wolfgang Boettcher, Frank Merkle, and Heinz-Hermann Weitkemper, "History of Extracorporeal Circulation: The Conceptual and Developmental Period," *Journal of the American Society of Extra-Corporeal Technology* 35 (2003): 172–183; Robert L. Hewitt and Oscar Creech, "History of the Pump Oxygenator," *Archives of Surgery* 93 (1966): 680–696; Heinz-Gerd Zimmer, "Perfusion of Isolated Organs and the First Heart-Lung Machine," *Canadian Journal of Cardiology* 17 (2001): 963–969.
61   *"In January":* Berg, p. 291.
62   *"After weeks of trial":* CAL to Durkin, May 20, 1966, CALPY; Edwards and Edwards, p. 91; Malinin, "Remembering Alexis Carrel," p. 130.
62   *"The oil flask was made":* CAL, "An Apparatus for the Culture of Whole Organs," *JEM* 62 (1935): 409–431.
63   *"Then he'd walk":* AOV, p. 136.
64   *"In June 1934":* Berg, p. 296.
64   *"Shortly after":* AOV, p. 18; Berg, p. 296; Ben Yagoda, *Will Rogers: A Biography* (New York: Knopf, 1993), p. 316.
64   *"Rogers, whose hugely popular":* Berg, p. 143; Milton, p. 152; Yagoda, pp. 242–243.
65   *"Because she had trouble":* Hertog, p. 239.
65   *"In reality":* Ibid., p. 219.
65   *" 'Oh, God,' "said Anne":* Berg, p. 296.

65 *"Bruno Richard Hauptmann"*: Fisher, *The Lindbergh Case*, pp. 184–190.

66 *"On December 3"*: Berg, p. 305; Hertog, p. 256.

66 *"Hauptmann's trial convened"*: Fisher, *The Lindbergh Case*, p. 272.

66 *"The biggest names"*: Berg, p. 308; Fisher, *The Lindbergh Case*, p. 270.

66 *"When Lindbergh"*: Berg, pp. 309, 315.

67 *"Here is a typical exchange"*: Fisher, *The Lindbergh Case*, pp. 329–330.

67 *"Roughly 60,000"*: Ibid., p. 284.

67 *Regarding Wilentz's summation:* Berg, pp. 331–332; Fisher, *The Lindbergh Case*, pp. 364–366.

67 *"A page in"*: CALPY.

70 *"On April 2"*: Ibid.

70 *"The exterior"*: CAL, "An Apparatus for the Culture of Whole Organs," *JEM* 62 (1935): 409–431.

70 *"The nutrient medium"*: TCOO, pp. 55–74; CAL, "An Apparatus."

70 *"When the oil flask"*: CAL, "An Apparatus"; Edwards and Edwards, p. 94; Malinin, "Remembering Alexis Carrel."

71 *"A cat lay"*: TCOO, pp. 11–12.

71 *"Carrel accepted"*: Ibid.

72 *"When finished"*: Ibid.

72 *"The left thyroid"*: Ibid.

72 *"Once inside that room"*: AC and CAL, "The Culture of Whole Organs, *Science* 81 (1935): 621–623.

72 *"There was a noticeable"*: Arthur Train Jr., "More Will Live," *Saturday Evening Post* 211 (July 23, 1938): 5–7, 67–70

74 *"Without saying"*: Ibid.

CHAPTER 6 Every Act of His Is Not a Fluke

75 *"Each of these tests"*: TCOO, pp. 11–12; AC and CAL, "The Culture of Whole Organs," *Science* 81 (1935): 621–623.

76 *"Organs cultured in"*: AC and CAL, "The Culture of Whole Organs."

76 *"In one especially"*: Ibid.

77 *"After those breakthroughs"*: Ibid.

77 *"ONE STEP"*: Gobind Behari Lal, "One Step Nearer to Immortality," *New York American*, June 21, 1935.

77 *" 'The development of' "*: William L. Laurence, "Carrel, Lindbergh Develop Device to Keep Organs Alive Outside Body," *NYT*, June 21, 1935.

77 *" 'Today it came' "*: AML, *Locked Rooms and Open Doors* (New York: Harcourt Brace Jovanovich, 1974), p. 278.

77 *"The movie starred"*: Internet, turnerclassicmovies.com.

78 *"Several tabloids"*: W. Sterling Edwards and Peter D. Edwards, *Alexis Carrel* (Springfield, Ill.: Charles C. Thomas, 1974), pp. 94–95.

78 *" 'Lindbergh is considered' "*: "Carrel Lauds Lindbergh," *NYT*, September 7, 1935.

78   *"When Carrel spoke"*: "Everybody Has Telepathic Power, Dr. Carrel Says after Research," *NYT*, September 18, 1935.

78   "*'I do not necessarily'*": Ibid.

78   "*'There is no escaping'*": Ibid.

79   *"Perhaps it would"*: Ibid.

79   "*'Who on Earth'*": David Le Vay, *Alexis Carrel* (Rockville, Md.: Kabel, 1996) pp. 231–232.

79   *"One autumn morning"*: CAL, "A Tribute to Alexis Carrel," in Robert W. Chambers and Joseph T. Durkin, eds., *Papers of the Alexis Carrel Centennial Conference* (Washington, D.C.: Georgetown University Press, 1974); CAL to Durkin, May 20, 1966, CALPY.

80   *"According to Sergeant"*: "Carrel Sees Lives Extended for Ages," *NYT*, December 13, 1935.

80   "*'The weak, the diseased'*": Ibid.; AC, "The Mystery of Death," in Eugene H. Pool, ed., *Medicine and Mankind* (New York: Appleton-Century, 1936), pp. 197–217; "A Flash in Eternity," *WP*, December 15, 1935.

81   *"Sent to Carrel"*: Joyce Milton, *Loss of Eden* (New York: HarperCollins, 1993), pp. 337–338.

81   "*'I have no desire'*": Le Vay, pp. 171–172.

82   "*'Had I seen'*": Herbert S. Gasser to AC, February 6, 1936, Malinin Collection, RAC; L. G. Walker Jr., "Alexis Carrel on Science and Pseudoscience," *Surgery, Gynecology, and Obstetrics* 168 (1989): 365–370.

82   *"Reporters and photographers"*: A. Scott Berg, *Lindbergh* (New York: Putnam, 1998), p. 337; Milton, p. 341.

82   *"LINDBERGH CASE REOPENED"*: Berg, p. 338.

82   *"The* New York Daily Mirror": Ibid.

83   "*'There are millions'*": Ibid., p. 339.

83   *"Another anonymous"*: Ibid.

83   *"One fall afternoon"*: *AOV*, p. 144.

83   *"In early December"*: Berg, p. 340.

83   *"Lindbergh chose England"*: *AOV*, p. 145.

84   "*'Dear Moynihan'*": AC to Lord Moynihan, December 16, 1935, CALPY.

84   *"He invited Lyman"*: Berg, p. 340.

84   *"At 10:30 p.m."*: Ibid.

84   *"While the text"*: Ibid., p. 341.

84   *"Monday's paper"*: Lauren D. Lyman, "Lindbergh Family Sails for England," *NYT*, December 23, 1935.

CHAPTER 7 Men of Genius Are Not Tall

86   *"Men of genius"*: *MTU*, p. 62

86   *"The feeble-minded"*: Ibid., p. 271.

86   *"The Caucasian"*: Ibid., p. 110.

86   *"Childless women"*: Ibid., p. 91.
86   *"The abnormal"*: Ibid., p. 318.
87   *"Erotic activity"*: Ibid., p. 143.
87   *"Certainty derived"*: Ibid., p. 122.
87   *"Clairvoyants can"*: Ibid., p. 161.
87   *"Democracies thwart"*: Ibid., p. 21
87   *"Natural selection"*: Ibid., p. 20.
87   *"Of all the errors"*: Alain Drouard, *Alexis Carrel (1873–1944): De la memoire à l'histoire* (Paris: L'Harmattan, 1995), p. 154, trans. for this project by Antony Shugaar.
87   *"And for this reason"*: *MTU*, p. 297.
87   *"But all children"*: Ibid., p. 309.
87   *"Those who murder"*: Ibid., pp. 318–319.
87   *"The same treatment"*: Ibid., p. 319.
87   *" 'Why,' Carrel asked"*: Ibid., p. 318.
87   *"The work of creating"*: Ibid., pp. 291–292.
87   *"Most of these experts"*: Laurent Mucchielli, "Elitist Utopia and Biological Myth: The Eugenics of Alexis Carrel," *Esprit* 238 (1997): 73–94. (Trans. Antony Shugaar.)
87   *"In the preface"*: *MTU*, p. ix.
88   *"Such men"*: Ibid., p. 291; Alain Drouard, "Entre les États-Unis et la France: Alexis Carrel et la 'science de l'homme,' " in Giuliana Gemelli, ed., *Big Culture: Intellectual Cooperation in Large Scale Cultural and Technical Systems* (Bologna: CLUEB, 1994), pp. 163–181. (Trans. Antony Shugaar.)
88   *"Through sheer"*: *MTU*, p. 296.
88   *"Carrel also called"*: The originator of the word "biocracy" was another French doctor, Édouard Toulouse, who coined the phrase in his book *Le progrès civique*, published in 1920.
88   *"But this can occur"*: Drouard, *Alexis Carrel*, p. 151.
88   *" 'This catastrophe' "*: Ibid.
89   *" 'Man cannot remake' "*: *MTU*, p. 274.
89   *"Within weeks of "*: *NYT*, November 25, 1935.
89   *"A year later"*: Alice Payne Hackett, *Seventy Years of Bestsellers* (New York: Bowker, 1967).
89   *"Carrel's book was translated"*: Andrés Horacio Reggiani, "Alexis Carrel, the Unknown: Eugenics and Population Research under Vichy," *French Historical Studies* 25 (2002): 331–356.
89   *"In France"*: Drouard, *Alexis Carrel*, 206.
89   *"Carrel's German publisher"*: David Le Vay, *Alexis Carrel* (Rockville, Md.: Kabel, 1996), p. 260.
90   *"According to the latter's"*: Frederic R. Coudert, "Reminiscences of Frederic R. Coudert," unpublished. Obtained by courtesy of Coudert Brothers.
90   *"After rescuing"*: Drouard, *Alexis Carrel*, p. 201; Le Vay, p. 235.

91   *"An excerpt published":* *Reader's Digest* 29 (September 1936): 113–125.

91   *" 'For probably' ":* Raymond Pearl, "Dr. Carrel Ponders the Nature and the Soul of Man," *NYTBR*, Sept. 29, 1935.

91   *"Reviewing Carrel in":* Sir Arthur Keith, "Life from a Laboratory Window," *British Medical Journal* 2 (November 30, 1935): 1057–1058.

91   "Scouting, *the journal":* Drouard, *Alexis Carrel*, pp. 176–177; Le Vay, p. 264.

91   *"The Reverend George A. Buttrick":* "Sermons Tomorrow in City Churches," *NYT*, February 29, 1936; "Carrel Book Is Praised," *NYT*, December 30, 1935.

92   *" 'For me, personally' ":* Drouard, *Alexis Carrel*, pp. 179–180

93   *"Huxley was so":* Aldous Huxley, *After Many a Summer Dies the Swan* (New York: Harper, 1939). See also Clive James, "Out of Sight: The Curious Career of Aldous Huxley," *New Yorker* (March 17, 2003), Internet.

93   *" 'His name was' ":* Le Vay, p. 261.

CHAPTER 8   A Tiny Puff of Smoke

95   *"The 'demand' ":* CAL to AC, April 16, 1936, CALPY.

95   *" 'I wouldn't take' ":* Joyce Milton, *Loss of Eden* (New York: HarperCollins, 1993), p. 205.

96   *"The dock in Liverpool":* Kenneth S. Davis, *The Hero* (Garden City, N.Y.: Doubleday, 1959), p. 360.

96   *"In truth, Long Barn":* A. Scott Berg, *Lindbergh* (New York: Putnam, 1998), p. 347.

96   *"Carrel gave Lindbergh":* Davis, p. 363.

97   *"Shortly after 8:40":* Jim Fisher, *The Lindbergh Case* (New Brunswick, N.J.: Rutgers University Press, 1998), p. 425–427.

97   *"The United Press":* Berg, p. 352.

98   *" 'There has never' ":* Ibid.

98   *"He tried to partner":* Theodore I. Malinin, *Surgery and Life* (New York: Harcourt Brace Jovanovich, 1979), p. 144.

98   *" 'I have missed' ":* CAL to Irene McFaul, May 13, 1936, CALPY.

99   *"He began this":* Malinin, p. 147.

100   *" 'I have been thinking' ":* CAL to AC, January 1, 1937, CALPY.

100   *"When Lindbergh heard":* Malinin, p. 147.

100   *" 'Mme. Carrel is my' ":* *FN*, p. 44.

100   *" 'Mme. Carrel took' ":* Berg, pp. 354–355.

101   *" 'Had Yogic masters' ":* *AOV*, p. 368.

101   *"Lindbergh asked":* Ibid.

101   *"Lindbergh was mesmerized":* Ibid.

102   *"Carrel urged":* AC to CAL, April 7, 1936, CALPY.

CHAPTER 9 The Most Interesting Place in the World Today

103   *"The invitation"*: Max Wallace, *The American Axis* (New York: St. Martin's, 2002), p. 112.
103   *"When Smith"*: A. Scott Berg, *Lindbergh* (New York: Putnam, 1998), p. 356.
103   *" 'Although I' "*: Walter S. Ross, *The Last Hero* (New York: Harper and Row, 1976), pp. 209–210.
104   *" 'Comparatively little' "*: Wallace, p. 113.
104   *"What Germany also"*: Ibid., p. 117.
105   *"The first words"*: Ibid., p. 114.
105   *" 'The neatness' "*: *FN*, p. 83.
105   *"Charles's days"*: Berg, p. 357.
105   *" 'The sound of drums' "*: *FN*, p. 84.
105   *" 'With the Smiths' "*: Ibid., pp. 95–96.
106   *"Though Anne had"*: Ibid., pp. 97–98.
106   *"The air minister, who'd recorded"*: Biography of Hermann Göring, Internet, firstworldwar.com.
106   *" 'Why don't you work here' "*: *FN*, p. 98
106   *" 'Just like a child' "*: Ibid.
107   *" 'I had intended' "*: CAL to AC, July 27, 1936, CALPY.
107   *" '[Hitler] is undoubtedly' "*: Wallace, p. 118.
107   *"Anne, too"*: *FN*, p. 100; Berg, p. 361.
107   *" 'There is no' "*: *FN*, p. 100.
108   *"Though Carrel supported"*: F. T. Rapaport, "Alexis Carrel, Triumph and Tragedy," *Transplantation Proceedings* 19 (1987): 3–8.
108   *"Even Anne"*: *FN*, p. 102.
108   *"The details of "*: "Lindbergh Adds to Fame as Scientist," *Literary Digest* (August 22, 1936): 16–17; "Lindbergh Active at Science Session," *NYT*, August 11, 1936.
109   *" 'Carrel and an assistant' "*: "Lindbergh Bows as Scientist, Exhibits Mechanical Heart," *WP*, August 12, 1936.
109   *" 'I immediately agreed' "*: Interview with Dr. Richard Bing, February 24, 2002. Bing has written about his experiences with Carrel and Lindbergh in several medical journals. See, for example, Richard J. Bing, "The Lone Eagle's Contribution to Cardiology," *Clinical Cardiology* 22 (1999): 494–495; Bing, "Evolution in Cardiology: Triumph and Defeat," *Perspectives in Biology and Medicine* 34 (1990): 1–15; and Bing, "Lindbergh and the Biological Sciences (A Personal Reminiscence)," *Texas Heart Institute Journal* 14 (1987): 231–237.
111   *"He also declined"*: David A. Cumache to AC, August 18, 1936, Malinin Collection, RAC.
112   *" 'My Nobel Prize was used' "*: AC to *Svenska Dagladet*, October 10, 1933, ACPG.

112 *"Saint-Gildas was a spectacular place"*: *AOV*, pp. 365–367; W. Sterling Edwards and Peter D. Edwards, *Alexis Carrel* (Springfield, Ill.: Charles C. Thomas, 1974), p. vii; David Le Vay, *Alexis Carrel* (Rockville, Md.: Kabel, 1996). p. 184; Robert Soupault, *Alexis Carrel* (Paris: Les Sept Couleurs, 1972), p. 155.

112 *"The Carrels' house"*: Soupault, p. 155.

113 *"It is not pleasant"*: CAL to Raymond Parker, October 14, 1936, CALPY.

113 *"You felt the forces"*: *AOV*, p. 367

114 *"He'd created 'beasts'"*: Joseph T. Durkin, *Hope for Our Time* (New York: Harper and Row, 1965), p. 41.

114 *"Charles's aura"*: *FN*, p. 108.

115 *"Median life expectancy"*: Theodore I. Malinin, *Surgery and Life* (New York: Harcourt Brace Jovanovich, 1979), p. 223.

115 *"A vexing conclusion"*: *MTU*, pp. 183–184.

116 *"An improvement in"*: AC to CAL, November 25, 1936, CALPY.

116 *"Bing has learned"*: AC to CAL, December 22, 1936, CALPY.

116 *"I am designing"*: CAL to Robert Goddard, September 14, 1936, CALPY.

116 *"The finished tank"*: CAL to AC, January 16, 1937, CALPY.

116 *"It is quite gratifying"*: AC to CAL, January 30, 1937, CALPY.

116 *"As you know, kidneys"*: AC to CAL, May 3, 1937, CALPY.

117 *"[Robert] Yerkes"*: AC to CAL, January 21, 1937, CALPY.

CHAPTER 10 The Exploration of This Realm Is a Great New Adventure

118 *"One realized how"*: *AOV*, p. 151.

118 *"The dark, servile"*: Ibid., p. 149.

118 *"Ragged, hungry"*: Ibid., p. 150.

119 *"Lindbergh saw what"*: Ibid., p. 157.

119 *"The ongoing supremacy"*: Ibid., p. 152.

119 *"Suppose, he asked"*: Ibid.

119 *"That frightening:"*: Ibid., p. 154.

119 *"As a child"*: Ibid., p. 153.

119 *"The American takes"*: Ibid.

119 *"The goal of"*: CAL to AC, March 7, 1937, CALPY.

119 *"Lindbergh's other areas"*: Ibid.

119 *"The Lindberghs attended"*: "Lindberghs Visit Parley," *NYT*, March 2, 1937; "Lindbergh Embarrassed," *NYT*, March 4, 1937.

119 *"What other man"*: "Still Modest," *NYT*, March 5, 1937.

120 *"He would not sit"*: CAL to AC, March 7, 1937, CALPY.

120 *"He classed his"*: Ibid.

120 *"There is probably"*: AC to CAL, June 15, 1937, CALPY.

120 *"We whites"*: *AOV*, p. 152.

121 *"Lindbergh drove"*: Ibid., p. 369.

122 *"I wondered whether"*: CAL to AC, December 6, 1939, CALPY.

122 *"I found that rectal"*: *AOV*, p. 369.

122 *"He strongly advised"*: AC to CAL, May 3, 1937, CALPY.

122 *"If it is possible"*: Theodore I. Malinin, *Surgery and Life* (New York: Harcourt Brace Jovanovich, 1979), p. 143.

123 *"In February 1937"*: "See Man as Whole, Dr. Carrel Urges," *NYT*, February 22, 1937.

124 *"In reality, Anne was"*: Susan Hertog, *Anne Morrow Lindbergh* (New York: Nan Talese/Doubleday, 1999), p. 307.

124 *"While circling"*: *AOV*, pp. 364–365.

125 *"A hack for the* Sunday Express*"*: Victor Burnett, "Lindbergh Seeks the Secret of Life," *Sunday Express*, July 11, 1937, ACPG.

126 *"I am here"*: The article in *Paris Soir* was summarized on July 14, 1937, in *NYT*.

126 *"Two of Lindbergh's"*: "Lindbergh Visits France," *NYT*, August 9, 1937; "Dr. Carrel and Lindbergh Work on New Experiment," *NYT*, September 18, 1937.

126 *"Those who have"*: AC, "The Making of Civilized Men," *Dartmouth Alumni Magazine* 30 (November 1937): 8–11; "Calls on Science to Save Humanity," *NYT*, October 12, 1937.

127 *"The businessmen"*: David Le Vay, *Alexis Carrel* (Rockville, Md.: Kabel, 1996) p. 179.

127 *"The quality of tissues"*: AC, "The Prolongation of Life," *Vital Speeches of the Day* 4 (January 15, 1938): 200–202; "Carrel Predicts Longer Life by Slowing 'Clock,'" *WP*, December 4, 1937.

127 *"Prolonging the life"*: AC, "The Prolongation of Life."

CHAPTER 11 The Reaction of a Man Would Probably Be Similar

128 *"Those events included"*: "Lindbergh Ignores Today's Tributes," *NYT*, May 20, 1937; "Lindbergh Flight Celebrated by All Save 'Lone Eagle,'" *Christian Science Monitor*, May 20, 1937; A. Scott Berg, *Lindbergh* (New York: Putnam, 1998), p. 366.

128 *"Colonel Ernst Udet"*: Berg, p. 214.

129 *"There he became"*: Ibid., p. 368.

129 *"I have never seen"*: CAL to Jack Allard, November 11, 1937, CALPY.

129 *"According to Smith"*: Max Wallace, *The American Axis* (New York: St. Martin's, 2002), pp. 157–158.

129 *"The size of Germany's"*: Ibid.

130 *"The link between"*: *FN*, p. 185.

130 *"The Germans and Italians"*: CAL to Harry Davison, October 28, 1937, CALPY.

130 *"Fanaticism and crime"*: CAL to AC, September 9, 1937, CALPY.

131 *"Within minutes of "*: "Lindberghs Arrive Home on Surprise Holiday Visit," *NYT*, December 6, 1937.
131 *"Lindbergh vented"*: Berg, p. 369.
131 *"One press mogul"*: Ibid.
131 *"One reason was"*: *FN*, pp. 200–201.
132 *"'Germany,' he wrote"*: Berg, p. 370.
133 *"In posing these"*: Charles Platt, "Life Unlimited," *Cryonics* (Fourth Quarter, 1997), Internet.
134 *"It seemed likely"*: *AOV*, pp. 369–370.
134 *"A letter mailed"*: CAL to William J. Ellis, January 31, 1938, CALPY.
134 *"'Colonel Lindbergh returned'"*: AC to Simon Flexner, January 1, 1938, Flexner Papers, APSL.
135 *"Madame Carrel, in New York"*: *FN*, p. 216.
136 *"On March 11, 1938"*: Ibid., p. 224; "What the Papers Said" (about *Bremen* and *Titanic*), Internet, www.southampton.gov.uk.

CHAPTER 12 Two Men Sitting on Two Rocks

138 *"'Our conversation ended'"*: Herbert S. Gasser to AC, March 11, 1938, Malinin Collection, RAC.
138 *"'[This] is the first'"*: AC to Harvey Cushing, March 17, 1938, ACPG.
139 *"'I am constantly'"*: CAL to Paul B. Hoeber Jr., April 22, 1938, CALPY.
140 *"'More than any'"*: "Men in Black," *Time* (June 13, 1938): 40–43.
140 *"'It makes an arresting'"*: Ibid.
140 *"Nine hundred perfusion"*: William L. Laurence, "Organs Live Days in Lindbergh 'Pump,'" *NYT*, April 22, 1938.
141 *"In an editorial"*: "Medical Engineering," *NYT*, April 23, 1938.
141 *"'History is being made'"*: Waldemar Kaempffert, "Carrel, at 65, Is Deep in New Studies of Man," *NYTM* (June 26, 1938): 3, 17–18.
141 *"'The machine was never'"*: "Carrel Discredits 'Guinea Pig' Tales," *NYT*, June 28, 1938.
142 *"'Dr. Carrel looked'"*: *FN*, p. 312.
142 *"'C[harles] says'"*: Ibid., p. 313.
142 *"'Dr. Carrel asks'"*: Ibid., p. 320.
142 *"A few weeks before"*: AC to CAL, April 5, 1938, CALPY.
142 *"Even better, there"*: CAL to AC, April 8, 1938, CALPY.
142 *"How silly it is'"*: *WJ*, p. 39.
143 *"'How strangely this'"*: Joseph T. Durkin, *Hope for Our Time* (New York: Harper and Row, 1965), p. 98; Leon Sokoloff, "Alexis Carrel and the Jews at the Rockefeller Institute," *Korot* 11 (1995): 66–81.
143 *"'You brought'"*: AC to Stephen S. Wise, June 14, 1938, Malinin Collection, RAC.
144 *"His commitment to"*: AC to Simon Flexner, August 25, 1938, Flexner Papers, APSL.

144 *"So on August 7":* AOV, p. 163.

144 *"A collective farm":* WJ, p. 64.

144 *"An airplane factory":* Ibid., p. 54.

144 *" 'They command' ":* Ibid., p. 61.

144 *" 'They preach' ":* Ibid., p. 63.

145 *"The first of these":* "Lindbergh Visits Benes," NYT, September 4, 1938.

145 *"But Lindbergh warned":* WJ, p. 67

145 *"On September 10":* AOV, p. 169; WJ, p. 70.

145 *"Every German he":* WJ, p. 131.

146 *"Typical of (footnote)":* Max Wallace, *The American Axis* (New York: St. Martin's, 2002), p. 200.

146 *"In his view":* AC to Attale Guigou, September 17, 1938, Malinin Collection, RAC.

147 *"As an alternative":* WJ, p. 32.

147 *"Many of the ideas":* David Le Vay, *Alexis Carrel* (Rockville, Md.: Kabel, 1996), pp. 331–334. See also Desmond Lee, "Translator's Introduction," in *Plato's The Republic*, 2nd ed., rev. (New York: Penguin, 1987); C. D. C. Reeve, "Plato, *Republic* (c. 380 BC): The Psycho-Politics of Justice," in Jorge J. E. Gracia, Gregory M. Reichberg, and Bernard N. Schumacher, eds., *The Classics of Western Philosophy: A Reader's Guide* (Oxford: Blackwell, 2003); Karl Popper, *The Open Society and Its Enemies*, Vol. 1, *The Spell of Plato* (London: Routledge, 1945); C. C. W. Taylor, "Plato's Totalitarianism," *Polis* 5 (1986): 4–29.

148 *" 'As I gain experience' ":* CAL to AC, April 17, 1937, CALPY.

149 *" 'As time passes' ":* CAL to Truman Smith, May 9, 1938, CALPY.

149 *"The future of":* WJ, p. 110.

149 *" 'Weakness is' ":* Durkin, *Hope*, p. 93.

150 *" '[How could I] spend' ":* AOV, p. 373.

150 *" 'I decided' ":* Ibid., p. 159.

CHAPTER 13 **By Order of the Führer**

151 *"In the months":* Wayne S. Cole, *Charles A. Lindbergh and the Battle against American Intervention in World War II* (New York: Harcourt Brace Jovanovich, 1974), p. 49.

152 *"After a farewell dinner":* WJ, p. 71.

152 *"At his meeting with":* Ibid., p. 72.

152 *" 'I feel certain' ":* Cole, p. 53.

152 *"On September 22":* Peter Collier and David Horowitz, *The Kennedys* (New York: Warner, 1985), p. 94; Susan Hertog, *Anne Morrow Lindbergh* (New York: Nan A. Talese/Doubleday, 1999), p. 332.

152 *"The Luftwaffe's bomber force":* A. Scott Berg, *Lindbergh* (New York: Putnam, 1998), p. 375; Cole, p. 37.

153 *"When the definitive'"*: Joseph Driscoll, "Gave Expert Advice; English Believed Him; Factor for Peace," *WP*, October 16, 1938.

153 *"We had to make'"*: Max Wallace, *The American Axis* (New York: St. Martin's, 2002), p. 173.

154 *"My greatest hope'"*: *AOV*, p. 186.

154 *"Beginning on October 5"*: Hertog, pp. 334–335; Wallace, pp. 182–183.

154 *"All over Germany"*: Wallace, p. 183.

155 *"It had been"*: *AOV*, pp. 199–101; Berg, p. 377; "Lindbergh Tours Reich, Wide Facilities Afforded Him to Survey Aviation Development," *NYT*, October 19, 1938.

155 *"Inside was"*: "Hitler Grants Lindbergh High Decoration after Bitter Attacks on Flier by Russians," *NYT*, October 20, 1938; Berg, pp. 377–378. It is unclear whether Lindbergh wore the medal. Accounts in the *New York Times* and *Newsweek* said he did; Colonel Smith said he did not.

156 *"The albatross'"*: Hertog, p. 337.

CHAPTER 14 **I'll Take a Rain Check**

157 *"I want to express'"*: CAL to Hermann Göring, October 25, 1938, CALPY.

157 *"The most attractive'"*: *WJ*, p. 111.

158 *"The officer phoned"*: Ibid.

158 *"What we do know"*: Leonard Mosley, *Lindbergh* (Garden City, N.Y.: Doubleday, 1976), p. 237.

158 *"There is nothing'"*: CAL to AC, October 28, 1938, CALPY.

158 *"This brutal mayhem"*: Peter Gay, "My German Question," *American Scholar* 67 (1998), Internet; "Kristallnacht," *Jewish Virtual Library*, Internet.

159 *"As the historian"*: Gay, "My German Question."

159 *"It seemed so"*: *WJ*, pp. 115–116.

159 *"The outrage"*: Anne O'Hare McCormick, "Nazi Day of Terror a Threat to All Civilization," *NYT*, November 12, 1938; "Dewey and Smith Lead Protest Here against Anti-Semitic Riots in Reich," *NYT*, November 12, 1938. "Lindbergh Said to Plan to Move to Berlin Because of Reich's Aviation Research Fame," *NYT*, November 16, 1938.

159 *"There is a great deal"*: AC to CAL, November 18, 1938, CALPY.

160 *"An official of"*: "Expects Nazi Propaganda," *NYT*, November 28, 1938.

160 *"TWA dropped"*: "Drops Lindbergh Slogan," *NYT*, December 6, 1938; "Ickes Hits Takers of Hitler Medals," *NYT*, December 19, 1938.

160 *"You [must] not'"*: AC to CAL, November 30, 1938, CALPY.

160 *"With confused emotions'"*: "The Talk of the Town," *New Yorker*, November 26, 1938.

160 *"I have found'"*: CAL to AC, December 10, 1938, CALPY.

161 *"In America"*: *WJ*, p. 166.

161 *"The strength'"*: CAL to William C. Bullitt, October 26, 1938, CALPY.

162 *"She is at least"*: WJ, p. 176.

162 *"This was the only"*: Ibid., p. xviii.

162 *"But as A. Scott Berg"*: A. Scott Berg, *Lindbergh* (New York: Putnam, 1998), p. 385.

162 *"The steward"*: Ibid., p. 386.

163 *"Carrel had important"*: Joyce Milton, *Loss of Eden* (New York: HarperCollins, 1993), p. 375.

163 *"Lindy Comes Home"*: *New York Daily Mirror*, April 15, 1939.

164 *"I went first"*: WJ, p. 183.

164 *"The first mission"*: Ibid., pp. 183–184.

164 *"The president 'leaned'"*: Ibid., pp. 186–187.

165 *"Unlike Lindbergh"*: Mosley, p. 255.

165 *"He was assigned"*: Berg, p. 388.

165 *"I am glad"*: AC to CAL, April 20, 1939, CALPY.

166 *"Those experiences"*: AOV, p. 134.

167 *"It might seem"*: "Carrel Explains Mechanical Heart," NYT, May 6, 1939.

167 *"On June 3"*: WJ, p. 207.

167 *"He, too"*: Robert Olby, "The Rockefeller University and the Molecular Revolution in Biology," in Darwin Stapleton, ed., *Creating a Tradition of Biomedical Research* (New York: Rockefeller University Press, 2004), pp. 271–282.

167 *"Lederle was soon"*: "New Vaccine Checks Disease of Horses," NYT, September 4, 1938.

168 *"To Wyckoff's surprise"*: WJ, p. 207.

168 *"The place has"*: Ibid., p. 213.

168 *"In an article"*: "Heroes: Press vs. Lindbergh," *Time* (June 19, 1939).

169 *"Everyone has agreed"*: WJ, p. 216.

169 *"[We] discussed"*: Ibid., p. 245.

169 *"Whenever the Jewish"*: Berg, p. 393.

170 *"'Colonel, I'm going'"*: Kenneth S. Davis, *The Hero* (Garden City, N.Y.: Doubleday, 1959), p. 386.

170 *"YES, 80'"*: WJ, p. 247.

170 *"After Lindbergh mentioned"*: Ibid.

CHAPTER 15 **Not Merely a Schoolboy Hero, but a Schoolboy**

171 *"Mr. Lewis, this is'"*: Kenneth S. Davis, *The Hero* (Garden City, N.Y.: Doubleday, 1959), p. 388.

171 *"By the time"*: WJ, p. 256.

172 *"Using Colonel Smith"*: Ibid., pp. 257–258.

172 *"Shortly after six p.m."*: Ibid., p. 258.

172 *"Just before 9:45 p.m."*: Ibid.

172 *"In times of"*: "Lindbergh's Appeal for Isolation," NYT, September 16, 1939.

173 *"Aviation, he wrote"*: CAL, "Aviation, Geography and Race," *Reader's Digest*

(November 1939): 64–67. This article is available today at an Internet site maintained by a former Ku Klux Klan leader, David Duke. See www.storm front.org.

173 *"The Republican senator":* WJ, p. 274.
173 *"'No one ... exert[s]'":* A. Scott Berg, *Lindbergh* (New York: Putnam, 1998), p. 395.
173 *"Ninety percent":* Ibid., pp. 397–398.
173 *"One American who":* "Tunney Hits at Lindbergh," *NYT*, October 20, 1939.
174 *"As for expected":* "Lindbergh Scored by Press in London," *NYT*, October 16, 1939; "London Audience Cheers Lampooning of Lindbergh," *NYT*, October 20, 1939; Charles W. Hurd, "Lindbergh Speech Assailed in Senate," *NYT*, October 15, 1939.
174 *"He liked [the Nazis]'":* "British Host Gives Lindbergh Excuse," *NYT*, October 22, 1939.
175 *"'Bitter criticism'":* WWW, pp. 65–66.
175 *"Anne's mother":* "Mrs. Morrow Differs with Col. Lindbergh on Arms Ban, Joins Group Seeking Repeal," *NYT*, October 22, 1939.
175 *"'I am speaking'":* AML, "A Prayer for Peace," *Reader's Digest* (January, 1940): 1–8.
176 *"Just as he'd":* SSL, pp. 261–262.

CHAPTER 16 For We All Know What Awaits Us

177 *"There was so":* Alain Drouard, *Alexis Carrel*, (Paris: L'Harmattan, 1995), pp. 201–202.
178 *"Instead of uniting":* David Le Vay, *Alexis Carrel* (Rockville, Md.: Kabel, 1996), p. 269.
178 *"As was his custom":* Ibid.
178 *"Working conditions":* Drouard, pp. 201–222; Le Vay, p. 269.
178 *"'The Germans have trained'":* Le Vay, p. 270.
178 *"One day there":* Ibid., pp. 270–271.
178 *"'We are like cells'":* "Dr. Carrel Warns French of Plight If They Lose" *NYT*, December 7, 1939.
179 *"'I am disgusted'":* Le Vay, p. 271.
179 *"'I believe we are'":* Drouard, p. 237.
179 *"To the reporters":* "Noted Surgeon Returns, Reports Brittany Calm," *NYT*, May 28, 1940.
179 *"Lindbergh learned":* WJ, p. 351; "Noted Surgeon Returns."
180 *"The 'countries which'":* CAL, "What Substitute for War," *Atlantic Monthly* (March 1940).
181 *"'Carrel is able'":* WJ, p. 351.
181 *"Instead, he would":* "Lindbergh's Sourland Estate Is Offered to New Jersey for Orphanage and Camp," *NYT*, May 30, 1940.

181 *"Let us not'"*: "The Text of President Roosevelt's Address at Charlottesville," *NYT*, June 11, 1940.

182 *"[Roosevelt's] speeches'"*: *WJ*, p. 356.

182 *"Most people'"*: Drouard, p. 139.

182 *"Carrel's conversation"*: *WJ*, p. 358.

183 *" 'But it's the Nazis"*: Joyce Milton, *Loss of Eden* (New York: HarperCollins, 1993), p. 389.

183 *" 'What do you think'"*: *WJ*, pp. 360–361.

183 *" 'It might be possible'"*: Ibid.

183 *"Lindbergh wasn't immune"*: A. Scott Berg, *Lindbergh* (New York: Putnam, 1998), p. 403.

184 *" 'Germany has demonstrated'"*: CAL to Truman Smith, May 23, 1940, CALPY.

184 *" 'But if [returning]'"*: *WJ*, pp. 360–361.

184 *"The Nazis' capture"*: "Nazi Sees 'New Era' Reforming Europe," *NYT*, June 17, 1940.

184 *"This was because"*: AC, "Married Love," *Reader's Digest* 35 (July 1939): 12–16; AC, "Do You Know How to Live?" *Reader's Digest* 35 (August 1939): 19–22; AC, "Work in the Laboratory of Your Private Life," *Reader's Digest* 37 (September 1940): 7–11.

CHAPTER 17 There Is Much I Do Not Like That Is Happening in the World

186 *"A war between'"*: "Lindbergh Urges We 'Cooperate' with Germany If Reich Wins War," *NYT*, August 5, 1940; "Text of Col. Lindbergh's Speech Appealing for Peace Plea to Europe," *NYT*, August 5, 1940.

187 *" 'From 1936 on'"*: "Text of Col. Lindbergh's Speech."

187 *Regarding Kurt Vonnegut*: Max Wallace, *The American Axis* (New York: St. Martin's, 2002), p. 275.

187 *"William Yandell Elliott"*: "Plea by Lindbergh Called Cowardice," *NYT*, August 8, 1940.

187 *"Walter Winchell"*: A. Scott Berg, *Lindbergh* (New York: Putnam, 1998), p. 409; Ralph Ingersoll, "Denouncing Charles A. Lindbergh," *PM*, August 6, 1940.

187 *" 'The Lone Eagle had flown'"*: Wallace, p. 260.

187 *"They were students"*: Berg, p. 411.

188 *"Soon the organization's"*: Ibid., pp. 411–412.

188 *" 'All the logic'"*: *WJ*, p. 402.

189 *"The thirty-minute address"*: Ibid., p. 411.

189 *" '[Your] generation'"*: "Proposes Restudy of Foreign Policy," *NYT*, October 31, 1940.

189 *"Lindbergh was received"*: *WJ*, p. 411.

189 *"This response was"*: Ibid., p. 409.

189 *"Johnson: 'You are not'"*: "Urges Neutrality, Aviator Testifies He Wants Neither Side to Win Conflict," *NYT*, January 24, 1941.

190 *"'The man who built'"*: "Pepper Denounces Plea by Lindbergh," *NYT*, August 6, 1940.

190 *"Cornelius Vanderbilt Jr."*: Berg, pp. 415–416.

191 *"When he thought it"*: Ibid., p. 436; Leonard Mosley, *Lindbergh* (Garden City, N.Y.: Doubleday, 1976), p. 293.

191 *"Thirty-five thousand"*: "British Seek Another A.E.F., Lindbergh Tells 10,000 Here," *NYT*, April 24, 1941.

192 *"Many of these"*: "Lindbergh to Lead Anti-Convoy Rally," *NYT*, April 23, 1941.

192 *"Birkhead's opinion"*: "The Text of Col. Lindbergh's Address at Rally of the America First Here," *NYT*, April 24, 1941.

192 *"Lindbergh's wife"*: WWW, p. 178.

192 *"The crowd was"*: Ibid.

192 *"After a columnist"*: "Lindbergh Praised in Nazi Newspaper," *NYT*, April 24, 1941.

192 *"So when, at a"*: "President Defines Lindbergh's Niche," *NYT*, April 25, 1941.

192 *"Lindbergh was so angry"*: CAL to President Roosevelt, April 28, 1941, CALPY; "Lindbergh Quits Air Corps," *NYT*, April 29, 1941.

193 *"On May 23"*: "500 Police Ready for Peace Rally," *NYT*, May 23, 1941.

193 *"After shouts of"*: Berg, p. 419.

193 *"That so-called election"*: "Lindbergh Joins in Wheeler Plea to U.S. to Shun War," *NYT*, May 24, 1941.

193 *"'This Knight'"*: "Lindbergh Called Nazi Tool by Ickes," *NYT*, July 15, 1941.

193 *"More than 200"*: Berg, p. 421; Wallace, p. 287.

194 *"The title had"*: Wayne S. Cole, *Charles A. Lindbergh and the Battle against American Intervention in World War II* (New York: Harcourt Brace Jovanovich, 1974) p. 157.

194 *"Roosevelt's vice president"*: Ibid., p. 160.

194 *"So was the state's"*: "Press Reaction to the Lindbergh Speech at Des Moines," American Jewish Committee, October 1941.

194 *"When Roosevelt's speech"*: WJ, p. 537.

195 *"'It is now two years'"*: "Who Are the War Agitators?" Internet, pbs.org.

196 *"Writing in his journal"*: WJ, p. 538.

196 *"He traveled east"*: Ibid.

196 *"'We have sustained'"*: Mosley, p. 302.

197 *"WE ARE NOT NAZIS"*: "Press Reaction to the Lindbergh Speech."

197 *"A study just completed"*: Berg, p. 429.

197 *"A survey by"*: Wallace, p. 292.

197 *"Wendell Willkie"*: Berg, p. 428.

197   *"Thomas E. Dewey"*: "Dewey Denounces Lindbergh's Talk," *NYT*, September 15, 1942.

197   *"Walter Winchell gloated"*: Berg, p. 429.

197   *"After a contentious"*: Cole, pp. 183–184.

198   *" 'It seems that' "*: WJ, p. 539.

198   *" 'They were* far *above' "*: *WJ*, p. 552.

198   *"As he hammered"*: Ibid., pp. 558–559.

198   *" 'What's all this' "*: Joyce Milton, *Loss of Eden* (New York: HarperCollins, 1993), p. 404.

198   *" 'All that I feared' "*: WJ, pp. 565–566.

199   *" 'From my reading' "*: Ibid., p. 580.

199   *" 'Mister Charlie Lindbergh' "*: From Woody Guthrie, *This Land Is Your Land*, Asch Recordings, Vol. 1 (Smithsonian Folkways, 1997).

CHAPTER 18   The Tissues, the Blood, and the Mind of Man

201   *"He is committing' "*: Alain Drouard, *Alexis Carrel* (Paris: L'Harmattan, 1995), p. 139.

202   *"That worry turned"*: David Le Vay, *Alexis Carrel* (Rockville, Md.: Kabel, 1996), p. 281.

202   *"According to an article"*: "Lindberghs' French Home Occupied by Nazi Troops," *NYT*, October 7, 1940.

202   *"A lone Wehrmacht sentry"*: WJ, p. 493.

203   *"Carrel also called"*: Le Vay, p. 272; Robert Soupault, *Alexis Carrel* (Paris: Les Sept Couleurs, 1972), p. 211.

203   *" 'The idea of' "*: Le Vay, p. 274.

203   *" 'I will go' "*: James Wood Johnson, "We Saw Spain Starving," *Saturday Evening Post* 213 (June 28, 1941): 12–13, 38, 41–42.

204   *"The French government"*: CAL to Madame Carrel, January 29, 1941, CALPY.

204   *" 'I believe Dr. Carrel' "*: WJ, p. 444.

204   *"Carrel and Johnson left"*: Drouard, p. 204.

204   *"According to the"*: "Dr. Carrel Sails for Survey in Europe; Will Study Effects of Cold and Malnutrition," *NYT*, February 2, 1941.

205   *" 'It is with mixed' "*: Soupault, p. 213.

205   *"Leaving that environment"*: Johnson, "We Saw Spain Starving."

205   *"The medicines would"*: Soupault, p. 216.

206   *" 'I was not sent' "*: Ibid., p. 217.

206   *"France fell"*: Ibid.

206   *" 'Man cannot remake' "*: MTU, p. 274.

206   *"The situation in"*: Soupault, p. 218.

207   *"These two men were"*: Drouard, p. 206; Le Vay, p. 285.

209   *"A witness to"*: Soupault, p. 227.

209 *"CARREL BELIEVED"*: *NYT*, May 16, 1941.
209 *"THANKS VERY"*: Soupault, p. 224.
209 *"The institute they"*: Drouard, p. 208.
209 *" 'Lord, make me' "*: Ibid., p. 218.
209 *"Carrel's 'Institute of Man' "*: Ibid., p. 227.
209 *" 'The aim of' "*: Le Vay, p. 293.
210 *"When Abetz"*: Drouard, p. 220; Le Vay, p. 288; Soupault, p. 239.
210 *"Other Frenchmen"*: Le Vay, p. 296.
210 *"In the end"*: Drouard, p. 211; Le Vay, p. 296.
210 *"Offers of special treatment"*: Le Vay, p. 287.
210 *"Pierre Laval"*: Drouard, p. 241.
211 *"Even the New York Times"*: "Hitler Demanding Showdown in Vichy," *NYT*, April 16, 1942.
211 *"As Gasser explained"*: Notes of Katherine Crutcher regarding meeting with Herbert S. Gasser, February 9, 1942, Malinin Collection, RAC.
211 *"The salaries"*: Soupault, p. 230.
211 *"By the summer"*: Drouard, p. 224.
212 *"The foundation's expenses"*: Ibid.
212 *"The pro-Vichy journalist"*: Le Vay, p. 295.
212 *" 'Measures to encourage' "*: Andrés Horacio Reggiani, "Alexis Carrel, the Unknown: Eugenics and Population Research under Vichy," *French Historical Studies* 25 (2002): 331–356.
212 *"It is 'a method' "*: Ibid.
213 *" 'We're interested' "*: Le Vay, p. 301.
213 *"Missenard, a civil engineer"*: Reggiani, "Alexis Carrel."
213 *"Researchers traveled"*: Ibid.
213 *"Serge Huard"*: Soupault, p. 225.
213 *"Carrel was criticized"*: Drouard, p. 214.
214 *"Though he published"*: Le Vay, p. 295.
214 *"Those pitiful sales"*: Drouard, p. 209; Le Vay, p. 304.
214 *"He had expected"*: Drouard, p. 210.
214 *"He experienced shortness"*: Soupault, p. 243.
214 *"The first strike"*: Drouard, pp. 212–213.
215 *"Just as Carrel was"*: Ibid., p. 215.
215 *"Premier Laval"*: Ibid., p. 207.
215 *"Carrel was now living"*: Soupault, p. 237.
215 *" 'The most difficult' "*: Le Vay, p. 304.
215 *" 'My Foundation is' "*: Drouard, p. 217.
216 *"After a while"*: Le Vay, p. 305.
216 *" 'This man does not' "*: Ibid.
217 *"While being visited"*: Soupault, p. 243.
217 *"Pasteur Vallery-Radot"*: Drouard, p. 218; Soupault, p. 246.
217 *"Radio programs"*: Le Vay, p. 306.

217   *"The bored policeman":* Ibid.

217   *" 'As you can see' ":* Ibid., p. 307.

218   *"FRENCH HEALTH CHIEF":* *NYT,* August 29, 1944.

218   *" 'I was living' ":* "Dr. Carrel Denies He Aided Germans," *NYT,* September 1, 1944.

218   *" 'This man of science' ":* Melville Cane, "Alexis Carrel," *Nation* (October 28, 1944): 512.

218   *"Frederic Coudert":* Soupault, p. 248.

218   *"Simon Flexner":* Flexner Papers, APSL.

218   *" 'I hope that you' ":* AC to Katherine Crutcher, August 30, 1944, Flexner Papers, APSL.

219   *"When one doctor asked":* Soupault, p. 248.

219   *" 'It is in the hour' ":* Ibid., p. 249.

219   *"At 5:15 a.m.":* Ibid.

CHAPTER 19   The World Was Never Clearer

220   *"Lindbergh learned":* "Dr. Alexis Carrel Dies in Paris at 71," *NYT,* November 6, 1944.

220   *" 'I cannot yet' ":* CAL to Irene McFaul, May 11, 1945, CALPY.

220   *"A radio report":* David Le Vay, *Alexis Carrel* (Rockville, Md.: Kabel, 1996), p. 307.

221   *"On March 24, 1942":* WJ, p. 608.

221   *"In August 1927":* AOV, p. 23; Patricia Zacharias, "Lucky Lindy and His Ties to Detroit," *Detroit News,* Internet.

221   *"On April 2, 1942":* WJ, pp. 612–613; Austin Weber, "Willow Run: A Historical Perspective," *Assembly* (August 1, 2001), Internet.

221   *"After taking one B-24":* WJ, p. 613.

222   *"After giving birth":* A. Scott Berg, *Lindbergh* (New York: Putnam, 1998), pp. 455–456.

222   *"On September 18":* WJ, p. 717.

222   *" 'He heard the devil' ":* Leonard Mosley, *Lindbergh* (Garden City, N.Y.: Doubleday, 1976), p. 318.

222   *"A new American fighter plane":* Berg, p. 447; www.aviation-history.com.

222   *"When Lindbergh read":* WJ, pp. 723, 729.

223   *"But before Boothby":* Ibid., p. 721.

223   *"Boothby's chamber":* Ibid., p. 719.

223   *"On September 24":* Ibid., p. 721.

224   *" 'Dr. Boothby is much' ":* Ibid., p. 723.

224   *"On October 1":* AOV, pp. 25, 729.

225   *"This knowledge saved":* OFAL, pp. 6–8.

225   *"This process began":* Berg, pp. 447–448.

225   *"Lindbergh jumped at":* www.aviation-history.com.

226 *"Because Lindbergh had experience"*: Berg, p. 448; Mosley, pp. 320–321; *WJ*, pp. 755–756.

226 *"On April 24, 1944": WJ*, p. 787.

226 *"Lindbergh's only luggage"*: Mosley, p. 321; *WJ*, p. 775.

226 *" 'Since I can carry' "*: *WJ*, p. 775.

226 *"After stops"*: Mosley, pp. 321–322.

227 *"Along with three"*: *WJ*, p. 814.

227 *"Four Corsairs"*: Ibid., pp. 815–818.

227 *" 'I hope' "*: Ibid., p. 816.

227 *" 'You didn't fire' "*: Ibid., p. 818.

228 *" 'You press a button' "*: Ibid., p. 835; CAL, "Thoughts of a Combat Pilot," *Saturday Evening Post* 272 (September–October 2000): 66–70. (Reprint of article originally published in October 1954.)

228 *" 'Lindbergh from Doakes' "*: Berg, p. 453.

229 *" '[He] flies directly' "*: *WJ*, pp. 888–889.

229 *" 'Lindbergh got a Jap' "*: Berg, p. 453; "Lindbergh Downed Zero Plane in 1944 over Borneo, Says Writer," *NYT*, November 30, 1945.

229 *" 'Want to go' "*: Internet, www.charleslindbergh.com; Charles MacDonald, "Lindbergh in Battle," *Colliers* 117 (February 16, 1946): 12–14.

229 *"They took off"*: *WJ*, p. 890.

230 *" 'Bandits, two o'clock' "*: Internet, www.charleslindbergh.com; MacDonald, "Lindbergh in Battle."

230 *" 'I bank right' "*: *WJ*, p. 892.

230 *" 'I hunch down' "*: Ibid.

231 *"Japanese bodies"*: Ibid., pp. 880, 882–884.

231 *" 'I see that' "*: Ibid., p. 882.

231 *"Lindbergh found"*: Ibid., p. 885.

231 *" 'I have never felt' "*: Ibid., p. 882.

231 *" 'We are . . . constantly' "*: Ibid., p. 847.

231 *"In a letter"*: Kenneth S. Davis, *The Hero* (Garden City, N.Y.: Doubleday, 1959), p. 422.

232 *" 'I have done' "*: Berg, p. 456.

232 *"On September 20, 1944"*: *WJ*, pp. 926–927.

CHAPTER 20 The Grandeur of His Life

233 *"A technical mission"*: A. Scott Berg, *Lindbergh* (New York: Putnam, 1998), p. 463.

233 *"Its objective"*: *AOV*, p. 344.

234 *" 'The lush green' "*: *WJ*, p. 934.

234 *" 'I stand looking' "*: Ibid., p. 936.

234 *" 'Six years away' "*: Ibid., p. 938.

235 *" 'He was innocent!' "*: *AOV*, pp. 375–376; *WJ*, pp. 940–941.

236 *"Parking his car"*: *AOV*, pp. 373–374.

236 *"He walked first"*: Ibid., p. 374.

236 *" 'It was a solemn' "*: Ibid., p. 375.

237 *"He thought of Lyon"*: Ibid., p. 374.

237 *" 'Carrel's body' "*: Ibid.

238 *"He'd written a letter"*: CAL to AC, October 28, 1938, CALPY.

238 *" 'It is a city' "*: *WJ*, p. 944.

238 *"Messerschmitt had recently"*: Berg, pp. 465–466; *WJ*, pp. 955–957.

238 *"In the fall of 1944"*: Internet, www.fighter-planes.com; www.militaryfactory.com; www.warbirdalley.com.

239 *"A Soviet interrogator"*: *WJ*, pp. 969–970.

239 *"In return for"*: Ibid.

239 *"They were 'disarmed' "*: Ibid., p. 948.

240 *"It was a death factory"*: Yves Béon, *Planet Dora: A Memoir of the Holocaust and the Birth of the Space Age* (Boulder, Colo.: Westview, 1997); Internet, www.v2rocket.com.

240 *" 'I felt I had woken' "*: *WJ*, p. 992.

240 *" 'Dozens of' "*: Ibid.

241 *"It was a smell"*: Ibid.

241 *"Moments later"*: Ibid.

241 *"Thousands of identification cards"*: Ibid., p. 993.

241 *"It was a 'low, small' "*: Ibid., p. 994.

242 *"When they arrived"*: Ibid.

242 *"On the floor"*: Ibid.

242 *"The night before"*: Ibid., p. 991.

242 *" 'He is hardly' "*: Ibid., p. 995.

242 *" 'Twenty-five thousand' "*: Ibid.

242 *" 'It covered' "*: Ibid.

242 *" 'It was terrible' "*: Ibid.

243 *"He was in front"*: Ibid., p. 996.

243 *"It was a knee"*: Ibid.

243 *"His mind was frozen"*: Ibid., p. 995.

243 *" 'Here was a place' "*: Ibid.

243 *" 'A man who' "*: Ibid., p. 949.

243 *"He remembered"*: Berg, p. 469.

243 *"Now, nine years later"*: *WJ*, p. 949.

244 *" 'As far back' "*: Ibid., p. 997.

244 *"Lindbergh thought of his"*: Ibid.

244 *"That pit, and"*: Ibid., pp. 997–998.

246 *" 'The weak and the fools' "*: AC, "The Mystery of Death," in Eugene H. Pool, ed., *Medicine and Mankind* (New York: Appleton-Century, 1936), pp. 197–217; "Carrel Sees Lives Extended for Ages," *NYT*, December 13, 1935.

246 *"It was in"*: *WJ*, p. 993.

CHAPTER 21 I Felt the Godlike Power

247 *"On August 6"*: Harry S. Truman, "Statement by the President of the United States," August 6, 1945, Internet, trumanlibrary.org.

247 "Enola Gay, *Lindbergh saw*": Internet, www.boeing.com.

247 *"History showed"*: CAL to Robert M. Hutchins, November 25, 1946, CALPY.

249 *"The next morning"*: CAL to Boris Bakhmeteff and Frederic Coudert, August 8, 1945, CALPY.

249 " *'A small hole' "*: "Experiment Made on the Mummy," October 1, 1925, ACPG.

249 *"The new home"*: A. Scott Berg, *Lindbergh* (New York: Putnam, 1998), pp. 478–479.

249 *"In August 1953 (footnote)"*: "Georgetown Gets Carrel Papers," *NYT*, August 20, 1953.

250 *"On December 17"*: "Lindbergh Urges Power-Backed UNO," *NYT*, December 18, 1945.

250 *"This slender book"*: *OFAL*, pp. v–vii.

251 " *'Few nations' "*: Ibid., p. 18.

251 " *'Scientific man' "*: Ibid., p. 42.

251 " *'I grew up' "*: Ibid., pp. 49–50.

251 " *'I have seen' "*: Ibid., p. 51.

252 " *'By worshipping science' "*: Ibid., pp. 52–53.

252 " *'Man has never' "*: Ibid., p. 56.

252 "Of Flight": Berg, p. 485.

252 *"The man who was"*: Ibid., p. 486.

252 *"Lindbergh wrote* Spirit": Ibid., p. 488.

253 *"In February"*: "Lindbergh Is Named a Brigadier General," *NYT*, February 16, 1954.

253 *"The trustees"*: Charles Grutzner, "'54 Pulitzer Play Is 'Teahouse'; Lindbergh Wins Biography Prize," *NYT*, May 4, 1954.

253 " *'Boom days' "*: Berg, p. 490.

253 *"His biggest traveling year"*: Ibid., p. 495.

254 *"A meditation on"*: Susan Hertog, *Anne Morrow Lindbergh* (New York: Nan A. Talese/Doubleday, 1999), p. 432.

254 "Gift from the Sea *sold"*: Berg, p. 498.

254 " *'There is a terrible' "*: Ibid., p. 499.

255 *"He would spend"*: Ibid., p. 495.

255 " *'This family is not' "*: Reeve Lindbergh, *Under a Wing* (New York: Simon & Schuster, 1998), p. 45.

255 *"He censored"*: Ibid., p. 14.

256 *"Scott's Cove was safe"*: Berg, p. 496.

256 *"When Anne repeated"*: Ibid.

256 *"That Anne was now discussing"*: Hertog, p. 423.

256 *"The two had met"*: Ibid., p. 422.

256 *"She rented"*: Berg, p. 508.

256 *"Almost fifty years"*: "Lindy's Secret Life," *People* (August 25, 2003).

257 *"She told her sister"*: Berg, p. 509.

CHAPTER 22 Only by Dying

258 *"The stranger"*: A. Scott Berg, *Lindbergh* (New York: Putnam, 1998), p. 521.

258 *Regarding Moral Re-Armament:* Internet, religiousmovements.lib.virginia.edu.

259 *"Until Newton briefed"*: Berg, p. 521.

259 *"The admission examination"*: Ibid.

259 *"One morning Lindbergh saw"*: AOV, p. 277.

260 *"The American watched"*: Ibid., p. 276.

260 *" 'You speak of' "*: Ibid., p. 272.

260 *"In February 1964"*: Leonard Mosley, *Lindbergh* (Garden City, N.Y.: Double-day, 1976), pp. 363–364.

261 *"The spotted beast"*: CAL, "Is Civilization Progress?" *Reader's Digest* (July 1964): 67–75.

261 *" 'What progress that' "*: Ibid.

262 *" 'When I watch' "*: AOV, p. 36.

262 *" 'Only by dying' "*: Ibid.

262 *" 'How much have I' "*: "Is Civilization Progress?"

263 *" 'Had not the primitive' "*: Ibid.

263 *" 'Living in the' "*: Ibid.

263 *"After returning"*: "Lindbergh Active in World Wildlife Fund," *WP*, April 28, 1965.

263 *Regarding CAL's environmental activities: AOV*, pp. 32–35; Berg, pp. 526–528, 539–544; John Chamberlain, "These Days . . . Lindbergh and the Blue Whale," *WP*, November 5, 1966; Susan M. Gray, *Charles A. Lindbergh and the American Dilemma* (Bowling Green, Ohio: Bowling Green State University Popular Press, 1988), pp. 86–93; Mosley, pp. 361–387; "Lindbergh in Jakarta Tracking Rhinoceros," *NYT*, May 21, 1967; "Lindbergh Urges Alaskans to Protect Their Resources," *NYT*, March 20, 1968; "Lindbergh, on Taiwan Visit, Urges Conservation," *NYT*, October 31, 1968; "Appeal by Lindbergh Spurs Manila's Aid for Wildlife," *NYT*, January 29, 1969; "Philippines Is Battling to Preserve Rare Wildlife," *NYT*, April 1, 1969; Alden Whitman, "Lindbergh Traveling Widely as Conservationist," *NYT*, June 23, 1969; Alden Whitman, "The Return of Charles Lindbergh," *NYTM*, May 23, 1971.

264 *"Twice she had to"*: Berg, p. 519.

265 *" 'What does the' "*: Ibid., p. 548.

265 *"His unexpected return"*: AOV, pp. 399–400; John W. Nelson, "The Lone Eagle as Medical Researcher," *Navy Medicine* 94 (November–December 2003): 18–22; Vernon P. Perry, "Charles A. Lindbergh, 1902–1974," *In Vitro* 11 (1975): 247–250.

265   " *'Three weeks ago'* ": Nelson, "The Lone Eagle."

266   *"So instead of leaving"*: Nate Haseltine, "Lindbergh Aids Study of Tissues," *WP*, March 14, 1967.

266   " *'His industry'* ": Perry, "Charles A. Lindbergh."

267   " *'If I were entering'* ": CAL, "The Wisdom of Wildness," *Life* (December 22, 1967): 8–10.

267   " *'I have never experienced'* ": Berg, p. 537.

268   " *'What motivates'* ": CAL, "Letter from Lindbergh," *Life* (July 4, 1969): 60A–60C.

268   " *'How mechanical'* ": Ibid.

269   " *'The cycle of life'* ": Ibid.

CHAPTER 23   No, Science Has Abandoned *Me*

270   *"The NBC News anchor"*: A. Scott Berg, *Lindbergh* (New York: Putnam, 1998), p. 537.

270   *"In the period"*: "Conservation Prizes Going to Lindbergh and Senator," *NYT*, January 13, 1969; Berg, p. 544.

270   *"The Samoan"*: Berg, p. 527.

270   *"President Marcos"*: Ibid., p. 540.

271   " *'We won'* ": *WJ*, p. xv.

271   *"A front-page story"*: Alden Whitman, "Lindbergh Says U.S. 'Lost' World War II," *NYT*, August 30, 1970.

271   *"The Pulitzer Prize–winning historian"*: Ibid.

271   *"An essay in"*: Jean Stafford, "Gooney Bird," *New York Review of Books* (October 8, 1970), Internet.

272   " *'Although Lindbergh makes'* ": Ibid.

272   " *'I was reminded'* ": Interview with Dr. Richard Bing, February 24, 2002.

274   " *'How time collapses'* ": Richard Bing, "The Lone Eagle's Contribution to Cardiology," *Clinical Cardiology* 22 (1999): 494–495.

275   *"Pryor offered"*: Berg, p. 534.

275   *"In October 1972"*: Ibid., p. 549.

276   *"One day he walked"*: Ibid.

276   *"Eight papers"*: Robert W. Chambers and Joseph T. Durkin, eds., *Papers of the Alexis Carrel Centennial Conference* (Washington, D.C.: Georgetown University Press, 1974).

277   *"Dr. Charles A. Hufnagel"*: "Cardiovascular Perfusion Timeline," in *Cardiovascular Perfusion Handbook,* Internet.

277   " *'I first met'* ": CAL, "A Tribute to Alexis Carrel," in Chambers and Durkin.

279   *"On August 14"*: Berg, p. 553.

279   *"When he was finished"*: Ibid.

280   *"After a few telephone calls"*: Ibid., p. 554.

280   *"One of Lindbergh's physicians"*: Ibid.

280 *"But you're abandoning'"*: Brendan Gill, *Lindbergh Alone* (St. Paul, Mn.: Minnesota Historical Society Press, 2002), p. 174.

281 *"He'd come to Maui"*: Berg, p. 556.

281 *" 'Now, doctor,' "*: Ibid., p. 560.

281 *"For the next week"*: Ibid., pp. 556–561.

282 *" 'You're going through' "*: Ibid.

282 *"He tried other valves"*: Gill, p. 174.

EPILOGUE All of Us Are Following

283 *"In a 1937"*: John H. Gibbon, Jr., "Artificial Maintenance of Circulation during Experimental Occlusion of Pulmonary Artery," *Archives of Surgery* 34 (1937): 1105–1131.

283 *"Sixteen years later"*: Lyman A. Brewer III, "Open Heart Surgery and Myocardial Revascularization: Historical Notes," *American Journal of Surgery* 141 (1981): 618–631; Adora Ann Fou, "John H. Gibbon: The First 20 Years of the Heart-Lung Machine," *Texas Heart Institute Journal* 24 (1997); John H. Gibbon Jr., "The Development of the Heart-Lung Apparatus," *Review of Surgery* 27 (1970): 231–244. Bernard J. Miller, "The Development of Heart Lung Machines," *Surgery, Gynecology, and Obstetrics* 154 (1982): 403–414; Harris B. Shumacker Jr., "Birth of an Idea and the Development of Cardiopulmonary Bypass," in Glenn P. Gravlee et al., eds., *Cardiopulmonary Bypass: Principles and Practice*, 2nd ed. (Philadelphia, Pa.: Lippincott Williams and Wilkins, 2000), pp. 22–34.

283 *"The first implantation"*: Denton A. Cooley et al., "Cardiac Transplantation as Palliation of Advanced Heart Disease," *Archives of Surgery* 98 (1969): 619–625; Denton A. Cooley, "The Total Artificial Heart," *Nature Medicine* 9 (2003): 108–111.

283 *"The first artificial heart"*: George Raine, "Artificial Heart Implant Begun in Salt Lake," *NYT*, December 2, 1982.

283 *"The patient survived"*: Lawrence K. Altman, "Barney Clark Dies on 112th Day with Permanent Artificial Heart," *NYT*, March 24, 1983; Lawrence K. Altman, "Dr. Clark's Death Laid to Failure of All Organs but Artificial Heart," *NYT*, March 25, 1983.

283 *"An updated version"*: Barnaby J. Feder, "Regulators Approve Artificial Heart," *NYT*, October 19, 2004.

284 *"These 'neobladders'"*: Anthony Atala et al., "Tissue-Engineered Autologous Bladders for Patients Needing Cystoplasty," *Lancet* 367 (2006): 1215–1216; Rick Weiss, "First Bladders Grown in Lab Transplanted," *WP*, April 4, 2006.

284 *"Tissue-engineering projects"*: Mrunal S. Chapekar, "Tissue Engineering: Challenges and Opportunities," *Journal of Biomedical Materials and Research*, 53 (2000): 617–620; Doug Garr, "The Human Body Shop," *Technology Review* (April 2001), Internet.

284 *" 'I give sixty' "*: Interview with Dr. Anthony Atala, June 29, 2006.

INDEX

—

Abetz, Otto, 210
Aero Club, 250
After Many a Summer Dies the Swan
    (Huxley), 93
aircraft, 17, 41, 42, 43, 48, 105, 129,
    165, 221–20, 232, 238, 260–61
Aldrich, Chester, 51
Aldrin, Edwin "Buzz," 268, 270
Alexis Carrel Centennial Conference,
    276–78
America First Committee, 187–89,
    191–98, 199, 202
American Heritage, 58
American Importer, 84, 96
American Legion, 19
American Philosophical Society,
    140–41
Anders, William, 267
animal-rights activists, 9–10, 11
anti-Semitism, *see* Jews
antiseptic agents, 8, 32, 71, 76–77, 237
Apollo 11 lunar mission, 267–69, 270
"Apparatus for the Culture of Whole
    Organs, An" (Lindbergh), 76

"Apparatus to Circulate Liquid under
    Constant Pressure in a Closed
    System" (Lindbergh), 35–36
arctic wolf, 264
argonauta shell, 254, 275
Armstrong, Alan J., 227
Armstrong, Neil A., 268, 270
Arnold, Henry H. "Hap," 164, 165,
    199, 252
Aquitania, 161–63
artificial heart, 3–4, 6–7, 16, 24, 160
    CAL as possessed of, 190, 193
    implantation of, 277, 283
    perfusion pump misinterpreted as,
        78, 139, 141
artificial life, 76, 77
artificial respiration, 3, 38
artificial skin, 28
astronauts, 267–68, 270
Atala, Anthony, 283–84
Atchley, Dana, 256–57, 279
Atlantic Monthly, 180
atomic bomb, 247–48, 255–56, 271
Aublant, Louis, 217

auras, human, 114
Australia, 119
Autobiography of Values (C. Lindbergh),
    20, 23, 24, 26, 45, 101, 118–19,
    120–21, 129, 134, 150, 154, 166,
    236, 279
"Aviation, Geography and Race"
    (C. Lindbergh), 173
aviation programs, national, 102, 103,
    105, 128–30, 132, 144–45, 148,
    151–53, 155, 158, 164–65, 188,
    233, 238–39

Baker, Lillian E., 76
Bakhmeteff, Boris, 30, 181, 249
Baldwin, Stanley, 151
Baruch, Bernard, 48
Battle Creek Sanatorium, 46
battlefield brutality, 180–81, 231, 244
Bäumker, Adolf, 155, 238, 239
Beebe, Eugene, 199
Beneš, Edvard, 145, 153
Benny, Jack, 67
Bent, Silas, 54
Berg, A. Scott, 162–63, 169–70, 271
Bernard, Claude, 9–10, 126
Bernhardt, Sarah, 5
Bing, Richard, 109–11, 116, 272–74
biocracy, 88
Birkhead, Leon M., 192
Bitz, Irving, 55
blood, 60–61
    clotting of, 7, 11
    red corpuscles of, 7, 49, 50–51, 59
    transfusion of, 10–11
    white cells of, 59
Boad Nelly, 55
Bonnet, Georges, 151
Boothby, Walter M., 222–25
Borah, William E., 173

Borman, Frank, 267
Bourdel, Maurice, 90
Bouteuil, Astrid, 256–57
Brackman, Robert, 131–32
Breckinridge, Henry, 55
Bremen, 136
Brewster, Kingman, 187–88, 189
Brinkley, David, 270
Bronx Home News, 55
Brown, Séquard, Charles-Édouard,
    60–61
Buchman, Frank, 258
Bullitt, William C., 161
Burnett, Victor, 125
Burns, James MacGregor, 271
Burrage, G. H., 18
Butler, Charles, 203
Buttrick, George A., 91

cancer, 12, 142, 275–82
Cane, Melville, 218
Canfield, Cass, 90–91
cannulas, glass, 34, 70, 72
Captiva Island, 254
CardioWest artificial heart, 283
Cardozo, Benjamin J., 30
Carlsberg Biological Institute, 101,
    109
Carrel, Alexis:
    appearance of, 4, 5, 32, 139–40
    body viewed as living machine by,
        16, 28, 38–40, 73, 115, 132–33,
        269, 283–84
    Cane's light verse about, 218
    death of, 219, 220, 233, 234–38, 248,
        278
    declining health of, 214–15, 216–19
    early medical career of, 6, 31–32
    forced retirement of, 137–41,
        142–43, 161, 165, 185, 201–2

French background of, 5, 6, 45, 108
great men in philosophy of, 6, 78–79,
   83
"latent life" concept of, 8, 120
light theories of, 14
Manhattan residence of, 29, 178,
   179–80, 202
as model for Huxley character, 93
new professional base sought by, 161,
   164, 167–69, 202
Nobel Prize of, 4, 7, 23, 28, 36, 92,
   112, 138, 140, 185, 237, 278
personality of, 6, 92–93, 108, 183,
   204, 219, 273, 277–78
physiognomy practiced by, 5, 211,
   217, 272
political views of, 30–31, 45–46,
   86–89, 110, 113–16, 117, 118,
   123, 130, 137–38, 146–47, 149,
   163, 165, 168, 172, 176, 189, 206,
   248, 252
resurrection and legal liability of, 39
at Rockefeller Institute, *see*
   Rockefeller Institute,
   Experimental Surgery Division of
surgical skills of, 10–11, 32–33,
   71–72, 166
on *Time* magazine covers, 92,
   139–40
Carrel, Anne de la Motte de la Mairie,
   29–30, 109, 112, 116, 124, 135,
   140, 163, 204, 211, 214, 215, 234,
   236
AC counseled by, 207–8
at AC's deathbed, 219, 220
in French army, 179, 183
Man, the Unknown defended by, 90
Nazi looting endured by, 202–3
occult interests of, 100–101, 104,
   114, 165, 178, 216

Carrel, Joseph, 178, 179, 202, 214, 215
Carrel-Dakin solution, 8, 32, 71, 237
Carrel Foundation, 207–19, 235, 277,
   278
   academic criticism of, 213–14
   coup organized against, 214–15
   employees of, 211–13, 214–15,
      217–18
   eugenics program of, 212–13
   funding of, 208, 215
   headquarters of, 210, 211, 214–15
   "high council of experts" in, 207–8,
      214
Carrel-Lindbergh relationship, 37, 57,
   69, 92, 108, 165–66, 235–38,
   272–74, 277–78
   CAL's move to England in, 83–84
   correspondence in, 98, 100, 101–2,
      107, 117, 119, 120, 122, 130, 158,
      159–60, 238
   first meeting in, 2, 4–16, 272
   Gasser's disapproval of, 138–39, 203
   long conversations in, 30–31, 32, 89,
      113–16, 142, 145–50, 151, 176,
      203–4, 237
   role reversal in, 183, 184, 186–87
   scientific partnership in, 24, 28
Cass Technical High School, 221
Castle, William R., 169
cell division, 28–29, 58
cell-mediated immunity, 59
centrifuges, 48–49, 59, 167
Century Association, 30, 143, 167, 203
Chamberlain, Neville, 151, 152, 153
Champlain, 179
Chase, Merrill, 58–59
Chaulnes, duchess de, 235
Chiang Kai-shek, 42
chick-heart tissue culture, 12–14, 49,
   79, 166, 272

chick-heart tissue culture (*cont.*)
  apparent immortality of, 13–14,
    28–29, 59, 73, 132
China, 42–45, 48, 50, 95, 271
  CAL's flood relief flights in, 42–44
  Western civilization as threatened by,
    44–45, 47, 161, 165, 176, 248
Chinese National Flood Relief
    Commission, 42–43
Churchill, Winston, 5, 153
civilizations, 14, 261–63
  ancient, destruction of, 118, 119, 260
  death as aid to, 80
  elite guidance of, 148
  great changes resisted in, 147–48
  as white, 118–19, 135
clairvoyance, 82, 87, 91, 94
Coli, François, 18–19
Collins, Michael, 268, 270
communism, 129, 143–44, 146, 148,
    180, 243
Compiègne, 180, 237
Concorde supersonic airplane, 260
Condon, James F. "Jafsie," 55, 65
Congress, U.S., 22, 32, 46, 47–48, 169,
    280
  CAL's testimony before, 189–91,
    264
Cooley, Denton, 283
Coolidge, Calvin, 2, 5, 18, 47
coronary artery bypass surgery, 4, 6
corpuscle washing machine, 49, 50–51,
    59
Coudert, Frederic, 30, 79, 90, 203, 205,
    211, 218, 234, 249
courage, 7, 18–19, 129–30
Crutcher, Katherine, 178, 203, 209,
    218–19
cryobiology, 133, 265–67
cryosurgery, 122

Culture of Organs, The (Carrel and
    Lindbergh), 125, 126, 139, 284
"Culture of Whole Organs, The"
    (Carrel and Lindbergh), 76–77
Curtiss P-36A fighter plane, 165
Cushing, Harvey, 138
Cyon, Élie de, 61
Cytology Congress, 101–2, 104,
    108–11, 112, 113, 272
Czechoslovakia, 145, 146, 151–54

Dakin, Henry J., 8, 32, 71, 237
Daladier, Édouard, 151, 153
Daniel Guggenheim Fund to Promote
    Aeronautics, 20
Dartmouth College, 126–27, 219
Darwin, Charles, 7, 45–46
Davis, Noel, 19
Davison, Harry, 107, 130
death, 16, 28, 33, 40, 60, 121, 122, 132,
    231, 274, 282
  as "accidental" phenomenon, 13, 39
  as AC's lecture subject, 79–80,
    81–82, 93–94, 137
  civilization as aided by, 80
  natural, 21–23, 24, 35, 66
  Nazism as pagan cult of, 108, 147,
    181, 247
  necessity of, 252, 262, 268–69
  of tissue cultures, 29
  visceral-organism experiment and,
    38
democracy, 106–7, 182, 255
  AC's rejection of, 30, 46, 78–79, 86,
    87, 88–89, 91, 147
  biocracy vs., 88
  CAL's view of, 107, 160, 176, 191
  meritocracy vs., 87, 88, 123
Dempsey, Jack, 174
Denmark, 101–2, 104, 108–11, 113

Des Moines Register, 194
DeVries, William C., 287
Dewey, George, 19
Dewey, Thomas E., 197
Diamond, Jack "Legs," 55
Donzelot, Édouard, 214, 216
Dora concentration camp, 240–46, 247, 253
  crematorium at, 241–43
  rodent-like stench of, 241, 242, 245–46, 249
Dornier DO 17 airplane, 129
Driscoll, Joseph, 153
Dr. Seuss, cartoons of, 187
du Moulin de Labarthète, Henri, 209
du Noüy, Pierre Lecomte, 100
Durkin, Father Joseph T., 249*n*, 266, 276

Earle, E. P., 81
Einstein, Albert, 79, 277
Eisenhower, Dwight D., 253
Elisabeth Morrow School, 83
elitism, 47, 87–88, 91, 93, 95–96, 113–14, 148, 248
Elliott, William Yandell, 187
Ellis, William, 81, 134
embryo juice, 29
England, *see* Great Britain
Engstrom, Martin, 26
Enola Gay, 247
eugenics, 45–47, 78–80, 86, 87–89, 108, 113, 126–27, 213, 245–46
  demographic survey in, 123, 212–13
  euthanasia in, 87
  in Nazi Germany, 89, 104–5, 111
  selective breeding and, 46, 47, 87, 114
  Spanish children and, 205, 220
  for U.S. air corps, 165

U.S. as in need of, 119, 123
  wars and, 46, 203, 205, 272
  see also "high council of experts"
evolution, 45, 261–63
  CAL's role in, 7
  natural selection in, 9, 30–31, 80, 87, 245–46, 262, 273
experimental medicine, field of, 9–10, 146–47

F4U Corsair fighter plane, 225–28
Fabre-Luce, Alfred, 212
Felmy, Helmuth, 152–53
Firestone, Harvey, 169
Fischer, Albert, 101–2, 109
Fitzgerald, F. Scott, 18
Flagg, Paluel, 2, 3–4
Flexner, Abraham, 98
Flexner, Simon, 81–82, 90, 98, 134–35, 144, 218
Flynn, John T., 188
Fontanne, Lynn, 67
Ford, Gerald, 187
Ford, Henry, 188, 221–22, 272
Ford Motor Company, 221–22, 225, 232
Ford Trimotor 40-AT-10 airplane, 17
France, 6, 19, 30, 37, 45, 60–61, 78, 89–90, 104, 106, 107, 123, 129, 135, 160, 222, 237, 260, 271
  AC's early medical career in, 31–32, 215
  CAL's celebrated landing in, 1
  declining birthrate of, 212–13
  deficiencies of, 148, 152, 165, 178–79, 180, 183, 187, 206, 209
  inbred gibbon colony in, 100
France, in World War II, 136, 145, 152, 169, 177–84, 201–19

France, in World War II (*cont.*)
AC as suspected collaborationist in, 205, 210–11, 214, 216, 217–19, 220, 234, 235, 272, 273
AC's personal mission to, 203–7, 220
AC's radio address to, 178–79
fall of, 179, 182–84, 201, 206
liberation of, 205, 216
resistance forces in, 201, 205, 217
Vichy government of, 202–3, 204, 205–7, 208–9, 210–11, 212, 216
Franco, Francisco, 205
Freud, Sigmund, 19
Friends of Democracy, 192

Galileo Galilei, 119
Gallavardin, Louis, 37
Galton, Francis, 45–46
Gannett, Frank E., 131, 197
Garbo, Greta, 256
Garden City Hotel, 111
Gasser, Herbert S., 81–82, 93–94, 137–39, 145, 161, 167, 203
Gay, Peter, 159
Geist, Raymond, 156
"General Estimate (of Germany's Air Power) of Nov. 1, 1937" (Smith and Lindbergh), 129–30
George VI, King of England, 169
Georgetown University, 266, 270
AC's materials at, 249*n*
Alexis Carrel Centennial Conference at, 276–78
Germany, Nazi, 103–8, 113–14, 136, 144, 145–50, 151–56, 176–85, 201–19, 251, 273
AC's anti-German sentiments and, 89, 108, 130, 146, 176, 179, 180–84
aviation program of, 102, 103, 105, 128–30, 132, 145, 148, 151–53, 155, 158, 164–65, 188, 233, 238–39
brutal excesses of, 148, 154, 180–81, 190, 240–46, 271
CAL as propaganda tool of, 160–61
CAL's admiration of, 102, 106–7, 129–30, 147–50, 157–61, 174–75, 245, 250, 274
CAL's postwar mission to, 233–46, 247, 248
as death of Western civilization, 108, 181, 182–83
expansion as "natural right" of, 180, 181
Jews persecuted by, 104–5, 111, 143, 145–46, 154–55, 158–59, 160, 184, 195, 196, 244
Lindberghs' proposed residence in, 157–58, 159–60
Man, the Unknown as published in, 89
Nuremberg laws of, 104–5, 111
as pagan death cult, 108, 147, 181, 247
rocket program of, 233, 239–46; *see also* Dora concentration camp
as savior of Western civilization, 121, 129–30, 147, 148–50, 151–52, 161, 165, 180–81, 186–87, 243, 248–49
see also France, in World War II; Hitler, Adolf; World War II, CAL's isolationist stance on
Gibbon, John H., 283
Gifford, George, 238
Gift from the Sea (A. M. Lindbergh), 254–55, 258, 275
Gillon, Jean-Jacques, 212, 214, 215, 219
Gish, Lillian, 188

Glory That Was Greece, The (Stobart), 162

Goddard, Robert H., 116, 267

Goebbels, Joseph, 202

gold certificates, 65–66

Gone with the Wind (Mitchell), 89

Göring, Hermann, 103–4, 108, 152–53, 157, 202, 274
  Lindberghs photographed with, 106, 271
  Nazi medal presented by, 155–56, 193, 238

Gow, Betty, 52, 53

Grant, J. B., 43–44

Great Britain, 2, 83–85, 95–102, 106, 113, 123–24, 129, 131, 135–36, 144, 145, 148, 152, 187, 189–90, 192, 238, 260, 271
  CAL's return to U.S. from, 161–64
  CAL's scientific research in, 96–100, 116–17, 121–22, 124
  CAL's stance criticized in, 174–75
  deficiencies of, 148, 165, 180, 187
  Long Barn residence in, 95, 96, 116, 121–22, 124, 128, 148, 174
  medical community of, 84, 97, 98, 101
  press of, 96, 125–26, 158
  in World War II, 136, 145, 152–53, 169, 174, 181–82, 189–90, 192, 195–96, 216

Gros, André, 215, 219, 277

Grynzspan, Herschel, 159

Guthrie, Woody, 199–200

Hamburger Fremdenblatt, 192

Hammerstein, Oscar, 191

Hand, Learned, 48

Harper publishers, 79, 90–91

Harris, Hayden B., 99

Harrison, Ross G., 11

Hauptmann, Bruno Richard, 65–67
  execution of, 82, 97–98
  trial of, 66–67, 82–83, 92n, 172

Hay, Robert, 28–29

Hayflick, Leonard, 28, 29

heads, severed, 15, 23, 61, 78

heart, 6, 8, 13, 62, 63, 265–67
  mitral valve of, 3, 34–35
  see also artificial heart; chick-heart tissue culture

Heinkel Corporation, 105

Heinkel, Ernst, 155

helicopter, German, 129

Hepburn, Katharine, 256

Hesshaimer, Brigitte, 257, 258

high-altitude flying, 222–25, 226, 232

"high council of experts," 87–88, 89, 91, 123
  of Carrel Foundation, 207–8, 214
  immortality and, 88, 114–16, 147

High Fields estate, 48, 50, 51–53, 54, 55, 61, 168
  as "Institute of Man" site, 168–69, 177, 181

Hinghwa, 43

Hitler, Adolf, 103, 105, 106–8, 136, 176, 194, 196, 274
  CAL's admiration of, 102, 106–7, 111, 147–48, 174
  CAL's appeasement efforts and, 145–46, 150, 151–54, 192
  as "cleansing wave," 147–48, 154–55, 184
  Göring's photograph of, 106, 271
  legacy of, 243–44, 248–49
  in Munich pact, 153, 154, 156
  Nazi medal bestowed by, 155–56, 157, 158, 159–61, 174, 193, 238
  Poland invaded by, 170, 171, 180

Hitler Youth, 105, 154
Hoeber, Paul B., Jr., 139
Hoffman, Harold S., 82
Hoffman, Heinrich, 106
Hoover, Herbert, 30, 169, 173, 199
Hoover, J. Edgar, 92
Hopf, Otto, 33–34, 58, 63, 67, 70, 71,
    76, 98, 133, 248
Howell, Milton, 279–81
Huard, Serge, 213
Hufnagel, Charles A., 277
Hull, Cordell, 151, 152
Huxley, Aldous, 93
Hyman, George, 275–76
hypothermia experiments (artificial
    hibernation), 101, 119, 121–22,
    124, 130, 142, 144, 222
  heat-regulating mechanism in,
    133–34, 224

Ickes, Harold, 160, 191, 193
Illiec island, 140, 141, 144, 145, 146,
    151, 158, 161–62, 183, 250
  house on, 135, 160, 236
  Nazi looting of, 201–2
  purchase of, 135–36
Illinois, University of, 123, 212
immortality, 60, 77, 78, 92, 126–27,
    140, 216, 274, 282
  apparent, of chick-heart tissue
    culture, 13–14, 28–29, 59, 73,
    132
  CAL's quest for, 21–24, 113, 247
  of chosen few, 47, 48, 80, 88, 93,
    114–16, 127, 147, 237
  genetic, 262–63
  replaceable body parts in, 16, 28,
    39–40, 63, 73, 115, 132–33, 269
  as ultimate research goal, 60, 63, 80,
    81–82, 94, 113, 115–16, 121, 130,

    132–33, 146, 149, 167, 237, 248,
    252, 262, 268–69, 278
  yogis and, 101, 119, 120, 121; *see also*
    hypothermia experiments
"Immortality of Animal Tissues and Its
    Significance, The" (Carrel), 46
immune system, 15, 59, 284
India, 101, 117, 118–21, 224, 248
  Parliament of Religions of, 119
  poverty observed in, 118–19
  see also yogis
Indonesia, 264, 267
infection, 7, 11, 12, 132
  antiseptics for, 8, 32, 71, 76–77, 237
  as perfusion problem, 15, 24, 33, 35,
    40, 62–64, 70, 73, 75, 77
Ingersoll, Ralph, 187
"Institute of Man," 114–16, 202, 206,
    207–19, 278
  AC's speaking campaign for, 114–15,
    123, 126–27, 137, 212, 219
  in France, *see* Carrel Foundation
  funding of, 177, 181, 258
  High Fields as proposed site of,
    168–69, 177, 181
  planning of, 167–69, 170
  see also "high council of experts"
intercontinental survey flights, 41–45,
    47, 50, 95
"Is Civilization Progress?" (C.
    Lindbergh), 261–63

Japan, 41–42, 50, 161, 165, 176
  in World War II, 198, 225–32, 244,
    247–48
Jarvik, Robert, 283
Jews, 91–92, 107, 137, 147, 272
  alleged U.S. "influence" of, 169,
    195–98
  Nazi persecution of, 104–5, 111, 143,

145–46, 154–55, 158–59, 160,
184, 195, 196, 244
as "race," 143–44, 184, 194
in *Wartime Journals*, 162–63, 169–70,
271
Johnson, James Wood, 203, 204–7, 209
Johnson, Luther A., 189–90
Johnson, Lyndon, 264
Journal of Experimental Medicine, 13, 76
*Journal of the American Medical
Association*, 9, 13, 38
Jovanovich, William, 162, 271, 279

Kaempffert, Waldemar, 141
Karloff, Boris, 77
Keith, Arthur, 91
Kellogg, John Harvey, 46
Kennedy, Joseph P., 151–52
Kenya, 259–63
Keyhoe, Donald E., 40
kidney, mechanical, 96–99, 113
kidney transplants, 14–15
Kimberling, Mark O., 97
Kipahulu, 275, 276
Kissinger, Henry, 187
Konchellah, Jilin ole "John," 258–63
Kristallnacht, 158–59

La Guardia, Fiorello, 128
Lambert, Adrian V. S., 10–11
Land, Charles, 27, 123, 166
Langendorff, Oscar, 61
Laplace, Madam, 216
Lardner, John, 19
Laurence, William L., 77
Laval, Pierre, 210, 211, 213, 215, 217
Le Gallois, Julien-Jean-César, 15, 60,
76, 77, 248
Lehman, Herbert, 128
Lend-Lease law, 189–90

"Letter from Lindbergh," 268–69
Lewis, Fulton, 169, 170, 171, 172
Libman, Emanuel, 143
Life, 267–69
Life and Death at Low Temperatures
(Luyet), 133
life expectancy, 115
life extension, 119, 120, 122, 133, 134,
166, 223, 224
life stream, 29, 262
Lilienthal Society for Aeronautical
Research, 128, 154
Lindbergh (Berg), 162–63, 271
Lindbergh, Anne Morrow, 40, 48, 77,
82, 84–85, 141, 142, 162, 165,
171, 174, 231–32, 278–82
as author, 50, 95, 114, 175–76, 194,
254–55, 258, 274, 275
Brackman portrait of, 131–32
CAL's first meeting with, 2–3
CAL's isolationist stance supported
by, 175–76, 192, 193
Charles Jr.'s kidnapping and, 51–53,
55, 56, 65
Darien residence of, 249–50, 253,
254–56, 261, 264, 276, 278–79
first flight described by, 3
in Germany, 102, 105–6, 107–8, 128,
154–55, 156, 157–58, 274
intercontinental survey flights
crewed by, 41–42, 50, 95
Lloyd Neck residence of, 167, 170,
173
Madame Carrel and, 100, 114, 124
marital relationship of, 253–57,
264–65, 274, 278
in Maui, 274–75, 276, 278, 279–82
pregnancies of, 2, 3, 36, 50, 55,
61–62, 118, 123–24, 221
psychiatrist consulted by, 256

Lindbergh, Anne Morrow (*cont.*)
  romantic affair of, 256–57, 279
  in Switzerland, 258–59
  wedding of, 3, 46, 253
  Westport residence of, 232, 249
Lindbergh, Anne Spencer "Ansy," 249,
  255
Lindbergh, C. A. (Charles's father),
  22–23, 26–27, 44, 110, 119, 166,
  235, 256, 281
  as congressman, 22, 32, 46, 191, 280
  death of, 32
Lindbergh, Charles A.:
  appearance of, 2, 35, 139, 163, 267
  in army air corps reserves, 164, 165,
    170, 171, 192–93, 198–99, 253
  bioengineering ideas of, 3–4, 5–7,
    16
  Brackman portrait of, 131–32
  childhood of, 6, 13, 20, 21–23,
    25–28, 44, 119, 135, 249, 253,
    280, 281
  death of, 282
  declining health of, 275–82
  early failures of, 20, 148, 188
  education of, 7, 20, 148–49, 188
  as environmentalist, 263–65, 267,
    269, 270, 274, 275
  family farm of, 6, 13, 21, 26, 47
  global travels of, 253–57, 258–65,
    274, 275
  as godlike, 20–21, 27, 176, 251–52
  grave planned by, 276, 281, 282
  hate mail received by, 82–83, 85
  Hauptmann trial testimony of,
    66–67, 82–83, 172
  human evil encountered by, 243–46
  Manhattan apartment of, 61–62, 64
  mechanical engineering skills of,
    25–28, 39, 40, 57–59

natural death as mystery to, 21–23,
    24, 35, 66
  nuclear war feared by, 247–48,
    255–56, 271
  as parent, 121, 135, 198, 250,
    255–56, 265
  personality of, 18, 26, 36–37, 57,
    110, 174–75, 191, 272
  popularity regained by, 252–53, 270
  press attention resented by, 17, 36,
    39, 50, 56, 82, 113, 126, 130, 131,
    135, 160–61, 168, 174
  Princeton residence of, 2, 24, 40
  privacy guarded by, 2, 36, 50, 57,
    160, 168
  public adulation as bewildering to,
    17–18
  Pulitzer Prize of, 253
  retrospective self-examination of,
    240–46, 247–53, 260, 261–63
  romantic affair of, 256–57, 258
  science as viewed by, 17, 23, 27, 35,
    47, 50, 246, 247–49, 250–52, 263,
    267, 268–69, 273–74, 280
  single-minded concentration of, 57,
    64, 266–67
  on *Time* magazine cover, 139–40, 168
  Utah desert epiphany of, 17–21, 23,
    35, 50, 166, 247, 262
  wealth of, 19–20
Lindbergh, Charles A., transatlantic
  flight of, 1–2, 17–20, 40, 46, 54,
    161–62, 221, 222, 247, 250, 261,
    266, 267, 280
  CAL's motives for, 268
  courage shown by, 7, 18–19
  financial rewards of, 19–20
  heroic stature bestowed by, 7, 17–19,
    36, 42, 49–50, 85, 168
  Ortieg Prize for, 18, 19

tenth anniversary celebration of, 128
twenty-fifth anniversary celebration
of, 252
Lindbergh, Charles A., Jr., 2, 36, 40,
41, 48, 50
kidnapping of, 51–53, 54–57, 65–67,
87, 107, 139, 146*n*, 174
body discovered, 55–56, 57, 61
ladder involved in kidnapping, 54,
66, 67
manhunt for kidnapper, 54–55
ransom paid for, 55, 65–66
see also Hauptmann, Bruno Richard
Lindbergh, Evangeline Land, 22–23,
26–27, 57, 83, 104, 221, 253, 256,
281
Lindbergh, Jon, 64, 82, 131, 135, 162,
165, 198, 278, 279, 280, 281
birth of, 61–62
career of, 255
in England, 95, 96, 121
on Illiec, 141, 144, 158
threats against, 83, 84, 85
Lindbergh, Land, 123–24, 131, 158,
162, 165, 198, 255, 278, 279, 281
Lindbergh, Reeve, 249, 255, 264–65
Lindbergh, Scott Morrow, 222, 255,
279, 281
Lippmann, Walter, 48
Little School, 83
Liu, J. Heng, 43–44
Lloyd George, David, 151
Lockheed Sirius airplane, 41, 42, 43, 48
L'Oiseau Blanc, 18–19
Longworth, Alice Roosevelt, 188
Lourdes, shrine of, 31–32
Lovell, James, 267
Ludwig, Carl, 61
Luftwaffe, 103, 105, 128–29, 152–53,
155, 158, 164, 238

Luyet, Basile, 133
Lyman, Lauren D., 84–85
Lyon, University of, 31–32, 215, 237

MacArthur, Douglas, 92
MacDonald, Charles, 229–30
Mackaye, Percy, 219
McCoy, Frank Ross, 132
McFaul, Irene, 98–99, 220
Malinin, Theodore I., 265–67, 277
Man, the Unknown (Carrel), 79, 86–94,
203, 217, 237
colleagues' response to, 92–93
Gasser's reaction to, 93–94
ideas expressed in, 86–89, 110,
114–16, 165, 206, 207–8, 214
Platonic ideas echoed by, 147
publishers of, 89–91
sequel to, 216
sermons on, 91–92, 143
success of, 89–91, 93, 95, 114
Time magazine cover story on,
92
writing of, 90
Marcos, Ferdinand, 263–64, 270
Marcos, Imelda, 278
Markey, Morris, 36–37
Marquand, John P., 188
Masai tribe, 258–63
Maui, 274–75, 276, 278, 279–82
Mayo Clinic, 222–25
media coverage, 35–37, 42, 46, 50, 80,
113, 119, 162, 179
of AC's mission to France, 204, 206,
209, 211
of AC's speeches, 123, 140–41
CAL's harassment by, 83
of CAL's isolationist stance, 173–75,
187, 190–91, 193, 194, 196–98
of CAL's Nazi medal, 156, 159–61

media coverage (*cont.*)
  of CAL's transatlantic flight, 1, 18, 19, 20, 49
  of Charles Jr.'s kidnapping, 54, 56, 139
  of chick-heart tissue culture, 13, 49
  of Hauptmann's appeals, 82–83
  of Hauptmann trial, 66–67
  interventionist sentiment of, 169, 172
  of Kristallnacht, 158–59
  of Lindberghs' arrival in England, 96
  of *Man, the Unknown*, 91
  of Nuremberg laws, 104–5
  of perfusion pump invention, 77–79, 82, 139–41
  of Saint-Gildas, 125–27
  of *Wartime Journals*, 271–72
melena neonatorum, 10–11
"Memorandum as to a Proposed Center of Integrated Scientific Research" (Carrel), 168
Messerschmitt, Willy E., 155, 238–39
metabolites, 13, 14
Methods of Tissue Culture (Parker), 49
Milch, Erhard, 155
Miller, Danforth, 230
Missenard, André, 90, 207–8, 211–13, 214, 215
"Mister Charlie Lindbergh" (Guthrie), 199–200
Mitsubishi Ki-51 fighter plane, 228–29
monkey-eating eagle, 263–65, 274
Mookerjee, H. C., 120
Moorhead, Paul, 28, 29
Moral Re-Armament (MRA), 258–59
Morgan, Aubrey, 64–65, 175
Morgan, Elisabeth Morrow, 14, 16, 47, 55, 83, 175
  death of, 66, 67

heart condition of, 3, 5, 6, 34–35, 56, 64–65, 66
Morgan, J. P., and Company, 48
Morgenthau, Henry, Jr., 163, 197
Morrow, Betty, 41, 50, 55, 64, 84, 107, 131, 175, 256, 257
Morrow, Constance, 55, 175, 257
Morrow, Dwight, 2, 174
  funeral of, 47–48
Morrow, Dwight, Jr., 55, 256
Morrow, Elisabeth, *see* Morgan, Elisabeth Morrow
mousery experiment, 9, 30–31, 66, 178, 245–46, 249
mummy-revivification experiment, 249
Munich pact, 153, 154, 156
Mussolini, Benito, 92, 153
"Mystery of Death, The" lecture (Carrel), 79–80, 81–82, 93–94, 137

Naidu, Sarojini, 119
Nanking, flooding of, 42–43, 47
NASA, 267–68, 270
Nation, 218
Navy Medical Research Institute, 265–67, 277
necrosis, 63, 116
nerve tissue culture, 115, 117
Neutrality Act, 169, 175
New Republic, 188
Newton, James, 168–69, 258–59
New York Academy of Medicine, lecture series of, 79–80, 81–82, 93–94, 137
New Yorker, 19, 36–37, 160
New York Herald Tribune, 36, 37, 39, 136, 196
New York Times, 1, 13, 42, 46, 48, 49, 54, 66, 77, 78–79, 80, 84, 91, 92*n*,

119, 131, 141, 160, 167, 179, 184, 192, 196, 202, 204, 209, 211, 218, 220, 252, 253, 271

New York World, 13, 18, 19

Next Day Hill, 3, 41, 47–48, 50, 51, 52, 56, 61, 64, 82, 83, 131, 132, 164, 165, 174

Nicolson, Harold, 96, 174–75

Nordhausen V-2 rocket factory, 240–46; *see also* Dora concentration camp

Nordhoff-Jung medal, 185

North to the Orient (A. M. Lindbergh), 50, 95

Nungesser, Charles, 18–19

Nuremberg laws, 104–5, 111

occultism, 21, 82, 86, 88, 109–10, 114, 119, 120, 165, 273
  clairvoyance in, 82, 87, 91, 94
  ESP in, 88n, 109, 277
  miraculous healings in, 31–32, 91
  pendulum dowsing, 100–101, 104, 216
  rune casting in, 178, 216
  table lifting in, 110

Ochs, Adolph, 48

Of Flight and Life (C. Lindbergh), 250–52

Olympic Games of 1936, 106, 243

"On the Permanent Life of Tissues Outside the Organism" (Carrel), 13

open heart surgery, 4, 283

organ culture research, *see* perfusion

organ reimplantation, 132–33, 140

organ transplantation, 4, 8–9, 14–16, 23–24, 28, 76
  immune response problem of, 15, 284

military emergency, 265–67
  modern, 283–84
  visceral-organism experiment and, 38–39
  see also perfusion

Ortieg Prize, 18

P-38 Lightning fighter plane, 228–30

P-47 Thunderbolt fighter plane, 222, 225

Pan American Airways, 20, 253–54, 258, 260–61, 264, 274, 275

parapsychology, *see* occultism

Parker, Raymond C., 49, 113

Paumier, Raymond, 217–18

Pearl, Raymond, 91

Pechin, Edwin and Jeannie, 280, 281

pendulum dowsing, 100–101, 104, 216

Pennsylvania Railroad, 20

Pepper, Claude, 190, 193

perfusion, 15–16, 23–24, 30, 33–35, 47, 58–59, 62–64, 67–74, 75–79, 83–84, 111, 116, 127, 139–41, 237, 248, 252, 272–73, 283
  book about, 125, 126, 139
  founding experiments of, 60–61, 76
  of human organs, 68, 81, 97, 132, 133, 134–35, 140, 142
  infection problem of, 15, 24, 33, 35, 40, 62, 63, 64, 70, 73, 75, 77
  military research on, 265–67
  of nerve tissue, 115, 117
  nutrient medium of, 33, 34, 58, 62, 70, 72, 73, 96–99, 116
  political ramifications of, 146–47
  of primate organs, 130, 132, 133

perfusion pumps, 24, 39–40, 48, 60, 106
  capillary lift, 58
  insufficient pressure of, 35, 37, 58

perfusion pumps (*cont.*)
 rocking-coil, 33–36, 37, 40, 58
 testing of, 34–35
perfusion pumps, gas-powered pulse,
 62–64, 67–74, 75–79, 80, 81, 82,
 84, 115, 125, 132, 166, 272–73
 control gas of, 62, 63, 70, 72
 Cytology Congress demonstration
 of, 101–2, 104, 108–11, 113, 272
 mechanical kidney for, 96–99, 113
 media coverage of, 77–79, 82,
 139–41
 military application of, 265–67,
 277
 oil flask of, 62–63, 70, 71, 72, 98
 pressure tank for, 116–17, 121–22,
 130, 133–34, 222, 224
 public's misinterpretation of, 78, 139,
 141
 as scientific breakthrough, 110–11
 testing of, 71–74, 75–76, 97–98, 99
 on *Time* magazine cover, 139–40,
 248
peritonitis, 31, 38
Perroux, François, 215
Perry, Vernon P., 265–67, 277
Pétain, Henri Philippe Omer, 203, 206,
 208–9, 210, 216
Philippine Islands, 263–64, 270
Philosophers Club, 30, 45, 46, 79, 86,
 181, 203, 205
Plato, 147, 152, 162, 252
Plon publishing house, 79, 89–90
PM, 187, 196
Poland, 159, 170, 171, 180
Port-Blanc, 124, 236
"Prayer for Peace, A," (A. M.
 Lindbergh), 175–76
preserved-tissue experiments, 8–9
Presse, Dom Alexis, 207–8, 219

primate research, 81, 99–100, 117, 130,
 132, 265–67
 hypothermia experiments in, 122,
 124, 133–34
Pryor, Sam, 274–75, 276
Publisher's Weekly, 89

Race Betterment Conference, 46
racism, 45–46, 80, 104–5, 111, 119,
 129–30, 143–44, 173, 176, 184,
 194, 205, 246
 see also eugenics; Jews; white race
Ramakrishna, Sri, 119
Reader's Digest, 91, 279
 AC's articles for, 184–85, 273
 AML's article in, 175–76
 CAL's articles in, 173, 261–63
reconstructive surgery, 27–28
Réflexions sur la conduite de la vie
 (Carrel), 216
rejuvenation, 23, 29
replaceable body parts, 16, 28, 39–40,
 63, 73, 115, 132–33, 269, 283–84
Republic (Plato), 147, 152, 252
resuscitation experiments, 8
Richthofen Geschwader, 105
Rickenbacher, Eddie, 188
Ringer's solution, 38
Robinson, John B., 234–35
Rockefeller, John D., 4, 25
Rockefeller, John D., Jr., 93
Rockefeller Institute, 215, 218–19
 AC's materials stored at, 248–49
 AC's office at, 203, 211, 236–37
 core mission of, 92, 139
 directors of, 81–82, 90, 93–94,
 137–39, 161
 library of, 60–61, 211
 mandatory retirement policy of,
 137–41, 142–43, 161, 185

scientists at, 58–59, 92–93, 142–43, 167

Rockefeller Institute, Experimental Surgery Division of, 4–16, 23–24, 25–40, 41, 48–49, 50–51, 57–64, 66, 67–74, 75–76, 89, 92, 98–99, 100, 111, 116, 176, 178, 215, 222, 252, 268–69, 272

AC's office at, 4–5

animal-rights activists' campaign against, 9–10, 11

black operating suite of, 14–16, 32–33, 71–72, 90, 122, 166

black surgical gowns of, 8, 12, 14, 32, 71, 108, 109

CAL's Christmas holiday return to, 124–25, 130–36, 138–39

CAL's toolbox at, 25, 58

discontinuation of, 138

eugenics and, 46

glassblowing station of, 33–34, 58, 63, 70, 76, 98, 133, 248

incubation chamber of, 11–14, 28–29, 71–74, 75–76; *see also* tissue culture experiments

laboratory animals of, 6, 8, 9–10

mandate of, 6, 93–94

mousery of, 9, 30–31, 166, 178, 245–46, 249

mummy-revivification experiment of, 249

organ-transplantation experiments of, 15–16; *see also* perfusion

preserved-tissue experiments of, 8–9

proposed hypothermia chamber for, 122

public interest in, 79–80, 81–82

resuscitation experiments of, 8

semen-refrigeration experiments of, 9

technician assistants of, 12, 32, 58, 64, 68, 70, 71–72, 74, 138, 166

visceral-organism experiment of, 37–39

Rogers, Will, 64, 65

Roosevelt, Franklin Delano, 65, 66, 151–52, 159, 172, 197, 198, 233

CAL's meeting with, 164–65

CAL's patriotism impugned by, 192, 199, 253

in "great debate," 191, 192–94

Henry Ford and, 221, 222

interventionist sentiment of, 161, 169, 181–82, 189, 195

personality of, 191

"shoot on sight" policy of, 194

Rosen, John, 256

Rosenberg, Alfred, 184

Rosenberger, Heinz, 24

Rosner, Morris "Mickey," 55

Rucart, Marc, 177–78

rune casting, 178, 216

Saarinen, Eero, 222

Sackville-West, Vita, 96, 174

Saint-Gildas island, 30, 84, 112–16, 135, 140, 141–43, 169, 177, 178, 186, 207–8, 214, 216

AC's grave on, 235–38

CAL's visits to, 113–16, 123–26, 141–42, 145–50, 151, 237

house on, 112, 236

Nazi looting of, 202–3

press stories on, 125–27

research laboratory built on, 142, 144

Samoa, 270

Saturday Evening Post, 40, 41, 205

Schmidt, Axel, 61

Schorr, David, 201
Schwarzkopf, H. Norman, 65, 97
Science, 35, 50–51, 76–77
semen, 9, 29
Shriver, Sargent, 187–88
Siboney, 224–25
sleeping sickness, vaccine for, 167
Smith, Kay, 105, 130
Smith, Meryl, 230
Smith, Truman, 102, 103–4, 105,
    129–30, 149, 155, 156, 170, 172,
    184
Soupault, Robert, 206
Soviet Union, 136, 145, 146, 169, 181,
    216, 239, 271
    aviation program of, 144
    in cold war, 233, 250, 255–56
    Western civilization threatened by,
        144, 148, 154, 161, 165, 180,
        243
Spain, 203, 205, 206, 220
Spanish-American War, 19
Spectator, 174
sperm banks, 9
Spirit of St. Louis, 2, 6, 19, 21, 161–62,
    221, 222, 228, 247
Spirit of St. Louis, The (C. Lindbergh),
    21, 252, 253, 271–72
Spitale, Salvatore "Salvy," 55
Stafford, Jean, 271–72
stem cell research, 16
Stewart, Potter, 187–88
Stimson, Henry L., 199, 218, 221
Stobart, J. C., 162
Strehler, Bernard, 28–29
Stuart, R. Douglas "Bob," 188, 194
Stürmer, Der, 146
Sudetenland, 145, 153
supersonic airplanes, 260–61
Syzygy, Yulian, 93

Taiwan, 264
tamarau, 264, 274
Taufa-ahau Tupou IV, 264
Thomas, Ambroise, 135, 183
Thomas, Norman, 188
Tiepolo, Giambattista, 162
Time, 92, 139–40, 168, 248
tissue banks, 9
tissue culture research, 11–14, 109, 113,
    142, 167, 237
    centrifuges for, 48–49, 59
    corpuscle washing machine for, 49,
        50–51, 59
    flasks for, 11, 57–58, 121
    infection of, 12
    limited life span of, 28–29
    metabolites in, 13, 14
    nerve, 115, 117
    nutrient medium of, 12, 29, 48–49,
        57–58
    waste products of, 12, 13
    see also chick-heart tissue culture
tissue-engineering projects,
    283–84
Titanic disaster, 136
Tonga, 264
Toynbee, Arnold, 263
Train, Arthur, Jr., 39
Train, Arthur, Sr., 39
Transcontinental Air Transport, 17
Trenchard, Thomas W., 67
Truman, Harry, 233, 247
Tufele-Faiaoga, 270
Tunney, Gene, 173–74
TWA airline, 17, 20, 160
Tyrode solution, 68–69

Udet, Ernst, 128–29, 155
Uellendahl, E. H., 239–40, 241–43
United Aircraft, 225–26, 232

V-2 rockets, 239–46
vaccine industry, 12, 167
Vallery-Radot, Pasteur, 217, 218, 219
Vanderbilt, Cornelius, Jr., 190–91
Van Vorst, Bessie, 59–60
vascular anastomosis technique, 4, 6,
    8–9, 10–11, 14–15, 31, 32, 38, 76,
    237, 278
Verdienstkreuz Deutscher Adler medal,
    155–56, 157, 158, 159–61, 238
viruses, 167–68
visceral-organism experiment, 37–39
Völkischer Beobachter, 196
vom Rath, Ernst, 159
von Braun, Wernher, 239
Vonnegut, Kurt, 187
Voyageurs National Park, 264

Walking Dead, The, 77
Wallace, DeWitt, 173, 279
Wallace, Henry A., 194
*Wartime Journals of Charles A.
    Lindbergh, The,* 270–72, 274,
    279
    Jews in, 162–63, 169–70, 271
Washington Post, 18, 153
We (C. Lindbergh), 20, 252
Western civilization, 30, 44–48, 86, 87,
    92, 107, 113, 114–16, 123, 127,
    135, 162, 205
    Asian barbarians as threat to, 44–45,
        46, 47, 118–19, 120–21, 144, 148,
        154, 161, 165, 172, 176, 180, 248
    immigrants as threat to, 45, 213
    Nazi Germany as death of, 108, 181,
        182–83
    Nazi Germany as savior of, 121,
        129–30, 147, 148–50, 151–52,
        161, 165, 180–81, 186–87, 243,
        248–49

Whately, Aloysius "Olly," 51
Whately, Elsie, 51, 52–53
"What Substitute for War?"
    (C. Lindbergh), 180
White, E. B., 160
white race, 30, 47, 86, 115, 130, 135,
    172, 176
    black race vs., 262–63
    faster-breeding racial inferiors vs.,
        45, 46, 88, 118–19, 120–21, 146,
        173, 248
    of Nazi Germany, 104–5, 106–7,
        113–14, 129–30
    see also Western civilization
"Who Are the War Agitators?"
    (C. Lindbergh), 194–96
wildlife conservation, 263–65, 267, 269,
    274, 275
    CAL's awards for, 270
Wilentz, David, 67
Willkie, Wendell, 193, 197
Willow Run, 221, 222, 225
Wilson, Hugh R., 155–56, 157, 159
Winchell, Walter, 187, 197
"Wisdom of Wildness, The"
    (C. Lindbergh), 267
Wise, Stephen S., 91–92, 143
Witkowski, Jan A., 29
Wolff, Kurt and Helen, 258
Wood, Louis E., 226
Wood, Robert E., 188, 197, 198
Woodbridge, Frederick, 30
Woodring, Harry Hines, 164
Wooster, Stanton, 19
World's Fair of 1939, 167
World War I, 18, 29, 32, 46, 104, 174,
    177, 188, 199
    AC as French army surgeon in, 8, 89,
        108, 130, 180–81, 237, 278
    eugenic damage created by, 46

World War I (*cont.*)
  Germans in, 89, 106, 108, 128, 129,
    145–46, 147, 180–81, 206
World War II, 129, 135–36, 145–46,
    158–59, 160, 169, 170, 171–232
  atomic bomb dropped in, 247–48
  CAL in Pacific theater of, 225–32,
    244
  as eugenic disaster, 203, 205, 272
  U.S. aircraft manufacturing in,
    221–26, 232
  U.S. entry into, 198–200
  see also Germany, Nazi; Hitler, Adolf
World War II, CAL's isolationist stance
    on, 129, 150, 161, 164, 165, 171–
    76, 179–84, 186–200, 245, 250
  CAL's official military service
    precluded by, 198–200, 201, 221,
    252
  cartoons about, 187, 194
  congressional testimony of, 189–91,
    264
  criticism of, 173–76, 182, 187, 189–
    200, 201, 221
  foreign aid rejected in, 179, 181–83,
    189–90, 201
  Guthrie's song about, 199–200
  Jews targeted in, 195–98

Nazi-U.S. treaty recommended in,
    186
  negotiated peace recommended in,
    180
  Roosevelt's "great debate" with, 191,
    192–94
  speaking campaign in, 170, 171–73,
    175, 179, 182, 186–87, 188–98,
    199, 202
  subsequent justification of, 271
World Wildlife Fund, 263
World's Work, 17
Wright, Orville, 128, 250
Wright, Wilbur, 250
Wyckoff, Ralph, 167–69, 170, 171

Yale University, 187–89
  CAL's speech at, 188–89
  Lindbergh Collection of, 67–68, 279
  Primate Biology Laboratories of, 117
Yerkes, Robert, 117
yogis, 119–20
  body temperature controlled by, 101,
    119, 121, 224
  breathing exercises of, 224
  claimed feats of, 120
Younghusband, Sir Francis, 101, 119,
    120